Environmental Biotechnology

This book approaches the topic of environmental biotechnology in a clear, integrated, and meaningful way, covering both the fundamentals and biochemical processes involved, as well as the technologies themselves within different areas of application. As part of the framework, it also provides a thorough description of the pollution and its control, and the role of microorganisms in a wide range of ecosystems and deterioration processes.

Features:

- Focuses on the role of microorganisms in a wide range of ecosystems and deterioration processes.
- Explains underlying concepts of environment, interlinks them from an ecological point of view, and describes the approaches for waste treatment.
- Describes the concepts and fate processes of environmental contaminants, contaminant patterns in soil, groundwater, and surface water.
- Includes novel research findings and applications of biosurfactants.
- Discusses biodegradation as a key process in the bioremediation of recalcitrant compounds.

This book is aimed at Primarily Senior Undergraduates including Graduate Students and Researchers in Biotechnology, Environmental Science/Engineering, Conservation Biology, Microbiology, Waste Management, and Ecology.

Sibi G received his Ph.D. degree in 2010 from Bangalore University, Bangalore. He is the recipient of research projects from University Grants Commission, New Delhi, and Karnataka State Council for Science and Technology, Bangalore. He has written two books and published more than 100 research articles in peer-reviewed journals. He has also participated in a range of forums on life sciences and presented various academic as well as research-based papers.

Emerging Materials and Technologies

Series Editor
Boris I. Kharissov

For more information about this series, please visit: www.routledge.com/Emerging-
Materials-and-Technologies/book-series/CRCEMT

Environmental Biotechnology
Fundamentals to Modern Techniques

Sibi G

CRC Press
Taylor & Francis Group
Boca Raton London New York

CRC Press is an imprint of the
Taylor & Francis Group, an **informa** business

Designed cover image: ©Shutterstock

First edition published 2023
by CRC Press
6000 Broken Sound Parkway NW, Suite 300, Boca Raton, FL 33487-2742

and by CRC Press
4 Park Square, Milton Park, Abingdon, Oxon, OX14 4RN

CRC Press is an imprint of Taylor & Francis Group, LLC

© 2023 Sibi G

ISBN: 9781032224497 (hbk)
ISBN: 9781032224503 (pbk)
ISBN: 9781003272618 (ebk)

DOI: 10.1201/9781003272618

Typeset in Times
by codeMantra

Contents

PART II   Bioremediation

PART III   Wastes and Waste Management

Preface

The book devotes detailed chapters to each of the four main parts of Environmental Biotechnology—Biogeochemical cycles, Pollution and pollution control, Bioremediation, Wastes and waste management—dealing with both the microbiological and biotechnological aspects.

With advancements in the area of Environmental Biotechnology, researchers are looking for new opportunities to improve quality standards and environment. Recent technologies have given impetus to the possibility of cost intensive and eco-friendly technology for producing high-quality products and efficient ways to recycle waste to minimize environmental pollution are the need of the hour. The use of bioremediation technologies through microbial communities is another viable option to remediate environmental pollutants such as heavy metals, pesticides, and dyes. It continues to revolutionize the understanding of basic life sustaining processes in the environment, identification and exploitation of molecules, and its use to provide clean technologies and deal with environmental problems.

Overall, the volume provides a comprehensive overview of the current issues in the field of environmental biotechnology which is believed as immediate value to environmentalists and policymakers, as well as to students in environmental science.

Sibi G
Bangalore India

Part I

Biogeochemical Cycles, Pollution and Pollution Control

BIOGEOCHEMICAL CYCLES

Carbon dioxide concentrations in the atmospheric exceeded 400 parts per million (ppm) for the first time in May 2013, increasing from 315 ppm in 1958, reports the U.S. National Oceanic and Atmospheric Administration. What has driven this increase? Carbon is cycled between organic forms such as sugar or other cellular building blocks and inorganic forms such as carbon dioxide. Vast amounts of organic matter are photosynthetically produced on Earth each year utilizing carbon dioxide from the atmosphere. Most of this material is consumed and degraded but a part of it, over the millennia, has been stored in permafrost, peat bogs, and as fossil fuels. A delicate global balance of organic and inorganic carbon has been maintained largely driven by microbial activity. Human use of stored organic carbon (fossil fuels and peat) and recent warming of permafrost have upset this balance in favor of the release of carbon dioxide to the atmosphere. Most scientists agree that this is having a major impact on global warming, and is an excellent example of a major perturbation of the carbon cycle—one that is occurring during our lifetimes. Cycling between organic and inorganic forms is not limited to carbon. All of the major elements found in biological organisms, as well as some of the minor and trace elements, are similarly cycled in predictable and definable ways.

Taken together, various element cycles are called the biogeochemical cycles. Understanding these cycles allows scientists to understand and predict the development of microbial communities and activities in the environment. There are many activities that can be harnessed in a beneficial way, such as for remediation of organic

DOI: 10.1201/9781003272618-1

and metal pollutants, or for recovery of precious metals such as copper or uranium from low-grade ores. There are detrimental aspects of the cycles that can cause global environmental problems, for example, the formation of acid rain and acid mine drainage, metal corrosion processes, and formation of nitrous oxide, which can deplete Earth's ozone layer. As these examples illustrate, the microbial activities that drive biogeochemical cycles are highly relevant to the field of environmental microbiology. Thus, the knowledge of these cycles is increasingly critical as the human population continues to grow, and the impact of human activity on Earth's environment becomes more significant.

GAIA HYPOTHESIS

In the early 1970s, James Lovelock theorized that Earth behaves like a super organism, and this concept developed into what is now known as the Gaia hypothesis. To quote "Living organisms and their material environment are tightly coupled. The coupled system is a super organism, and as it evolves there emerges a new property, the ability to self-regulate climate and chemistry." The basic tenet of this hypothesis is that Earth's physicochemical properties are self-regulated so that they are maintained in a favorable range for life. As evidence for this, consider that the sun has heated up by 30% during the past 4–5 billion years. Given Earth's original carbon dioxide-rich atmosphere, the average surface temperature of a lifeless Earth today would be approximately 290°C. In fact, when one compares Earth's present-day atmosphere with the atmospheres found on our nearest neighbors Venus and Mars, one can see that something has drastically affected the development of Earth's atmosphere. According to the Gaia hypothesis, this is the development and continued presence of life. Microbial activity, and later the appearance of plants, has changed the original heat-trapping carbon dioxide-rich atmosphere to the present oxidizing, carbon dioxide-poor atmosphere. This has allowed Earth to maintain an average surface temperature of 13°C, which is favorable to the life that exists on Earth.

How do biogeochemical activities relate to the Gaia hypothesis? These biological activities have driven the response to the slow warming of the sun resulting in the major atmospheric changes that have occurred over the last 4–5 billion years. When Earth was formed 4–5 billion years ago, a reducing (anaerobic) atmosphere existed. The initial reactions that mediated the formation of organic carbon were abiotic, driven by large influxes of ultraviolet (UV) light. The resulting reservoir of organic matter was utilized by early anaerobic heterotrophic organisms. This was followed by the development of the ability of microbes to fix carbon dioxide photosynthetically. Evidence from stromatolite fossils suggests that the ability to photosynthesize was developed at least 3.5 billion years ago.

The evolution of photosynthetic organisms tapped into an unlimited source of energy, the sun, and provided a mechanism for carbon recycling, i.e., the first carbon cycle. This first carbon cycle was maintained for approximately 1.5 billion years. Geologic evidence then suggests that approximately 2 billion years ago, photosynthetic microorganisms developed the ability to produce oxygen. This allowed oxygen to accumulate in the atmosphere, resulting, in time, in a change from reducing to oxidizing conditions. Further, oxygen accumulation in the atmosphere created an ozone layer, which reduced the influx of harmful UV radiation, allowing the development of higher forms of life to begin. At the same time that the carbon cycle evolved, the nitrogen cycle emerged because nitrogen was a limiting element for microbial growth. Although molecular nitrogen was abundant in the atmosphere, microbial cells could not directly utilize nitrogen as N_2 gas. Cells require organic nitrogen compounds or reduced inorganic forms of nitrogen for growth. Therefore, under the reducing conditions found on early Earth, some organisms developed a mechanism for fixing nitrogen using the enzyme nitrogenase. Nitrogen fixation remains an important microbiological process, and to this day the majority of nitrogenase enzymes are totally inhibited in the presence of oxygen. When considered over this geologic time scale of several billion years, it is apparent that biogeochemical activities have been unidirectional. This means that the predominant microbial activities on Earth have evolved over this long period of time to produce changes and to respond to changes that have occurred in the atmosphere, i.e., the appearance of oxygen and the decrease in carbon dioxide content. Presumably, these changes will continue to occur, but they occur so slowly that we do not have the capacity to observe them.

One can also consider biogeochemical activities on a more contemporary time scale, that of tens to hundreds of years. On this much shorter time scale, biogeochemical activities are regular and cyclic in nature, and it is these activities that are addressed in this chapter. On the one hand, the presumption that Earth is a superorganism that can respond to drastic environmental changes is heartening when one considers that human activity is causing unexpected changes in the atmosphere, such as ozone depletion and buildup of carbon dioxide. However, it is important to point out that the response of a superorganism is necessarily slow (thousands to millions of years), and as residents of Earth we must be sure not to overtax Earth's ability to respond to change by artificially changing the environment in a much shorter time frame.

POLLUTION: CAUSES, EFFECTS, AND CONTROL

The carbon dioxide concentration in the atmosphere has increased by nearly 30% since the beginning of the Industrial Revolution; more atmospheric nitrogen is now fixed by humanity than by all natural terrestrial sources combined; more than half of all accessible surface fresh water is put to use by humanity. The rates, scales, kinds, and combinations of changes occurring now are fundamentally different from those at any other time in history; we are changing Earth more rapidly than we are understanding it.

Almost any substance, synthetic or natural, can pollute, but it is synthetic and other industrial chemicals that most concern people. The word "waste" differs from pollutant, although waste can pollute. Waste often refers to garbage or trash. Examples include the garbage discarded by households or restaurants, or the construction debris discarded by builders, or material that has reached the end of its useful life. It is difficult to imagine a modern society without the benefits of chemicals and the chemical industry. Pharmaceuticals, petrochemicals, agrochemicals, industrial and consumer chemicals all contribute to our modern lifestyles. However, with the rise of chemical manufacture and use has come increasing public awareness and concern regarding the presence of chemicals in the environment. There is an important distinction between the presence of chemicals in the environment (contamination) and pollution. Although these terms tend to be used in similar ways in everyday speech and journalism, in scientific areas, there is a broad consensus that the term "contamination" should be used where a chemical is present in a given sample with no evidence of harm and "pollution" used in cases where the presence of the chemical is causing harm. Pollutants therefore are chemicals causing environmental harm.

Despite the fact that any chemical can be a pollutant, certain chemicals have been identified in regulation or by international agreement as being "priority chemicals for control." Such chemicals have generally been selected based on the following criteria:

- the chemicals are frequently found by monitoring programs
- they are toxic at low concentrations
- they bioaccumulate
- they are persistent
- they are carcinogens

Government, regulatory agencies, and industrial initiatives have recognized a hierarchy of approaches to priority pollutant control:

- replace: use another, more environmentally friendly chemical
- reduce: use as little of the priority pollutants as possible
- manage: use in a carefully managed way to minimize accidental or adventitious loss and waste

1 Biogeochemical Cycles

CARBON CYCLE

CARBON RESERVOIRS

A reservoir is a sink or source of an element such as carbon. There are various global reservoirs of carbon, some of which are immense in size and some of which are relatively small. The largest carbon reservoir is carbonate rock found in Earth's crust. This reservoir is four orders of magnitude larger than the carbonate reservoir found in the ocean, and six orders of magnitude larger than the carbon reservoir found as carbon dioxide in the atmosphere. If one considers these three reservoirs, it is obvious that the carbon most available for photosynthesis, which requires carbon dioxide, is in the smallest of the reservoirs, the atmosphere. Therefore, it is the smallest reservoir that is most actively cycled. It is small, actively cycled reservoirs such as atmospheric carbon dioxide that are subject to perturbation from human activity. In fact, since global industrialization began in the late 1800s, humans have affected several of the smaller carbon reservoirs. Utilization of fossil fuels (an example of a small, inactive carbon reservoir) and deforestation (an example of a small, active carbon reservoir) are two activities that have reduced the amount of fixed organic carbon in these reservoirs and added to the atmospheric carbon dioxide reservoir.

The increase in atmospheric carbon dioxide has not been as great as expected. This is because the reservoir of carbonate found in the ocean acts as a buffer between the atmospheric and sediment carbon reservoirs through the equilibrium equation shown below:

$$H_2CO_3 \rightarrow HCO_2 \rightarrow CO_2$$

Thus, some of the excess carbon dioxide that has been released has been absorbed by the oceans. However, there has still been a net efflux of carbon dioxide into the atmosphere of approximately 7×10^9 metric tons/year. The problem with this imbalance is that because atmospheric carbon dioxide is a small carbon reservoir, the result of a continued net efflux over the past 100 years or so has been a 28% increase in atmospheric carbon dioxide from 0.026% to 0.033%. A consequence of the increase in atmospheric carbon dioxide is that it contributes to global warming through the greenhouse effect. The greenhouse effect is caused by gases in the atmosphere that trap heat from the sun and cause Earth to warm up. This effect is not solely due to carbon dioxide; other gases such as methane, chlorofluorocarbons (CFCs), and nitrous oxide all contribute to the problem.

CARBON FIXATION AND ENERGY FLOW

The ability to photosynthesize allows sunlight energy to be trapped and stored. In this process, carbon dioxide is fixed into organic matter. Photosynthetic organisms,

DOI: 10.1201/9781003272618-2

5

also called primary producers, include plants and microorganisms such as algae, cyanobacteria, some bacteria, and some protozoa. The efficiency of sunlight trapping is very low; less than 0.1% of the sunlight energy that hits Earth is actually utilized. As the fixed sunlight energy moves up each level of the food chain, up to 90% or more of the trapped energy is lost through respiration. Despite this seemingly inefficient trapping, photoautotrophic primary producers support most of the considerable ecosystems found on Earth. Productivity varies widely among different ecosystems depending on the climate, the type of primary producer and whether the system is a managed one. For example, one of the most productive natural areas is the coral reefs. Managed agricultural systems such as corn and sugarcane systems are also very productive, but it should be remembered that a significant amount of energy is put into these systems in terms of fertilizer addition and care. The open ocean has much lower productivity, but covers the majority of Earth's surface, and so is a major contributor to primary production. In fact, aquatic and terrestrial environments contribute almost equally to global primary production. Plants predominate in terrestrial environments, but with the exception of immediate coastal zones, microorganisms are responsible for most primary production in aquatic environments. It follows that microorganisms are responsible for approximately one-half of all primary production on Earth.

Carbon Respiration

Carbon dioxide that is fixed into organic compounds as a result of photoautotrophic activity is available for consumption or respiration by animals and heterotrophic microorganisms. The end products of respiration are carbon dioxide and new cell mass. An interesting question to consider is the following: if respiration were to stop, how long would it take for photosynthesis to use up all of the carbon dioxide reservoir in the atmosphere? Based on estimates of global photosynthesis, it has been estimated that it would take 30–300 years. This illustrates the importance of both legs of the carbon cycle in maintaining a carbon balance.

The following sections discuss the most common organic compounds found in the environment and the microbial catabolic activities that have evolved in response. These include organic polymers, humus, and C_1 compounds such as methane (CH_4). It is important to understand the fate of these naturally occurring organic compounds because degradative activities that have evolved for these compounds form the basis for degradation pathways that may be applicable to organic contaminants that are spilled in the environment. But before looking more closely at the individual carbon compounds, it should be pointed out that the carbon cycle is actually not quite as simple as depicted in Figure 1.1. This simplified figure does not include anaerobic processes, which were predominant on early Earth and remain important in carbon cycling even today. A more complex carbon cycle containing anaerobic activity is shown in Figure 1.1.

Under anaerobic conditions, which predominated for the first few billion years on Earth, some cellular components were less degradable than others (e.g., octane, naphthalene, and cyclopentane). This is especially true for highly reduced molecules such as cellular lipids. These components were therefore left over and buried with

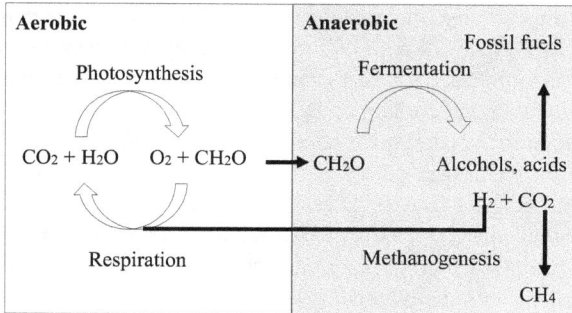

FIGURE 1.1 The carbon cycle.

sediments over time, and became the present-day fossil fuel reserves. Another carbon compound produced under anaerobic conditions is methane. Methane is produced in soils as an end product of anaerobic respiration. Methane is also produced in significant quantities under the anaerobic conditions found in ruminants such as cows as well as termite guts. It is also produced within landfills.

Organic Polymers

Organic polymers include plant polymers, fungal and bacterial cell wall polymers, and arthropod exoskeletons. Because these polymers constitute the majority of organic carbon, they are the basic food supply available to support heterotrophic activity. The three most common polymers are the plant polymers: cellulose, hemicellulose, and lignin. The various other polymers produced include starch (plants), chitin (fungi, arthropods), and peptidoglycan (bacteria). These various polymers can be divided into two groups on the basis of their structures: the carbohydrate-based polymers, which include the majority of the polymers found in the environment, and the phenylpropane-based polymer, lignin.

Carbohydrate-Based Polymers

Cellulose is not only the most abundant of the plant polymers, but it is also the most abundant polymer found on Earth. It is a linear molecule containing β-1,4 linked glucose subunits. Each molecule contains 1,000–10,000 subunits with a resulting molecular weight of up to 1.8×10^6. These linear molecules are arranged in microcrystalline fibers that help make up the woody structure of plants. Cellulose is not only a large molecule; it is also insoluble in water. How then do microbial cells get such a large, insoluble molecule across their walls and membranes? The answer is that they have developed an alternative strategy, which is to synthesize and release enzymes called extracellular enzymes, that can begin the polymer degradation process outside the cell. There are two extracellular enzymes that initiate cellulose degradation. These are β-1,4-endoglucanase and β-1,4-exoglucanase. Endoglucanase hydrolyzes cellulose molecules randomly within the polymer, producing smaller and smaller cellulose molecules. Exoglucanase consecutively hydrolyzes two glucose subunits from the reducing end of the cellulose molecule, releasing the disaccharide cellobiose. A third enzyme, known as β-glucosidase or cellobiase, then hydrolyzes cellobiose to glucose. Cellobiase can be found as both an extracellular and an intracellular enzyme. Both cellobiose and glucose can be taken up by many bacterial and fungal cells.

Hemicellulose is the second most common plant polymer. This molecule is more heterogeneous than cellulose, consisting of a mixture of several monosaccharides including various hexoses and pentoses as well as uronic acids. In addition, the polymer is branched instead of linear. An example of a hemicellulose polymer is the pectin molecule, which contains galacturonic acid and methylated galacturonic acid. Degradation of hemicelluloses is similar to the process described for cellulose except that, because the molecule is more heterogeneous, many more extracellular enzymes are involved. In addition to the two major plant polymers, several other important organic polymers are carbohydrate based. One of these is starch, a polysaccharide synthesized by plants to store energy. Starch is formed from glucose subunits and can be linear (α-1,4 linked), a structure known as amylose, or can be branched (α-1,4 and α-1,6 linked), a structure known as amylopectin. Amylases (α-1,4-linked exo- and endoglucanases) are extracellular enzymes produced by many bacteria and fungi. Amylases produce the disaccharide maltose, which can be taken up by cells and mineralized. Another common polymer is chitin, which is formed from β-1,4-linked subunits of N-acetylglucosamine. This linear, nitrogen-containing polymer is an important component of fungal cell walls and of the exoskeleton of arthropods. Finally, there is peptidoglycan, a polymer of N-acetylglucosamine and N-acetylmuramic acid, which is an important component of bacterial cell walls.

LIGNIN

Lignin is the third most common plant polymer, and is strikingly different in structure from all of the carbohydrate-based polymers. The basic building blocks of lignin are the two aromatic amino acids tyrosine and phenylalanine. These are converted to phenylpropene subunits such as coumaryl alcohol, coniferyl alcohol, and sinapyl alcohol. Then, 500–600 phenylpropene subunits are randomly polymerized, resulting in the formation of the amorphous aromatic polymer known as lignin. In plants, lignin surrounds cellulose microfibrils and strengthens the cell wall. Lignin also helps make plants more resistant to pathogens. Biodegradation of lignin is slower and less complete than degradation of other organic polymers. Lignin degrades slowly because it is constructed as a highly heterogeneous polymer, and in addition contains aromatic residues rather than carbohydrate residues. The great heterogeneity of the molecule precludes the evolution of specific degradative enzymes comparable to those for cellulose. Instead, a nonspecific extracellular enzyme, H_2O_2-dependent lignin peroxidase, is used in conjunction with an extracellular oxidase enzyme that generates H_2O_2. The peroxidase enzyme and H_2O_2 system generate oxygen-based free radicals that react with the lignin polymer to release phenylpropene residues. These residues are taken up by microbial cells and degraded. Biodegradation of intact lignin polymers occurs only aerobically, which is not surprising because reactive oxygen is needed to release lignin residues. However, once residues are released, they can be degraded under anaerobic conditions.

Phenylpropene residues are aromatic in nature, similar in structure to several types of organic pollutant molecules such as the BTEX (benzene, toluene, ethylbenzene, xylene) and polyaromatic hydrocarbon compounds found in crude oil as well as gasoline and creosote compounds found in wood preservatives. These naturally occurring aromatic biodegradation pathways are of considerable importance in the field of bioremediation. Lignin is degraded by a variety of microbes including fungi, actinomycetes, and bacteria. The best studied organism with respect to lignin degradation is the white rot fungus *Phanerochaete chrysosporium*. This organism is also capable of degrading several pollutant molecules with structures similar to those of lignin residues.

Humus

Humus forms in a two-stage process that involves the formation of reactive monomers during the degradation of organic matter, followed by the spontaneous polymerization of some of these monomers into the humus molecule. Although the majority of organic matter that is released into the environment is respired to form new cell mass and carbon dioxide, a small amount of this carbon becomes available to form humus. To understand this spontaneous process, consider the degradation of the common organic polymers found in soil, which were described in the preceding sections. Each of these polymers requires the production of extracellular enzymes that begin the polymer degradation process. In particular, for lignin, these extracellular enzymes are nonspecific and produce hydrogen peroxide and oxygen radicals. It is not surprising, then, that some of the reactive residues released during polymer degradation might repolymerize and result in the production of humus. In addition, nucleic acid and protein residues that are released from dying and decaying cells contribute to the pool of molecules available for humus formation. Considering the wide array of residues that can contribute to humus formation, it is not surprising that humus is even more heterogeneous than lignin.

Humus is the most complex organic molecule found in soil, and as a result it is the most stable organic molecule. The turnover rate for humus ranges from 2% to 5% per year, depending on climatic conditions. Humus provides a very slowly released source of carbon and energy for indigenous autochthonous microbial populations. The release of humic residues most likely occurs in a manner similar to release of lignin residues. Because the humus content of most soils does not change, the rate of formation of humus must be similar to the rate of turnover.

METHANE

Methanogenesis

The formation of methane, methanogenesis, is predominantly a microbial process, although a small amount of methane is generated naturally through volcanic activity. Methanogenesis is an anaerobic process and occurs extensively in specialized environments including water-saturated areas such as wetlands and paddy fields, anaerobic niches in the soil, landfills, the rumen, and termite guts. Methane is an end product of anaerobic degradation, and as such is associated with petroleum, natural gas, and coal deposits. At present, a substantial amount of methane is released to the atmosphere as a result of energy harvesting and utilization. A second way in which methane is released is through landfill gas emissions. Although methane makes a relatively minor carbon contribution to the global carbon cycle, methane emission is of concern from several environmental aspects. First, like carbon dioxide, methane is a greenhouse gas and contributes to global warming. In fact, it is the second most common greenhouse gas emitted to the atmosphere. Further, it is 22 times more effective than carbon dioxide at trapping heat. Second, localized production of methane in landfills can create safety and health concerns. Methane is an explosive gas at concentrations as small as 5%. Thus, to avoid accidents, the methane generated in a landfill must be managed in some way. If methane is present in concentrations higher than 35%, it can be collected and used for energy. Alternatively, the methane can be burned off at concentrations of 15% or higher. However, most commonly, it is simply vented to the atmosphere to prevent it from building up in high enough concentrations to ignite. Although venting landfill gas to the atmosphere does help prevent explosions, it clearly adds to the global warming problem.

The organisms responsible for methanogenesis are a group of obligately anaerobic archaebacteria called the methanogens. The basic metabolic pathway used by the methanogens is:

$$4H_2 + CO_2 \rightarrow CH_4 + 2H_2O \, \Delta G^0 = -130.7 \, kJ/mol$$

This is an exothermic reaction where CO_2 acts as the TEA and H_2 acts as the electron donor providing energy for the fixation of carbon dioxide. Methanogens that utilize CO_2/H_2 are therefore autotrophic. In addition to the autotrophic reaction shown in the above equation, methanogens can produce methane during heterotrophic growth on a limited number of other C_1 and C_2 substrates including acetate, methanol, and formate. Since there are very few carbon compounds that can be used by methanogens, these organisms are dependent on the production of these compounds by other microbes in the surrounding community. As such, an interdependent community of microbes typically develops in anaerobic environments. In this community, the more complex organic molecules are catabolized by populations that ferment or respire anaerobically, generating C_1 and C_2 carbon substrates as well as CO_2 and H_2 that are then used by methanogens.

Methane Oxidation

Clearly, methane as the end product of anaerobiosis is found extensively in nature. As such, it is an available food source, and a group of bacteria called the methanotrophs have developed the ability to utilize methane as a source of carbon and energy. The methanotrophs are chemoheterotrophic and obligately aerobic. They metabolize methane as shown in the below equation:

$$CH_4 + O_2 \rightarrow CH_3OH \rightarrow HCHO \rightarrow HCHOOH \rightarrow CO_2 + H_2O$$

The first enzyme in the biodegradation pathway is called methane monooxygenase. Oxygenases in general incorporate oxygen into a substrate and are important enzymes in the initial degradation steps for hydrocarbons. However, methane mono-oxygenase is of particular interest because it was the first of a series of enzymes isolated that can co-metabolize highly chlorinated solvents such as trichloroethene (TCE). Until this discovery, it was believed that biodegradation of highly chlorinated solvents could occur only under anaerobic conditions as an incomplete reaction. The application of methanogens for co-metabolic degradation of TCE is a strategy under development for bioremediation of groundwater contaminated with TCE. This is a good illustration of the way in which naturally occurring microbial activities can be harnessed to solve pollution problems.

CARBON MONOXIDE AND OTHER C₁ COMPOUNDS

Bacteria that can utilize C_1 carbon compounds other than methane are called methylotrophs. There are a number of important C_1 compounds produced from both natural and anthropogenic activities. One of these is carbon monoxide. The annual global production of carbon monoxide is $3-4 \times 10^9$ metric tons/year. The two major carbon monoxide inputs are abiotic. Approximately, 1.5×10^9 metric tons/year results from atmospheric photochemical oxidation of carbon compounds such as methane, and 1.6×10^9 metric tons/ year results from burning of wood, forests, and fossil fuels. A small proportion, 0.2×10^9 metric tons/year, results from biological activity in ocean and soil environments. Carbon monoxide is a highly toxic molecule because it has a strong affinity for cytochromes, and in binding to cytochromes, it can completely inhibit the activity of the respiratory electron transport chain. Destruction of carbon monoxide can occur abiotically by photochemical reactions in the atmosphere. Microbial processes also contribute significantly to its destruction, even though it is a highly toxic molecule. The destruction of carbon monoxide seems to be quite efficient because the level of carbon monoxide in the atmosphere has not risen significantly since industrialization began, even though CO emissions have increased. The ocean is a net producer of carbon monoxide and releases CO to the atmosphere. In contrast, the terrestrial environment is a net sink for carbon monoxide and absorbs approximately 0.43×10^9 metric tons/year. Key microbes found in terrestrial environments that can metabolize carbon monoxide include both aerobic and anaerobic organisms. Under aerobic conditions, *Pseudomonas carboxydoflava* is an example of an organism that oxidizes carbon monoxide to carbon dioxide:

$$CO + H_2O \rightarrow CO_2 + H_2$$

$$2H_2 + O_2 \rightarrow 2H_2O$$

This organism is a chemoautotroph and fixes the CO_2 generated in the above equation into organic carbon. The oxidation of the hydrogen produced provides energy for CO_2 fixation (see the following equation). Under anaerobic conditions, methanogenic bacteria can reduce carbon monoxide to methane:

$$CO + 3H_2 \rightarrow CH_4 + H_2O$$

A number of other C_1 compounds support the growth of methylotrophic bacteria. Many, but not all, methylotrophs are also methanotrophic. Both types of bacteria are widespread in the environment because these C_1 compounds are ubiquitous metabolites. In response to their presence, microbes have evolved the capacity to metabolize them under either aerobic or anaerobic conditions.

NITROGEN CYCLE

In contrast to carbon, elements such as nitrogen, sulfur, and iron are taken up in the form of mineral salts and cycle oxidoreductively. For example, nitrogen can exist in numerous oxidation states, from 23 in ammonium (NH_4^+) to 15 in nitrate (NO_3^-). These element cycles are referred to as the mineral cycles. The best studied and most complex of the mineral cycles is the nitrogen cycle. There is great interest in the nitrogen cycle because nitrogen is the mineral nutrient most in demand by microorganisms and plants. It is the fourth most common element found in cells, making up approximately 12% of cell dry weight, and includes the microbially catalyzed processes of nitrogen fixation, ammonium oxidation (aerobic nitrification and anaerobic anammox), assimilatory and dissimilatory nitrate reduction, ammonification, and ammonium assimilation. Similar to the carbon cycle, the global nitrogen cycle is currently undergoing major changes due to the ever-increasing demand for nitrogen in both agriculture and industry, fossil fuel burning and land use changes.

NITROGEN RESERVOIRS

Nitrogen in the form of the inert gas, dinitrogen (N_2), has accumulated in Earth's atmosphere since the planet was formed. Nitrogen gas is continually released into the atmosphere from volcanic and hydrothermal eruptions, and is one of the major global reservoirs of nitrogen. A second major reservoir is the nitrogen that is found in Earth's crust as bound, non-exchangeable ammonium. Neither of these reservoirs is actively cycled; the nitrogen in Earth's crust is unavailable and the N_2 in the atmosphere must be fixed before it is available for biological use. Nitrogen fixation is an energy-intensive and relatively slow process carried out by a limited number of microorganisms. Consequently, the amount of fixed nitrogen available is a limiting factor in primary production, and the nitrogen cycle is closely coupled to the carbon cycle. The pool of fixed nitrogen available can be divided into small reservoirs including the organic nitrogen found in living biomass and in dead organic matter and soluble inorganic nitrogen salts.

These small reservoirs tend to be actively cycled, particularly because nitrogen is often a limiting nutrient in the environment. For example, soluble inorganic nitrogen salts in terrestrial environments can have turnover rates greater than once per day. Nitrogen in plant biomass turns over approximately once a year, and nitrogen in organic matter turns over once in several decades.

NITROGEN FIXATION

Ultimately, all fixed forms of nitrogen, NH_4, NO_3, and organic N, come from atmospheric N_2. Approximately two-thirds of the N_2 fixed annually is microbial, half from terrestrial environments, including both natural systems and managed agricultural systems, and half from marine ecosystems. The remaining third comes from the manufacture of fertilizers. Recall that nitrogen fixation is energy intensive whether microbial or manufactured, and as a result, fertilizer prices are tied to the price of fossil fuels. As fertilizers are expensive, management alternatives to fertilizer addition have become attractive. These include rotation between nitrogen fixing crops such as soybeans and non-fixing crops such as corn.

Nitrogen is fixed into ammonia (NH_3) by over 100 different free-living bacteria, both aerobic and anaerobic, as well as some actinomycetes and cyanobacteria. For example, *Azotobacter* (aerobic), *Beijerinckia* (aerobic), *Azospirillum* (facultative), and *Clostridium* (anaerobic) can all fix N_2. Because fixed nitrogen is required by all biological organisms, nitrogen-fixing organisms occur in most environmental niches. The amount of N_2 fixed in each niche depends on the environment. Free-living bacterial cells that are not in the vicinity of a plant root fix small amounts of nitrogen (1–2 kg N/ha/year). Bacterial cells associated with the nutrient-rich rhizosphere environment can fix larger amounts of N_2 (2–25 kg N/ha/year). Cyanobacteria are the predominant N_2-fixing organisms in aquatic environments, and because they are photosynthetic, N_2 fixation rates are one to two orders of magnitude higher than those for free living non-photosynthetic bacteria. An evolutionary strategy developed collaboratively by plants and microbes to increase N_2 fixation efficiency was to enter into a symbiotic or mutualistic relationship to maximize N_2 fixation. The best studied of these symbioses is the *Rhizobium*-legume relationship, which can increase N_2 fixation to 200–300 kg N/ha/year. This symbiosis irrevocably changes both the plant and the microbe involved but is beneficial to both organisms.

As various transformations of nitrogen are discussed in this section, the objective is to understand how they are interconnected and controlled. As already mentioned, N_2 fixation is limited to bacteria and is an energy-intensive process. Therefore, it does not make sense for a microbe to fix N_2 if sufficient amounts are present for growth. Thus, one control on this part of the nitrogen cycle is that ammonia, the end product of N_2 fixation, is an inhibitor for the N_2-fixation reaction. A second control in some situations is the presence of oxygen. Nitrogenase is extremely oxygen sensitive, and some free-living aerobic bacteria fix N_2 only at reduced oxygen tension. Other bacteria such as *Azotobacter* and *Beijerinckia* can fix N_2 at normal oxygen tension because they have developed mechanisms to protect the nitrogenase enzyme.

Ammonia Assimilation (Immobilization) and Ammonification (Mineralization)

The end product of N_2 fixation is ammonia. In the environment, there exists an equilibrium between ammonia (NH_3) and ammonium (NH_4) that is driven by pH. This equilibrium favors ammonium formation at acid or near neutral pH. Thus, it is generally the ammonium form that is assimilated by cells into amino acids to form proteins, cell wall components such as N-acetylmuramic acid and purines and pyrimidines to form nucleic acids. This process is known as ammonium assimilation or immobilization. Nitrogen can also be immobilized by the uptake and incorporation of nitrate into organic matter, a process known as assimilatory nitrate reduction.

Because nitrate must be reduced to ammonium before it is incorporated into organic molecules, most organisms prefer to take up nitrogen as ammonium if it is available. The process that reverses immobilization, the release of ammonia from dead and decaying cells, is called ammonification or ammonium mineralization. Both immobilization and mineralization of nitrogen occur under aerobic and anaerobic conditions.

AMMONIA ASSIMILATION (IMMOBILIZATION)

There are two pathways that microbes use to assimilate ammonium. The first is a reversible reaction that incorporates or removes ammonium from the amino acid glutamate. This reaction is driven by ammonium availability. At high ammonium concentrations (>0.1 mM or >0.5 mg N/kg soil), in the presence of reducing equivalents (reduced nicotinamide adenine dinucleotide phosphate, $NADPH_2$), ammonium is incorporated into α-ketoglutarate to form glutamate using the GOGAT pathway. However, in most soil and many aquatic environments, ammonium is present at low concentrations. Therefore, microbes have a second ammonium uptake pathway that is energy dependent. This reaction is driven by ATP and two enzymes, glutamine synthase and glutamate synthetase. The first step in this reaction adds ammonium to glutamate to form glutamine, and the second step transfers ammonium molecule from glutamine to α-ketoglutarate resulting in the formation of two glutamate molecules.

AMMONIFICATION (MINERALIZATION)

Mineralization reactions can also occur extracellularly. Microorganisms release a variety of extracellular enzymes that initiate degradation of plant polymers. Microorganisms also release a variety of enzymes including proteases, lysozymes, nucleases, and ureases that can initiate degradation of nitrogen-containing molecules found outside the cell including proteins, cell walls, nucleic acids, and urea. Some of these monomers are taken up by the cell and degraded further, but some are acted upon by extracellular enzymes to release ammonium directly into the environment.

Which of these two processes, immobilization or mineralization, predominates in the environment? This depends on whether nitrogen is the limiting nutrient. If nitrogen is limiting, then immobilization will become the more important process. For environments where nitrogen is not limiting, mineralization will predominate. Nitrogen limitation is dictated by the carbon/nitrogen (C/N) ratio in the environment. Generally, the C/N ratio required for bacteria is 4–5 and for fungi is 10. However, one must take into account that only approximately 40% of the carbon in organic matter is actually incorporated into cell mass (the rest is lost as carbon dioxide). Thus, the C/N ratio must be increased by a factor of 2.5 to account for the carbon lost as carbon dioxide during respiration. Note that nitrogen is cycled more efficiently than carbon and there are essentially no losses in its uptake. In fact, a C/N ratio of 20 is not only the theoretical balance point but also the practically observed one. When organic amendments with C/N ratios less than 20 are added to soil, net mineralization of ammonium occurs. In contrast, when organic amendments with C/N ratios greater than 20 are added, net immobilization occurs.

There are numerous possible fates for ammonium that is released into the environment as a result of ammonium mineralization. It can be taken up by plants or microorganisms and incorporated into living biomass, or it can become bound to non-living organic matter such as soil colloids or humus. In this capacity, ammonium adds to the cation-exchange capacity of the soil. Ammonium can become fixed inside clay minerals, which essentially trap the molecule and remove the ammonium from active cycling. Also, because ammonium is volatile, some mineralized ammonium can escape into the atmosphere. Finally, ammonium can be utilized by chemoautotrophic microbes in a process known as nitrification.

AMMONIUM OXIDATION

Nitrification

Nitrification is the microbially catalyzed conversion of ammonium to nitrate (Figure 1.2). This is predominantly an aerobic chemoautotrophic process, but some methylotrophs can use the methane monooxygenase enzyme to oxidize ammonium and a few heterotrophic fungi and bacteria can also perform this oxidation. The autotrophic nitrifiers are a closely related group of bacteria. The best studied nitrifiers are from the genus *Nitrosomonas*, which oxidizes ammonium to nitrite, and *Nitrobacter*, which oxidizes nitrite to nitrate. The oxidation of ammonium is shown in the below equation:

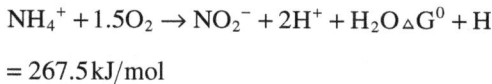

$$NH_4^+ + 1.5O_2 \rightarrow NO_2^- + 2H^+ + H_2O \, \Delta G^0 + H$$

$$= 267.5 \, kJ/mol$$

Nitrification is a two-step energy-producing reaction, and the energy produced is used to fix carbon dioxide. In the first step, 34 moles of ammonium are required

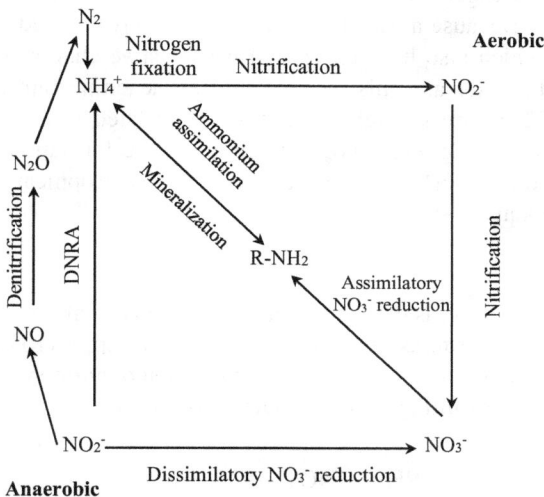

FIGURE 1.2   The nitrogen cycle.

to fix 1 mole of carbon dioxide which is mediated by the enzyme ammonium monooxygenase.

In the second step of nitrification, approximately 100 moles of nitrite are required to fix 1 mole of carbon dioxide as shown in the below equation:

$$NO_2^- + 0.5O_2 \rightarrow NO_3^-$$

$$\Delta G^0 = -87\,kJ/mol$$

These two types of nitrifiers, i.e., those that carry out the reactions shown in the above equations, are generally found together in the environment. As a result, nitrite does not normally accumulate in the environment. Nitrifiers are sensitive populations and the optimum pH for nitrification is 6.6–8.0. Nitrification rates are slowed in the environments with pH 6.0, and are completely inhibited when the pH drops below 4.5. Heterotrophic microbes that oxidize ammonium gain no energy from nitrification, so it is unclear why they carry out the reaction. Although the measured rates of autotrophic nitrification in the laboratory are an order of magnitude higher than those of heterotrophic nitrification, some data for acidic forest soils have indicated that heterotrophic nitrification may be more important in such environments (Pepper et al., 2014).

Nitrate does not normally accumulate in natural, undisturbed ecosystems. There are several reasons for this. One is that nitrifiers are sensitive to many environmental stresses. But perhaps the most important reason is that natural ecosystems do not have much excess ammonium. However, in agricultural systems that have large inputs of fertilizer, nitrification can become an important process resulting in the production of large amounts of nitrate. Other examples of managed systems that result in increased nitrogen inputs into the environment are feedlots, septic tanks, and landfills. The nitrogen released from these systems also becomes subject to nitrification processes. Because nitrate is an anion (negatively charged), it is very mobile in soil systems, which also have an overall net negative charge. Therefore, nitrate moves easily with water and this results in nitrate leaching into groundwater and surface waters. There are several health concerns related to high levels of nitrate in groundwater, including methemoglobinemia and the formation of nitrosamines. High levels of nitrate in surface waters can also lead to eutrophication and the degradation of surface aquatic systems.

Anammox

A relative recent discovery is that ammonium oxidation can also occur under anaerobic conditions using nitrite as the terminal electron acceptor (Kuenen, 2008). This process, known as anammox, is the bacterial oxidation of ammonium using nitrite as the electron acceptor resulting in the production of nitrogen:

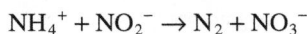

$$NH_4^+ + NO_2^- \rightarrow N_2 + NO_3^-$$

$$\Delta G^0 = -356\,kJ/mol$$

Anammox was first discovered in a bench-scale wastewater treatment reactor. Since then, it has been observed in a wide variety of both aquatic and terrestrial ecosystems with naturally low levels of oxygen. These include marine and freshwater sediments and anoxic water columns, anoxic terrestrial environments, as well as a number of managed systems including wastewater treatment plants, aquaculture, and landfill leachate treatment systems. All anammox bacteria identified are associated with the phylum Planctomycetes. This phylum is characterized by extremely slow growth rates and in addition has internal membrane bound structures. In anammox bacteria, one such structure is the "anammoxosome," the organelle where the ammonium oxidation and energy generation reactions take place. Anammox bacteria have not yet been isolated in pure culture. Anammox bacteria require habitats where both ammonium and nitrite are present. This occurs in the vicinity of aerobic–anaerobic interfaces. At these interfaces, ammonium is released both from the degradation of organic matter (either aerobically or anaerobically) and from dissimilatory nitrate reduction to ammonium (DNRA), an anaerobic process. Nitrite can be produced from nitrate reduction (anaerobic) and also from ammonium that has diffused to an oxic region and then undergone nitrification. Thus, competition for anammox substrates occurs at the aerobic-anaerobic interface making this an extremely complex activity.

An immediate and useful application for annamox that was recognized early is the removal of excess ammonium in wastewater treatment. Traditionally, this was done in a two-step process using nitrifying bacteria to oxidize the ammonium to nitrate under aerobic conditions and then switching to anoxic conditions to allow denitrification to reduce the nitrate to N_2. This is an energy-intensive process and can be replaced by an anammox process. Anammox requires that the first step of nitrification take place (using ammonia-oxidizing bacteria) converting some of the ammonium to nitrite. After the first step of nitrification occurs, anammox substrates, nitrite and ammonium, are all present. Anammox bacteria can then proceed (at approximately a 1:1 substrate ratio) resulting in the production of N_2. Research has shown that these steps can take place together in a reactor at one-third to one-half the cost of traditional ammonium removal.

A second discovery was made after anammox was first described. It is now thought that a significant portion (30%–50%) of the nitrogen loss in ocean environments that had been attributed to denitrification is actually due to anammox. In particular, in ocean zones where oxygen is less than 0.64 mg/L, regions termed oxygen minimum zones, anammox is now thought to dominate as the main cause of fixed nitrogen loss. This is of concern because scientists have suggested that ocean oxygen minimum zones are expanding as a result of climate change. This may in turn cause increased losses of fixed nitrogen through anammox and a reduction in this very important nutrient in ocean environments.

NITRATE REDUCTION

What are the possible fates of nitrate in the environment? Nitrate leaching into groundwater and surface waters is one possible fate. In addition, nitrate can be taken up and incorporated into living biomass by plants and microorganisms. The uptake of nitrate is followed by its reduction to ammonium, which is then incorporated into biomass. This process is called assimilatory nitrate reduction or nitrate immobilization. Finally, microorganisms can utilize nitrate as a terminal electron acceptor in anaerobic respiration to drive the oxidation of organic compounds. There are two separate pathways for this dissimilatory process, one called dissimilatory nitrate reduction to ammonium, where ammonium is the end product, and one called denitrification, where a mixture of gaseous products including N_2 and N_2O is formed.

ASSIMILATORY NITRATE REDUCTION

Assimilatory nitrate reduction refers to the uptake of nitrate, its reduction to ammonium, and its incorporation into biomass. Most microbes utilize ammonium preferentially, when it is present, to avoid having to reduce nitrate to ammonium, a process requiring energy. So, if ammonium is present in the environment, assimilatory nitrate reduction is suppressed. Oxygen does not inhibit this activity. In contrast to microbes, for plants that are actively photosynthesizing and producing energy, the uptake of nitrate for assimilation is less problematic in terms of energy. In fact, because nitrate is much more mobile than ammonium, it is possible that in the vicinity of the plant roots, nitrification of ammonium to nitrate makes nitrogen more available for plant uptake. Because this process incorporates nitrate into biomass, it is also known as nitrate immobilization.

DISSIMILATORY NITRATE REDUCTION

Dissimilatory Nitrate Reduction to Ammonium

There are two separate dissimilatory nitrate reduction processes both of which are used by facultative chemoheterotrophic organisms under microaerophilic or anaerobic conditions. The first process, called dissimilatory nitrate reduction to ammonium (DNRA), uses nitrate as a terminal electron acceptor to produce energy to drive the oxidation of organic compounds. The end product of DNRA is ammonium:

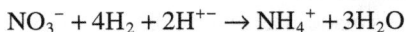

$$NO_3^- + 4H_2 + 2H^{+-} \rightarrow NH_4^+ + 3H_2O$$

$$\Delta G^0 = -603 \, kJ/8e^- \, transfer$$

The first step in this reaction, the reduction of nitrate to nitrite, is the energy-producing step. The further reduction of nitrite to ammonium is catalyzed by an NADH-dependent reductase. This second step provides no additional energy, but it does conserve fixed nitrogen and also regenerates reducing equivalents through the reoxidation of $NADH_2$ to NAD. These reducing equivalents are then used to help in the oxidation of carbon substrates. In fact, it has been demonstrated that under carbon-limiting conditions, nitrite accumulates (denitrification predominates), while under carbon-rich conditions, ammonium is the major product (DNRA predominates). A second environmental factor that selects for DNRA is low levels of available electron acceptors. It is therefore not surprising that this process is found predominantly in saturated, carbon-rich environments such as stagnant water, sewage sludge, some high organic matter sediments, and the rumen.

DENITRIFICATION

The second type of dissimilatory nitrate reduction is known as denitrification. Denitrification refers to the microbial reduction of nitrate, through various gaseous inorganic forms, to N_2. This is the primary type of dissimilatory nitrate reduction found in soil, and as such, is of concern because it cycles fixed nitrogen back into N_2. This process removes a limiting nutrient from the environment. Further, some of the gaseous intermediates formed during denitrification, e.g., nitrous oxide (N_2O), can cause depletion of the ozone layer and can also act as a greenhouse gas contributing to global warming. The overall reaction for denitrification is:

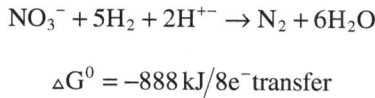

$$NO_3^- + 5H_2 + 2H^{+-} \rightarrow N_2 + 6H_2O$$

$$\Delta G^0 = -888\,kJ/8e^-\,transfer$$

Denitrification, when calculated in terms of energy produced for every eight-electron transfer, provides more energy per mole of nitrate reduced than DNRA. Thus, in a carbon-limited, electron acceptor-rich environment, denitrification will be the preferred process because it provides more energy than DNRA.

The first step, reduction of nitrate to nitrite, is catalyzed by the enzyme nitrate reductase. This is a membrane-bound molybdenum-iron-sulfur protein that is found not only in denitrifiers but also in DNRA organisms. Both the synthesis and the activity of nitrate reductase are inhibited by oxygen. Thus, both denitrification and DNRA are inhibited by oxygen. The second enzyme in this pathway is nitrite reductase, which catalyzes the conversion of nitrite to nitric oxide. Nitrite reductase is unique to denitrifying organisms and is not present in the DNRA process. It is found in the periplasm and exists in two forms, a copper-containing form and a heme form, both of which are distributed widely in the environment. Synthesis of nitrite reductase is inhibited by oxygen and induced by nitrate. Nitric oxide reductase, a membrane-bound protein, is the third enzyme in the pathway, catalyzing the conversion of nitric oxide to nitrous oxide. The synthesis of this enzyme is inhibited by oxygen and induced by various nitrogen oxide forms. Nitrous oxide reductase is the last enzyme in the pathway, and converts nitrous oxide to dinitrogen gas. This is a periplasmic copper-containing protein. The activity of the nitrous oxide reductase enzyme is inhibited by low pH, and is even more sensitive to oxygen than the other three enzymes in the denitrification pathway. Thus, nitrous oxide is the final product of denitrification under conditions of high oxygen (in a relative sense, given a microaerophilic niche) and low pH. In summary, both the synthesis and activity of denitrification enzymes are controlled by oxygen. Enzyme activity is more sensitive to oxygen than enzyme synthesis.

The amount of dissolved oxygen in equilibrium with water at 20°C and 1 atm pressure is 9.3 mg/L. However, as little as 0.5 mg/L or less inhibits the activity of denitrification enzymes. Therefore, nitrous oxide reductase is the most sensitive denitrification enzyme, and it is inhibited by dissolved oxygen concentrations of less than 0.2 mg/L. Whereas the denitrification pathway is very sensitive to oxygen, neither it nor the DNRA pathway is inhibited by ammonium as is the assimilatory nitrate reduction pathway. However, the initial nitrate level in an environmental system can help determine the extent of the denitrification pathway. Low nitrate levels tend to favor production of nitrous oxide as the end product. High nitrate levels favor production of N_2 gas, a much more desirable end product.

Organisms that denitrify are found widely in the environment and display a variety of different characteristics in terms of metabolism and activities. In contrast to DNRA organisms, which are predominantly heterotrophic using fermentative metabolism, the majority of denitrifiers are also heterotrophic, but use respiratory pathways of metabolism. However, some denitrifiers are autotrophic, some are fermentative, and some are associated with other aspects of the nitrogen cycle, for example, they can fix N_2.

SULFUR CYCLE

Sulfur is the tenth most abundant element in Earth's crust. It is an essential element for biological organisms, making up approximately 1% of the dry weight of a bacterial cell. Sulfur is not generally considered a limiting nutrient in the environment except in some intensive agricultural systems with high crop yields. Sulfur is cycled between oxidation states of +6 for sulfate (SO_4^{2-}) and −2 for sulfide (S^{2-}). In cells, sulfur is required for synthesis of the amino acids cysteine and methionine, and is also required for some vitamins, hormones, and coenzymes. In proteins, the sulfur-containing amino acid cysteine is especially important because the formation of disulfide bridges between cysteine residues helps govern protein folding and hence activity. All of these compounds contain sulfur in the reduced or sulfide form. Cells also contain organic sulfur compounds in which the sulfur is in the oxidized state. Examples of such compounds are glucose sulfate, choline sulfate, phenolic sulfate, and two ATP-sulfate compounds that are required in sulfate assimilation, and can also serve to store sulfur for the cell. Although the sulfur cycle is not as complex as the nitrogen cycle, the global impacts of the sulfur cycle are extremely important, including the formation of acid rain, acid mine drainage, and corrosion of concrete and metal.

SULFUR RESERVOIRS

Sulfur is outgassed from Earth's core through volcanic activity. The sulfur gases released, primarily sulfur dioxide (SO_2) and hydrogen sulfide (H_2S), become dissolved in the ocean and aquifers. Here, the hydrogen sulfide forms sparingly soluble metal sulfides, mainly iron sulfide (pyrite), and sulfur dioxide forms metal sulfates with calcium, barium, and strontium as shown in the below equations:

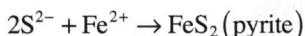

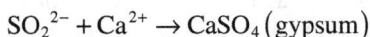

$$2S^{2-} + Fe^{2+} \rightarrow FeS_2 \,(\text{pyrite})$$

$$SO_2^{2-} + Ca^{2+} \rightarrow CaSO_4 \,(\text{gypsum})$$

This results in a substantial portion of the outgassed sulfur being converted to rock. Some of the gaseous sulfur compounds find their way into the upper reaches of the ocean and the soil. In these environments, microbes take up and cycle the sulfur. Finally, the small portions of these gases that remain after precipitating and cycling find their way into the atmosphere. Here, they are oxidized to the water-soluble sulfate form, which is washed out of the atmosphere by rain. Thus, the atmosphere is a relatively small reservoir of sulfur. Of the sulfur found in the atmosphere, the majority is found as sulfur dioxide. Currently, one-third to one-half of the sulfur dioxide emitted to the atmosphere is from industrial and automobile emissions due to the burning of fossil fuels. A smaller portion of the sulfur in the atmosphere is present as hydrogen sulfide, and is biological in origin.

The largest reservoir of sulfur is found in Earth's crust and is composed of inert elemental sulfur deposits, sulfur-metal precipitates such as pyrite (FeS_2) and gypsum ($CaSO_4$), as well as sulfur associated with buried fossil fuels. A second large reservoir that is slowly cycled is the sulfate found in the ocean, where it is the second most common anion (Dobrovolsky, 1994). Smaller and more actively cycled reservoirs of sulfur include sulfur found in biomass and organic matter found in the terrestrial and ocean environments. Two recent practices have caused a disturbance in the global sulfur reservoirs. The first is strip mining, which has exposed large areas of metal sulfide ores to the atmosphere, resulting in the formation of acid mine drainage. The second is the burning of fossil fuels, a sulfur reservoir that was quite inert until recently. This has resulted in sulfur dioxide emissions into the atmosphere with the resultant formation of acid rain (Figure 1.3).

ASSIMILATORY SULFATE REDUCTION AND SULFUR MINERALIZATION

The primary soluble form of inorganic sulfur found in soil is sulfate. Whereas plants and most microorganisms incorporate reduced sulfur (sulfide) into amino acids or other sulfur-requiring molecules, they take up sulfur in the oxidized sulfate form and then reduce it internally (Widdel, 1988). This is called assimilatory sulfate reduction.

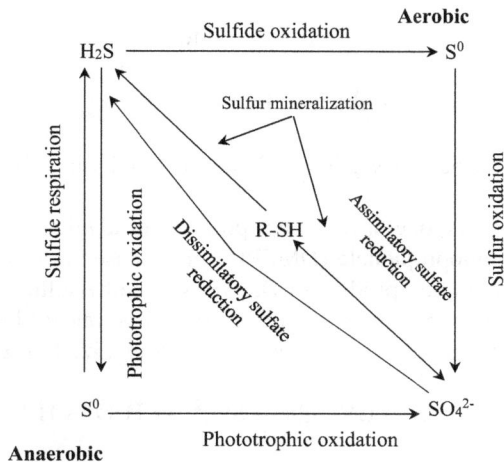

FIGURE 1.3   The sulfur cycle.

Cells assimilate sulfur in the form of sulfate because it is the most available sulfur form and also because sulfide is toxic. Sulfide toxicity occurs inside the cell when sulfide reacts with metals in cytochromes to form metal sulfide precipitates, destroying cytochrome activity. However, under the controlled conditions of sulfate reduction inside the cell, the sulfide can be removed immediately and incorporated into an organic form. Although this process does protect the cell from harmful effects of the sulfide, it is an energy-consuming reaction. After sulfate is transported inside the cell, ATP is used to convert the sulfate into the energy-rich molecule adenosine 5'-phosphosulfate (APS) (Eq. 1.1). A second ATP molecule is used to transform APS to 3'-phosphoaden osine-5'-phosphosulfate (PAPS) (Eq. 1.2). This allows the sulfate to be reduced to sulfite and then to sulfide in two steps (Eqs. 1.3 and 1.4). Most commonly, the amino acid serine is used to remove sulfide as it is reduced, forming the sulfur-containing amino acid cysteine. The release of sulfur from organic forms is called sulfur mineralization. The release of sulfur from organic molecules occurs under both aerobic and anaerobic conditions. The enzyme serine sulfhydrylase can remove sulfide from cysteine in the reverse of the reaction shown in Eq. (1.5), or a second enzyme, cysteine sulfhydrylase, can remove both sulfide and ammonia as shown in Eq. (1.6):

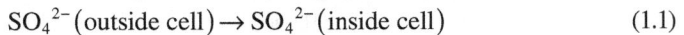

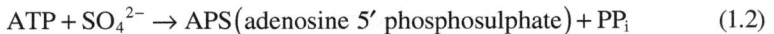

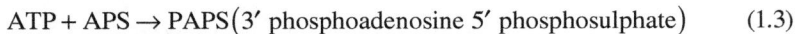

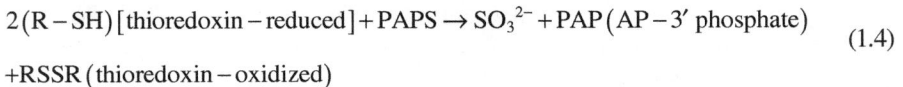

$$SO_4^{2-} \text{(outside cell)} \rightarrow SO_4^{2-} \text{(inside cell)} \tag{1.1}$$

$$ATP + SO_4^{2-} \rightarrow APS \text{(adenosine 5' phosphosulphate)} + PP_i \tag{1.2}$$

$$ATP + APS \rightarrow PAPS \text{(3' phosphoadenosine 5' phosphosulphate)} \tag{1.3}$$

$$2(R-SH) \text{[thioredoxin - reduced]} + PAPS \rightarrow SO_3^{2-} + PAP \text{(AP-3' phosphate)}$$
$$+ RSSR \text{(thioredoxin - oxidized)} \tag{1.4}$$

$$SO_3^{2-} + 3NADPH \rightarrow S^{2-} + 3NADP^+ \tag{1.5}$$

$$\text{O-acetyl L-serine} + S^{2-} \rightarrow \text{L-cysteine} + \text{acetate} + H_2O \tag{1.6}$$

Processing of Sulfate for Uptake into Bacteria

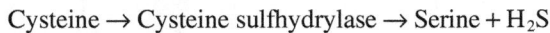

$$\text{Cysteine} \rightarrow \text{Cysteine sulfhydrylase} \rightarrow \text{Serine} + H_2S$$

In marine environments, one of the major products of algal metabolism is the compound dimethylsulfoniopropionate (DMSP), which is used in osmoregulation of the cell. The major degradation product of DMSP is dimethylsulfide (DMS). Both H_2S and DMS are volatile compounds and therefore can be released to the atmosphere. Once in the atmosphere, these compounds are photooxidized to sulfate:

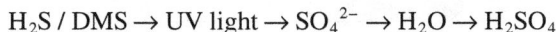

$$H_2S \text{ / DMS} \rightarrow \text{UV light} \rightarrow SO_4^{2-} \rightarrow H_2O \rightarrow H_2SO_4$$

Normal biological release of reduced volatile sulfur compounds results in the formation of approximately $1\,kg\ SO_4^{2-}$/ha/year. The use of fossil fuels, which all contain organic sulfur compounds, increases the amount of sulfur released to the atmosphere to up to 100 kg SO_4^2/ha/year in some urban areas. Exacerbating this problem is the fact that reserves of fossil fuels that are low in sulfur are shrinking, forcing the use of reserves with higher sulfur content. Burning of fossil fuels produces sulfite as shown in the following equation:

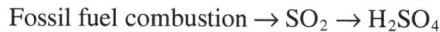

$$\text{Fossil fuel combustion} \rightarrow SO_2 \rightarrow H_2SO_4$$

Thus, increased emission of sulfur compounds to the atmosphere results in the formation of sulfur acid compounds. These acidic compounds dissolve in rainwater and can decrease the rainwater pH from neutral to as low as pH 3.5, a process also known as the formation of acid rain. Acid rain damages plant foliage, causes corrosion of stone and concrete building surfaces, and can affect weakly buffered soils and lakes.

Sulfur Oxidation

In the presence of oxygen, reduced sulfur compounds can support the growth of a group of chemoautotrophic bacteria under strictly aerobic conditions and a group of photoautotrophic bacteria under strictly anaerobic conditions. In addition, a number of aerobic heterotrophic microbes, including both bacteria and fungi, oxidize sulfur to thiosulfate or to sulfate. The heterotrophic sulfur oxidation pathway is still unclear, but apparently no energy is obtained in this process. Chemoautotrophs are considered the predominant sulfur oxidizers in most environments. However, because many chemoautotrophic sulfur oxidizers require a low pH for optimal activity, heterotrophs may be more important in some aerobic, neutral to alkaline soils. Further, heterotrophs may initiate sulfur oxidation, resulting in a lowered pH that is more amenable for chemoautotrophic activity.

Chemoautotrophic Sulfur Oxidation

Of the chemoautotrophs, most oxidize sulfide to elemental sulfur, which is then deposited inside the cell as characteristic granules:

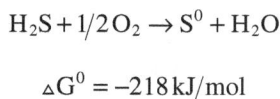

$$H_2S + 1/2 O_2 \rightarrow S^0 + H_2O$$

$$\Delta G^0 = -218 \, kJ/mol$$

The energy provided by this oxidation is used to fix CO_2 for cell growth. In examining this equation, it is apparent that these organisms require both oxygen and sulfide. However, reduced compounds are generally abundant in areas that contain little or no oxygen. So, these microbes are microaerophilic; they grow best under conditions of low oxygen tension. Characteristics of marsh sediments that contain these organisms are their black color due to sulfur deposits and their "rotten egg" smell due to the presence of H_2S. Most of these organisms are filamentous and can easily be observed by examining a marsh sediment under the microscope and looking for small white filaments. Some chemoautotrophs, most notably *Acidithiobacillus thiooxidans*, can oxidize elemental sulfur:

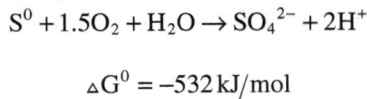

$$S^0 + 1.5O_2 + H_2O \rightarrow SO_4^{2-} + 2H^+$$

$$\Delta G^0 = -532\,\text{kJ/mol}$$

This reaction produces acid, and as a result, *A. thiooxidans* is extremely acid tolerant with an optimal growth pH of 2. It should be noted that there are various *Acidithiobacillus* species, and these vary widely in their acid tolerance (Baker and Banfield, 2003). However, the activity of *A. thiooxidans* in conjunction with the iron-oxidizing, acid-tolerant, chemoautotroph *Acidithiobacillus ferrooxidans* is responsible for the formation of acid mine drainage, an undesirable consequence of sulfur cycle activity. It should be noted that the same organisms can be harnessed for the acid leaching and recovery of precious metals from low-grade ore, also known as biometallurgy. Thus, depending on one's perspective, these organisms can be very harmful or very helpful.

Although most of the sulfur-oxidizing chemoautotrophs are obligate aerobes, there is one exception, *Acidithiobacillus denitrificans*, a facultative anaerobic organism that can substitute nitrate as a terminal electron acceptor for oxygen as shown in the following equation:

$$S^0 + NO_3^- + CaCO_3 \rightarrow CaSO_4 + N_2$$

In the above equation, the sulfate formed is shown as precipitating with calcium to form gypsum. *A. denitrificans* is not acid tolerant, and has an optimal pH for growth of 7.0.

Photoautotrophic Sulfur Oxidation

Photoautotrophic oxidation of sulfur is limited to green and purple sulfur bacteria. This group of bacteria evolved on early Earth when the atmosphere contained no oxygen. These microbes fix carbon using light energy, but instead of oxidizing water to oxygen, they use an analogous oxidization of sulfide to sulfur:

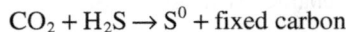

$$CO_2 + H_2S \rightarrow S^0 + \text{fixed carbon}$$

These organisms are found in mud and stagnant water, sulfur springs, and saline lakes. In each of these environments, both sulfide and light must be present. Although the contribution to primary productivity is small in comparison with aerobic photosynthesis, these organisms are important in the sulfur cycle. They serve to remove sulfide from the surrounding environment, effectively preventing its movement into the atmosphere and its precipitation as metal sulfide.

SULFUR REDUCTION

There are three types of sulfur reduction (Widdel, 1988). Assimilatory sulfate reduction occurs under either aerobic or anaerobic conditions. In contrast, there are two dissimilatory pathways, both of which use an inorganic form of sulfur as a terminal electron acceptor. In this case, sulfur reduction occurs only under anaerobic conditions. The two types of sulfur that can be used as terminal electron acceptors are elemental sulfur and sulfate. These two types of metabolism are differentiated as sulfur respiration and dissimilatory sulfate reduction. *Desulfuromonas acetoxidans* is an example of a bacterium that grows on small carbon compounds such as acetate, ethanol, and propanol, and it uses elemental sulfur as the terminal electron acceptor as shown in the below equation:

$$CH_3COOH + 2H_2O + 4S^0 \rightarrow 2CO_2 + 4S^{2-} + 8H^+$$

However, the use of sulfate as a terminal electron acceptor seems to be the more important environmental process. The following genera, all of which utilize sulfate as a terminal electron acceptor, are found widely distributed in the environment, especially in anaerobic sediments of aquatic environments, water-saturated soils, and animal intestines: *Desulfobacter, Desulfobulbus, Desulfococcus, Desulfonema, Desulfosarcina, Desulfotomaculum,* and *Desulfovibrio.* Together these organisms are known as the sulfate-reducing bacteria (SRB). They can utilize H_2 as an electron donor to drive the reduction of sulfate as shown in the below equation:

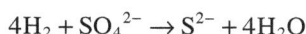

$$4H_2 + SO_4{}^{2-} \rightarrow S^{2-} + 4H_2O$$

Thus, SRB compete for available H_2 in the environment, as H_2 is also the electron donor required by methanogens. It should be noted that this is not usually a chemoautotrophic process because most SRB cannot fix carbon dioxide. Instead, they obtain carbon from low molecular weight compounds such as acetate or methanol. The overall reaction for utilization of methanol is shown in the below equation:

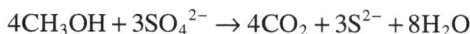

$$4CH_3OH + 3SO_4{}^{2-} \rightarrow 4CO_2 + 3S^{2-} + 8H_2O$$

Both sulfur and sulfate reducers are strict anaerobic chemoheterotrophic organisms that prefer small carbon substrates such as acetate, lactate, pyruvate, and low molecular-weight alcohols. Where do these small carbon compounds come from in the environment? They are byproducts of fermentation of plant and microbial biomass that occurs in anaerobic regions. Thus, the sulfate reducers are part of an anaerobic consortium of bacteria including fermenters, sulfate reducers, and methanogens, which together act to completely mineralize organic compounds to carbon dioxide and methane. More recently, it has been found that some SRB can also metabolize more complex carbon compounds including some aromatic compounds and some longer chain fatty acids. These organisms are being looked at closely to determine whether they can be used in remediation of contaminated sites that are highly anaerobic and that would be difficult to oxygenate.

The end product of sulfate reduction is hydrogen sulfide. It can be taken up by chemoautotrophs or photoautotrophs and reoxidized; it can be volatilized into the atmosphere; or it can react with metals to form metal sulfides. In fact, the activity of sulfate reducers and the production of hydrogen sulfide are responsible for the corrosion of underground metal pipes. In this process, the hydrogen sulfide produced reacts with ferrous iron metal to make more iron sulfide.

IRON CYCLE

IRON RESERVOIRS

Iron is the fourth most abundant element in Earth's crust. Iron generally exists in three oxidation states: 0, 12, and 13 corresponding to metallic iron (Fe^0), ferrous iron (Fe^{2+}), and ferric iron (Fe^{3+}). In the environment, iron is actively cycled between the 12 and 13 forms. Under aerobic conditions, iron is usually found in its most oxidized form (Fe^{3+}), which has low aqueous solubility. Under reducing or anaerobic conditions, Fe^{3+} is reduced to the ferrous form, Fe^{2+}, which has higher solubility. Iron is an essential but minor element for biological organisms, making up approximately 0.2% of the dry weight of a bacterial cell. Although the amount of iron in a cell is low, it has a very important function as a part of enzymes that are used in respiration and photosynthesis, both processes that require electron transfer.

Iron in Soils and Sediments

Iron is generally not a limiting nutrient in soil due to its high abundance in Earth's crust. However, even though iron abundance is high, the bioavailability of most iron minerals is quite limited. Thus, microorganisms have developed strategies to obtain iron from its mineral form, usually from iron oxides or iron oxyhydroxides. The best-studied strategy is the use of iron chelators known as siderophores. Siderophores are synthesized and released from the cell, where they bind Fe^{3+}, which helps keep this low solubility form of iron in solution. The soluble iron-siderophore complex is recognized by and binds to siderophore-specific receptors on the cell surface. The iron is released from the complex, reduced form and taken up into the cell as Fe^{2+}, its more soluble form.

Iron in Marine Environments

Unlike terrestrial and sediment environments, iron is considered a limiting nutrient in the modern marine environment (Raiswell, 2006). In fact, the extent of this limitation has been the focus of a vigorous debate in recent years. Recent data seem to indicate that for one-third of Earth's oceans, those that are more nutrient rich and support higher numbers of phytoplankton, iron is the limiting nutrient to growth (Boyd et al., 2007). How does iron enter the ocean environment? The largest flux of iron entering the oceans is fluvial in nature, meaning that it enters as suspended particles or dissolved iron carried by rivers, or is associated with glacial sediments (Jickells et al., 2005). For the most part, this iron is deposited in the sediments of near coastal areas, and does not reach the open ocean. Therefore, in the open ocean, the major pathway of iron entry is eolian or as dust that is carried through the atmosphere mainly from desert and other arid environment land surfaces.

IRON OXIDATION

Chemoautotrophs

Under aerobic conditions, ferrous iron tends to oxidize to the ferric form. Ferrous iron will autoxidize or spontaneously oxidize under aerobic conditions at pH > 5.

Iron oxidation also occurs biotically. In fact, reduced iron is an important source of energy for several specialized genera of chemoautotrophic bacteria, the iron oxidizers (Ehrlich, 1996):

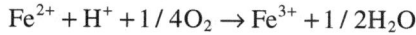

$$Fe^{2+} + H^+ + 1/4O_2 \rightarrow Fe^{3+} + 1/2H_2O$$

$$\Delta G^0 = -40\,kJ/mol$$

Note that compared to ammonia or sulfur oxidation, the yield of energy in iron oxidation is quite low. Nevertheless, this is an exploitable niche and an important biological reaction, which can have considerable environmental consequences. Iron oxidation is most often associated with acidic environments. For example, the acidophilic thermophile *Sulfolobus* is an archaean that was isolated from acidic hot springs. But iron oxidation is perhaps best studied in association with acid mine drainage and the acidophilic bacterium *A. ferrooxidans*. Interestingly, although *A. ferrooxidans* is best studied, nonculture-based analysis suggests that other iron oxidizers may actually play more important roles in the creation of acid conditions in mine tailings and the formation of acid mine drainage (Baker and Banfield, 2003). These include Bacteria such as *Leptospirillum ferrooxidans* and *Leptospirillum ferriphilum, Sulfobacillus acidophilus* and *Sulfobacillus thermosulfidooxidans*, and *A. ferrooxidans*, as well as some Archaea, e.g., *Ferroplasma acidiphilum*.

Iron oxidation has also been described in several marine genera at neutral pH. This process is problematic at neutral pH because (1) the energy yield is very low and (2) spontaneous oxidation of Fe^{2+} to Fe^{3+} (in the form of iron oxyhydroxides) occurs rapidly in the presence of oxygen and competes with the biological reaction. Neutrophilic iron oxidizers overcome these problems with very specific niche requirements. They position themselves in regions with low O_2 tension and high and constant Fe^{2+} concentrations. One marine environment that meets these requirements occurs in hydrothermal vents on the sea floor. Here, anoxic vent fluids charged with Fe^{2+} rapidly come in contact with the cold, oxygenated ocean water. Similar conditions can occur in municipal and industrial water pipelines.

The best-studied neutrophilic iron oxidizers are *Gallionella* and *Leptothrix*. *Gallionella* is capable of chemoautotrophic growth using Fe^{2+} as sole energy source under microaerobic conditions. *Leptothrix* is a sheath forming chemoheterotrophic organism that oxidizes both Fe^{2+} and Mn^{2+}, depositing an iron–manganese encrusted coating on its sheath. These microbes can deposit copious amounts of iron (and manganese) minerals on their surfaces. This can have serious economic consequences in some instances. For example, this process can cause extensive biofouling and corrosion of water pipelines.

Photoautotrophs

As shown below, some members of the purple and green bacteria can use Fe^{2+} as an electron donor to carry out anaerobic photosynthesis coupled to photoautotrophic growth (i.e., growth involving photosynthesis and oxidation):

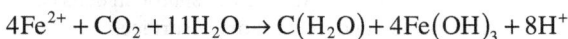

$$4Fe^{2+} + CO_2 + 11H_2O \rightarrow C(H_2O) + 4Fe(OH)_3 + 8H^+$$

It has been proposed that iron-based photosynthesis represents the transition between anaerobic photosynthesis that developed on early Earth and aerobic photosynthesis that began approximately 2 billion years ago (Jiao and Newman, 2007). In fact, it is thought that Fe^{2+} was the most widespread source of reducing power from 1.6 to 3.8 billion years ago, where under reducing conditions, Fe^{2+} was favored over Fe^{3+}. Under these conditions, the amount of iron in the marine environment was considerably higher (0.1–1 mmol/L) than it is today (0.03–1 nmol/L) (Bosak et al., 2007).

IRON REDUCTION

Iron is microbially reduced for two purposes, assimilation and energy generation. Assimilatory iron reduction is the reduction of Fe^{3+} for uptake and incorporation into cell constituents. This usually involves the release of siderophores, which complex Fe^{3+} in the environment exterior to the cell. The iron–siderophore complex then delivers Fe^{3+} to the cell, which is reduced to Fe^{2+} as it is taken up. Dissimilatory iron reduction or iron respiration is the use of Fe^{3+} as a terminal electron acceptor for the purpose of energy generation during anaerobic respiration. Due to its abundance in Earth's crust, Fe^{3+} found in iron oxides and oxyhydroxides serves as an important terminal electron acceptor for anaerobic heterotrophic bacteria. Because iron respiration has been an important activity during the evolution of Earth, there is a wide diversity among the bacteria and archaeans capable of carrying out this activity. The problem for the iron reducers is that most of the iron in environment is relatively unavailable. So, microorganisms have developed some very interesting strategies to solve this problem.

As shown in Figure 1.4 a and b, the first is to make direct contact with an iron oxide surface. In this case, the iron reductase is a membrane-bound enzyme allowing direct access of the enzyme with the substrate. A second strategy is to use an electron shuttle that can act as an intermediate in transferring electrons from the cell to the iron oxide surface. Possible electron shuttles in the environment include humic acids or other molecules that contain quinone-like structures that are reduced to a corresponding hydroquinone form. Humic acids are considered an exogenous electron shuttle source or one that is obtained from the environment outside the cell. Alternatively, some

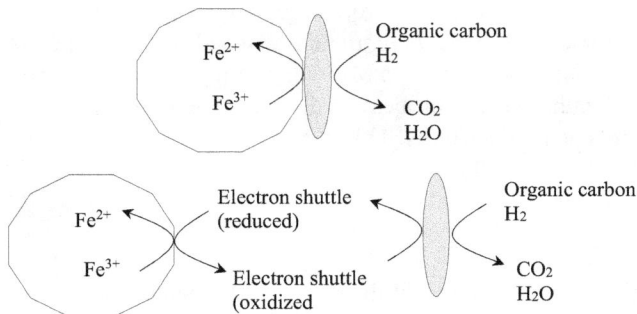

FIGURE 1.4   (a) Strategies used in iron respiration to facilitate electron transfer between microbial cells and iron oxide surfaces. (b) An electron shuttle in cells can be used to mediate the electron transfer between the cell and the iron oxide surface.

bacteria can make and release endogenous electron shuttles. For example, *Geothrix fermentans* releases a quinone-like electron shuttle during growth on lactate (electron donor) and poorly crystalline iron oxides (electron acceptor) (Nevin and Lovley, 2002).

CYCLING OF MANGANESE AND SELENIUM

Some chemolithotrophic bacteria can obtain energy from the reduction of Mn^{4+} to Mn^{2+} or from the oxidation of Mn^{2+} to Mn^{4+}. The oxidation of manganese by bacteria is associated with three different activities: (1) oxidation of soluble manganese, (2) oxidation of immobilized manganese, and (3) production of hydrogen peroxide resulting in the oxidation of manganese. Oxidation of soluble Mn^{2+} has been attributed to enzymes produced by *Leptothrix, Pseudomonas, Citrobacter, Bacillus, Arthrobacter,* and *Hypomicrobium.*

If electrons from the oxidation of soluble manganese are shuttled to molecular oxygen, cells can grow because sufficient energy is released (Figure 1.5). A second mechanism for manganese oxidation occurs if manganese is bound onto clay or insoluble manganese oxides. The enzyme associated with this activity is considered to be at the surface of the cell and bacteria that may be associated with this activity, including *Vibrio, Arthrobacter,* and *Oceanospirillum:*

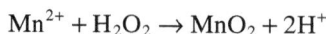

$$Mn^{2+} + H_2O_2 \rightarrow MnO_2 + 2H^+$$

Manganese oxides may consist of manganese dioxide (MnO_2), hausmannite (Mn_3O_4), or manganite (MnOOH), and microbial oxidation occurs more rapidly with amorphous manganese oxides than with crystalline forms. While several genera of bacteria are known to reduce manganese, the systems of *Geobacter* and *Shewanella* are highly characterized. Several additional reactions can account for the reduction of manganese; these may be attributed to microbial activity. Microorganisms may produce formic acid, H_2S, or Fe^{2+}, and these chemicals will react with MnO_2 to produce Mn^{2+}. This underscores the potential for interactions between biogeochemical cycles.

As a result of bacteria-oxidizing manganese, there are two important environmental observations: the formation of desert varnish (Kuhlman et al., 2006) and manganese nodules on the ocean floor (Hlawatsch et al., 2002). The manganese nodules may reach a diameter of over 25 cm, and because they frequently contain oxides of iron as well as manganese, they are referred to as *ferromanganese nodules.* The formation

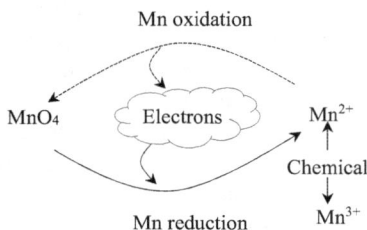

FIGURE 1.5 Microbial interactions with manganese (Barton, 2005).

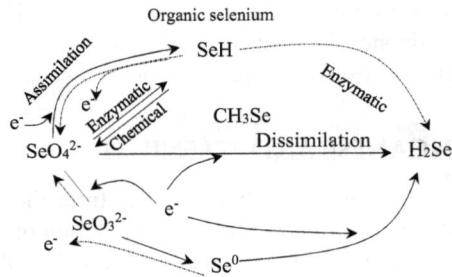

FIGURE 1.6 The selenium cycle.

of manganese nodules from marine waters is attributed to manganese-oxidizing bacteria and bacteria secreting a carbohydrate matrix to immobilize the manganese.

In terms of chemistry, selenium is similar to sulfur, with oxidation states ranging from -2 to $+6$ (see Figure 1.6). Selenium is required by microorganisms at relatively low levels for the formation of selenomethionine (Se^{2-}), where selenium replaces sulfur in methionine. Several dehydrogenases require selenomethionine for enzymatic activity. By nature, selenium is commonly found in aquatic systems as selenate (SeO^{2-}_4) or selenite (SeO^{2-}_3), and bacteria have specific uptake transport systems for these two selenium anions. At elevated levels, selenium is toxic to cells, and many environmental bacteria will reduce selenite to either elemental selenium (Se^0) or methylselenide (Se^{2-}) (Oremland et al., 2004). Thus, microbial activities contribute to the global cycling of selenium.

CYCLING OF HYDROGEN

Although little consideration is given to an official hydrogen cycle, the role of hydrogen in microbial systems is undisputed, and for this reason, there is merit in discussing the cycling of hydrogen in the microbial world. The principal reservoir for hydrogen is water with ionization of water, yielding protons plus a hydroxyl ion:

$$H_2O \leftrightarrow OH^- + H^+$$

Near neutrality, there is little ionization with few protons; however, if those protons are used in reactions, there is a continuous shift in equilibrium of the reaction to promote continuous ionization. Protons are extremely important in bacterial metabolism because respiratory processes pump protons out of the cell across the plasma membrane. A separation of protons and hydroxyl ions across the plasma contributes to the charge that may vary from -100 to $-250\,mV$. Proton movement across the membrane of bacteria is important for ATP synthesis, flagellar movement, nutrient uptake, and solute export. Water is the electron donor in aerobic photosynthesis, and as indicated in the following reaction, protons are released:

$$H_2O \rightarrow O_2 + 2H^+ + 2e^-$$

Through appropriate hydrogen carriers, these protons are transferred to CO_2 in the formation of carbohydrates. As various organic compounds are produced from carbohydrates, hydrogen atoms are a part of the building structures of cells. When cells use sugars for energy with the release of CO_2, hydrogen is released as protons, and these protons unite with O_2 to produce water.

There is another microbial activity involving hydrogen cycling, and it is associated with hydrogenase. Anaerobic heterotrophic bacteria will produce a variety of end products of fermentation, and H_2 is one of these. The formation of H_2 is a highly efficient means for bacteria to release electrons, and this reaction is catalyzed by hydrogenase:

$$2^+ + 2e^- \rightarrow H_2$$

Hydrogen gas could be released from the fermentation mixture as bubbles; however, many anaerobic bacteria can use hydrogen as an electron donor to energize growth. This uptake of hydrogen is also catalyzed by hydrogenase to produce the following reaction:

$$H_2 \rightarrow 2H^+ + 2e^-$$

With the same enzyme involved in two different activities, the direction of the hydrogenase reaction will be regulated by thermodynamics of the reaction concerned with the release or consumption of electrons. To underscore the role that hydrogen plays in microbial energetics, there are two separate reactions involving anaerobic systems. There is the interspecies H_2 transfer that occurs between a producer of H_2 and a consumer of H_2. An example of partners in hydrogen syntrophism would be *Syntrophomonas wolfei*, a H_2 producer, and a methanogen, a H_2 consumer. In this microbial partnership, consumption of H_2 by a methanogen enables *S. wolfei* to grow from fermentation of crotonic acid even though the thermodynamics of this reaction is unfavorable. Hydrogen cycling has been proposed to function in certain strains of SRB where protons from the cytoplasm are converted to H_2 in the periplasm and membrane hydrogenase directs protons to the cytoplasm (Odom and Peck, 1981). The hydrogen cycling in certain *Desulfovibrio* would benefit the bacterium by providing energy to the cell. There is also the production of H_2 by some nitrogen-fixing bacteria in an apparent release of excessive reducing activity.

TRANSFORMATION OF MERCURY

One of the most toxic metals is mercury, and microorganisms have unique processes that enable cells to grow in mercury-contaminated environments. As shown in Figure 1.7, mercury is moved through the biosphere, and some of these transformations are attributed to microorganisms.

Aerobic bacteria, anaerobic bacteria, and a few fungi convert Hg^{2+} to methyl mercury or dimethyl mercury. In aquatic environments, the concentrations of methyl mercury and Hg^{2+} are amplified as they move through the food chain, with concentration of mercury in fish exceeding safe limits of humans. Both methyl mercury and

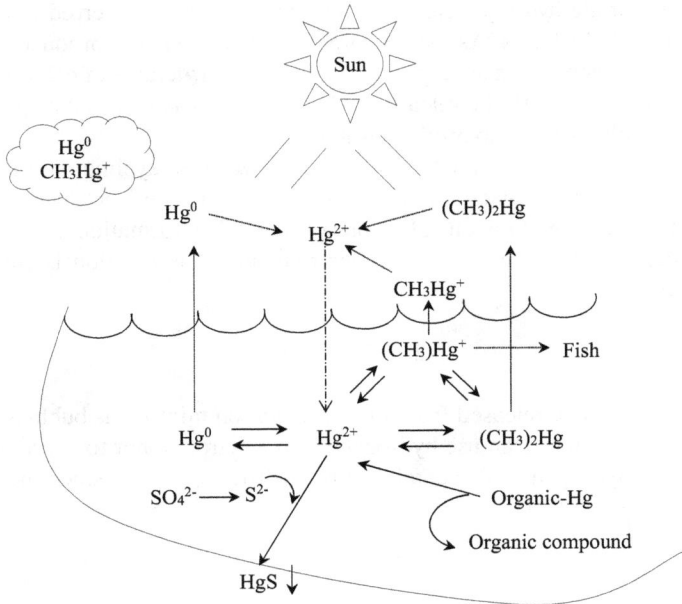

FIGURE 1.7   Role of microorganisms in global cycling of mercury (Silver and Misra, 1988).

dimethyl mercury may be released from the immediate environment into the atmosphere because of the high volatility of these compounds. Various Gram-positive and Gram-negative bacteria are capable of reducing Hg^{2+} to elemental mercury (Hg^0) using the *mer* operon. Mercury metal (Hg^0) is unusual because it is volatile at room temperature. Solar-mediated activities in the atmosphere are responsible for oxidization of Hg^0 to Hg^{2+} and conversion of methylated forms of mercury to Hg^{2+}. In addition to mercury entering the atmosphere from microbial activity, over 50,000 tons of Hg^0 and methyl mercury are released from coal-burning factories. The *mer* operon also encodes a mercury lyase that releases Hg^{2+} from organic mercury compounds.

REFERENCES

Baker BJ, Banfield JF. Microbial communities in acid mine drainage. *FEMS Microbiol. Ecol.* 2003;44:139–152.

Barton LL. *Structural and Functional Relationships in Prokaryotes.* New York: Springer; 2005.

Bosak T, Greene SE, Newman DK. A likely role for anoxygenic photosynthetic microbes in the formation of ancient stromatolites. *Geobiology.* 2007;5:119–126.

Boyd PW, Jickells T, Law CS, et al. Mesoscale iron enrichment experiments 1993–2005: synthesis and future directions. *Science.* 2007;315:612–617.

Dobrovolsky VV. *Biogeochemistry of the World's Land.* Boca Raton, FL: CRC Press; 1994.

Ehrlich HL. *Geomicrobiology.* New York: Marcel Dekker; 1996.

Hlawatsch S, Neumann T, Van Den Berg CMG, Kersten M, Harff J, Suess E. Fast-growing, shallow-water ferro-manganese nodules from the western Baltic Sea origin and modes of trace element incorporation. *Marine Geol.* 2002; 182:373–387.

Jiao Y, Newman DK. The pio operon is essential for phototrophic Fe(II) oxidation in *Rhodopseudomonas palustris* TIE-1. *J. Bacteriol.* 2007;189:1765–1773.

Jickells TD, An ZS, Andersen KK, et al. Global iron connections between desert dust, ocean biogeochemistry, and climate. *Science.* 2005;308:67–71.

Kuenen JG. Anammox bacteria: from to discovery to application. *Nat. Rev. Microbiol.* 2008;6:320–326.

Kuhlman KR, Fusco WG, La Duc MT, et al. Diversity of microorganisms within rock varnish in the Whipple Mountains, California. *Appl. Environ. Microbiol.* 2006;72:1708–1715.

Nevin KP, Lovley DR. Mechanisms for accessing insoluble Fe(III) oxide during dissimilatory Fe(III) reduction by *Geothrix fermentans*. *Appl. Environ. Microbiol.* 2002;68:2294–2299.

Odom JM, Peck HD Jr. Hydrogen cycling as a general mechanism for energy coupling in the sulphate-reducing bacteria, *Desulfovibrio* sp. *FEMS Microbial Lett.* 1981;12:47–50.

Oremland RS, Herbel MJ, Switzer Blum J, et al. Structural and spectral features of selenium nanospheres produced by Se-respiring bacteria. *Appl. Environ. Microbiol.* 2004;70:52–60.

Pepper IL, Gerba CP, Gentry TJ. *Environmental Microbiology*. Academic Press, Elsevier; 2014.

Raiswell R. Towards a global highly reactive iron cycle. *J. Geochem. Explor.* 2006;88:436–439.

Silver S, Misra TK. Plasmid-mediated heavy metal resistances. *Annu. Rev. Microbiol.* 1988;42:717–743.

Widdel F. Microbiology and ecology of sulfate- and sulfur reducing bacteria. In: Zehnder AJB (ed.) *Biology of Anaerobic Microorganisms*. New York: John Wiley & Sons; 1988:469–585.

2 Water Pollution

Water is one of our most important natural resources, and there are many conflicting demands upon it. Skillful management of our water bodies is required if they are to be used for such diverse purposes as domestic and industrial supply, crop irrigation, transport, recreation, sport and commercial fisheries, power generation, land drainage and flood protection, and waste disposal. An important objective of most water management programs is the preservation of aquatic life, partly as an end in itself and partly because water which sustains a rich and diverse fauna and flora is more likely to be useful to us, and less likely to be a hazard to our health, than one which is not so endowed. To meet this objective, it is necessary to maintain within certain limits factors such as water depth and flow regimes, temperature, turbidity and substratum characteristics, and the many parameters which contribute to the chemical quality of water. In waters which receive waste discharges, whether by design or by accident, one or more of these variables may come to lie outside the limits which can be tolerated by one or more of the species which live there, and consequently, the biological characteristics of water are altered. The biologist's role in the monitoring and control of water pollution is to detect accurately and to describe these alterations, to elucidate as precisely as possible the mechanisms by which they are brought about, and to seek to understand the qualitative and quantitative relationships between pollution and its biological consequences. He or she may also need to be aware of the application of biological processes in the control or amelioration of pollution, and of the serious consequences for public health which water pollution threatens. Finally, he or she must be able to offer constructive advice to other specialists—chemists, engineers, administrators, and legislators—who share the responsibility for managing our water resources (Abel, 1996).

Water pollution, like other environmental concerns, has been the focus of widespread public interest, and this interest seems to be increasing. While this has many obvious benefits, it sometimes can appear that the public perception of water pollution, as manifested for example by political debate and the activities of pressure groups, does not always accord with the scientific reality. This can lead to ill-advised or cost-ineffective actions, including legislation and regulation, in an attempt to deal with perceived or publicized problems which may, in fact, be less serious than others which are less well publicized or less easy for the educated layman to understand. It is therefore important to gain some idea of the real nature and extent of water pollution.

SOURCES AND EFFECTS OF WATER POLLUTANTS

There are hundreds perhaps thousands of pollutants whose effects are of actual or potential concern. Their numbers increase annually, as new compounds and formulations are synthesized. A substantial minority of these find commercial applications

DOI: 10.1201/9781003272618-3

and become significant pollutants of water during their manufacture and in subsequent use be made. Many pollutants also enter water through fallout from the atmosphere. Historically, control and prevention of water pollution have concentrated on point sources as these are more obvious, easily identifiable, and in theory, easier to regulate at the point of origin. As awareness has increased of the significance of diffuse sources of pollution, control strategies have been under development but are based more on the application of good practices designed to reduce pollutant impact rather than on regulation of specific sources of input.

Most effluents are complex mixtures of a large number of different harmful agents. These include toxic substances of many kinds, extreme levels of suspended solids, and dissolved and particulate putrescible organic matter. In addition, many effluents are hot, of extreme pH value, and normally contain high levels of dissolved salts. Some representative values for treated sewage effluent are given in Table 2.1 (Bond and Straub, 1974).

Most effluents also vary in their strength and composition on a seasonal, diurnal, or even hourly basis. Most sewage treatment plants report regular diurnal peaks and troughs in their output according to patterns of water use. Sometimes storm water drains are connected to the sewerage system, so the strength of the sewage effluent will vary with rainfall. Alterations in the strength and composition of sewage also influence the efficiency of the sewage treatment process, so that dilution of the influent does not necessarily cause an improvement in the quality of the effluent. In industrial plants, variations in the quality of the raw materials, or changes in specification of the finished product, will require changes in the operating conditions of the plant and lead to changes in the composition of the effluent. Many industrial processes are

TABLE 2.1
Analysis of Sewage Effluents after
Primary and Secondary
Treatment

1. Analysis	Range (mg/L)
Total solids	640–1170
Suspended solids	15–51
Biological oxygen demand	2–70
Chemical oxygen demand	31–155
Organic carbon	13–20
Anionic detergents	0.75–1.4
Ammonia	1.9–2.2
Nitrate	0.25–38
Nitrite	0.2–1.8
Chloride	69–300
Sulfate	61–270
Phosphate	6.2–9.6
Sodium	144–243
Potassium	20–26

"batch" rather than continuous processes, so that some effluent discharges will be intermittent rather than continuous.

Nevertheless, it is often possible to generalize about the effects of different kinds of effluent on their receiving waters. Broadly speaking, the effects of sewage effluent, kraft pulp mill effluent, coal mine effluent, and so on are consistent wherever they occur. However, a detailed understanding of the effects of individual components of an effluent and of the consequences of variation in the composition of an effluent is essential for pollution control. Wastewater treatment is expensive, and in order to devise cost-effective treatment processes, it is necessary to identify those components of the effluent which cause the greatest damage to the environment. This is because it is usually impossible to devise an economically feasible process which will be equally effective against all the components of a complex effluent. Frequently, treatment of an effluent to remove one component will exacerbate the problem of removing another; it is notoriously difficult, for example, to treat satisfactorily effluents which contain both cyanides and phenolic compounds, although many basic industrial processes produce just such an effluent. Therefore, in order to devise the optimum pollution control strategy, it is necessary to study, in the laboratory and in the field, the effects of effluents and of their individual components.

THE ENVIRONMENTAL REQUIREMENTS OF AQUATIC ORGANISMS

The effects of pollutants in aquatic ecosystems cannot be understood without some knowledge of the ecophysiology and basic biology of the aquatic biota. In order to survive, a living organism must spend its life in an environment which meets its needs: a suitable physical habitat which provides space, shelter, and a sufficient supply of food, oxygen and other metabolic requirements; and which is not subject to extremes of temperature or other physical variables which lie outside the range which the organism can tolerate. Obviously, different habitats have very different physical characteristics, and organisms have evolved a fascinating array of adaptations which have enabled them to colonize every part of the Earth. To a greater or lesser extent, every living organism is adapted—in its morphology, physiology, and behavior—to the environment it normally inhabits. Some are remarkably specialized, that is they are adapted to specific places and/or modes of life in which they are very successful, but are excluded from living in most habitats. Others are more generalized in their adaptations, perhaps being nowhere particularly abundant but able to survive reasonably well in a wider range of habitats. Few organisms, however, are universally distributed.

Any of a living organism's many individual requirements may be a limiting factor preventing the establishment, survival, or reproduction of a species in a particular habitat. Aquatic plants, for example, are commonly limited in their distribution and abundance by the availability of nutrients, such as phosphorus or nitrogen. An abundance of phosphorus is of no use to the plant if it has no nitrogen and vice versa. Further, the nutrients must be present in a form which the plant can use. Photosynthetic plants, of course, also require light, an important limiting factor in most aquatic habitats. The non-photosynthetic flora (fungi, bacteria) are more likely to be influenced by the levels of dissolved or particulate organic material present. Animals, in turn, are greatly influenced by the quality and quantity of the aquatic flora, because many animals rely upon plants for food, shelter, as a repository for eggs, and so on. Animals are influenced by the physical environment—current speed, nature of the substratum, and temperature—but in addition are generally much more susceptible than plants and bacteria to the prevailing levels of oxygen.

Although some aquatic animals are air breathers, the majority have to obtain their oxygen from water. Oxygen is not very soluble in water, and water is a rather dense and viscous medium. Moving through water requires a great deal of energy expenditure and therefore a high oxygen consumption. To obtain from water the meager amount of oxygen dissolved therein requires that the respiratory surfaces be moved through water or water be moved over the respiratory surfaces. Further, as water temperature increases, the solubility of oxygen in water decreases, while the oxygen requirement of the animal actually increases. Thus, the survival of animals in water is crucially dependent upon the extent to which their oxygen demand can be matched to the availability of oxygen from the environment.

ORGANIC POLLUTION

The discharge of excessive quantities of organic matter is undoubtedly the oldest, and even today, the most widespread form of water pollution. The major sources of organic pollution are sewage and domestic wastes; agriculture (especially runoff from inadequately stored animal wastes and silage); various forms of food processing and manufacture; and numerous industries involving the processing of natural materials such as textile and paper manufacture. Most organic waste waters contain a high proportion of suspended matter, and in part, their effects on the receiving water are similar to those of other forms of suspended solid. However, the most important consequences of organic pollution can be traced to its effect on the dissolved oxygen concentration in water and sediments. In an unpolluted water, the relatively small amount of dead organic matter is readily assimilated by the fauna and flora. Some is consumed by detritivorous animals and incorporated into their biomass. The remainder is decomposed by bacteria and fungi, which are themselves consumed by organisms at higher trophic levels. The activity of microorganisms results in the breakdown of complex organic molecules to simple, inorganic substances such as phosphate and nitrate, carbon dioxide, and water. During these metabolic processes, oxygen is consumed. However, where the organic load is light, the oxygen removed from water is readily replaced by photosynthesis and by reaeration from the atmosphere.

Where the input of organic material exceeds the capacity of the system to assimilate it, a number of changes take place. How far the sequence of changes proceeds depends upon the severity of the organic load and the physical characteristics of the receiving water? Initially, the enhanced level of organic matter will stimulate increased activity of the aerobic decomposer organisms. When their rate of oxygen consumption exceeds the rate of reaeration of water, the dissolved oxygen concentration in water will fall. This alone may be sufficient, as argued earlier, to eliminate some species, which may or may not be replaced by others with less rigorous demands for oxygen. If the drop in oxygen concentration is very severe, the aerobic decomposers themselves will no longer be able to function, and anaerobic organisms will become predominant.

The biochemical reactions involved in the breakdown of organic matter, and the microorganisms involved, are described. The composition of organic waste varies according to its source, and in particular, according to the relative abundance of material of plant, animal, or microbial origin. Most effluents, of course, also contain other materials, in particular toxic matter, derived from various sources. To illustrate the effects of the breakdown of organic matter on the receiving water, proteins may be used as an example. The first stage of decomposition of proteins is usually their breakdown by hydrolysis to their constituent amino acids. A typical amino acid is alanine and its breakdown under aerobic conditions may be summarized:

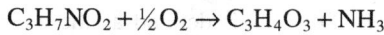

$$C_3H_7NO_2 + \tfrac{1}{2}O_2 \rightarrow C_3H_4O_3 + NH_3$$

$$\text{Alanine} \rightarrow \text{Pyruvate}$$

Pyruvic acid is an important substance in the metabolism of most living organisms. It is produced in normal metabolism during the glycolytic (anaerobic) phase of the breakdown of carbohydrates and, as in this case, from the breakdown of excess amino acids. In aerobic organisms, the pyruvate enters the citric acid cycle, the primary means by which compounds are broken down to release energy, carbon dioxide, and water. Under aerobic conditions, therefore, proteins will be broken down ultimately to these relatively innocuous compounds, while providing a source of metabolic energy for the organisms responsible for the catabolism. Ammonia is also a common end product of the metabolism of nitrogenous compounds (such as amino acids) and in aquatic organisms is generally excreted as such. Normally, the ammonia diffuses rapidly into the environment, but where the level of organic enrichment is high, it can create difficulties for living organisms as it is very toxic. Typically, therefore, organic wastes contain high levels of ammonia; the eventual fate of the ammonia is very relevant to the effects of water pollution.

Under anaerobic conditions, the breakdown of amino acids takes place through different metabolic pathways. Some amino acids, such as cysteine, contain sulfur as well as nitrogen, and its breakdown is used as an example in the following sequence of reactions which are catalyzed by acid-producing and methanogenic bacteria:

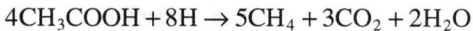

$$4CH_3H_7O_2NS + 8H_2O \rightarrow 4CH_3COOH + 4CO_2 + 4NH_3 + 4H_2S + 8H$$

$$4CH_3COOH + 8H \rightarrow 5CH_4 + 3CO_2 + 2H_2O$$

In this case, the products of decomposition include (in addition to ammonia, carbon dioxide, and water) acetic acid, hydrogen sulfide, and methane. These compounds are very toxic to most forms of aquatic life and, in addition, they are esthetically undesirable by virtue of their unpleasant odors. The fate of the ammonia largely depends upon the level of oxygen present. Under aerobic conditions, nitrifying bacteria predominate and the ammonia is converted to nitrite (e.g., by bacteria of the genus *Nitrosomonas*) and subsequently to nitrate (e.g., by *Nitrobacter* spp.). Thus, the toxic ammonia is oxidized to the less toxic nitrite and to the relatively innocuous nitrate. Since, however, both ammonia and nitrate are important plant nutrients, problems related to eutrophication can arise as a consequence of organic inputs to water. Under anaerobic conditions, denitrification of the nitrate typically takes place under the influence of other bacteria such as *Thiobacillus denitrificans* and some *Pseudomonas* species. These cause the reduction of nitrate to elemental nitrogen, which readily displaces the less-soluble oxygen from solution and contributes still further to the deoxygenation of water. Note that the gaseous phases of the cycle are usually of limited significance in the aquatic environment. These chemical changes, combined with the blanketing effect of fine organic particles on the substratum, lead to the deoxygenation of water and substratum, and readily bring about profound changes in the fauna and flora of the receiving water.

The strength of an organic effluent is frequently expressed in terms of its biological oxygen demand (BOD). This is defined as the quantity of oxygen utilized, expressed in mg/L, by the effluent during the microbial degradation of its organic content. BOD is typically measured by taking a sample of water or effluent and aerating it until it is saturated with oxygen. The dissolved oxygen concentration in part of the aerated sample is then determined by one of the well-known standard procedures. Another part of the aerated sample is incubated, in a sealed bottle of known volume for a period of (typically) 5 days, at a controlled temperature which is usually 15°C or 20°C. Usually, the incubation is carried out in the dark to eliminate oxygen production by any photosynthetic organisms which may be present. At the end of the incubation period, the concentration of dissolved oxygen remaining in the sample is determined. The difference between the initial and final dissolved oxygen concentrations is used to determine the BOD of the sample, which is expressed as milligrams of oxygen consumed per liter of sample. The basic assumption of the method is that oxygen is mainly consumed by aerobic microorganisms during the metabolism of organic matter. This is not necessarily true, since many effluents contain chemically reduced compounds which undergo oxidation by purely chemical reactions.

For this reason, it is often necessary to carry out other determinations (for example, of chemical oxygen demand or of permanganate value) in addition to BOD in order to characterize precisely the likely effects of an effluent or correctly interpret the significance of a BOD value. Within these limitations, BOD values are generally useful as indicators of the organic loading of water. They typically range from 1 or 2 mg/L in unpolluted water to 10,000 mg/L or more in raw wastes, untreated effluents, or severely polluted receiving waters. The wide range of expected values means that the precise details of the method of determination have to be frequently modified according to the circumstances; raw samples, for example, often need to be diluted before the determination is carried out. The incubation period of 5 days (rather than some shorter period) in practice compensates for the fact that some samples will have, initially, very few microorganisms present, while others will contain a large inoculum.

NUTRIENT POLLUTION

Plant growth in water may be limited by any of several factors, including light and the physical characteristics of the habitat. In many cases, however, the limiting factor is the availability of inorganic nutrients, particularly phosphate (Moss, 1988). Increased input of nutrients can therefore trigger increased plant growth which, if excessive, leads to changes in the biological characteristics of the receiving water.

The discharge of organic matter to water is an important source of plant nutrients, since the aerobic decomposition of organic matter results in the release of phosphate, nitrate, and other nutrients. Domestic sewage typically contains high levels of phosphate because detergent washing powder formulations normally contain high levels of phosphate. For example, the level of phosphate typically found in treated sewage effluent may be compared with the levels normally found in unpolluted waters, which range from about 0.001 to 1 mg/L. Food processing effluents are often high in nitrate and phosphate, and in agricultural areas, runoff from land carries nutrients into water, especially if artificial fertilizers are used. Many agricultural and forestry practices lead to increased soil erosion, carrying plant nutrients from land to water. Intensive rearing of livestock contributes significant nutrient loads to surface waters.

Increased plant growth can sometimes be considered beneficial, especially in oligotrophic waters where primary productivity is nutrient limited. Moderately increased plant growth can provide increased productivity of herbivorous and detritivorous animals, leading to increased overall productivity. It is not unknown, for example, for fishermen deliberately to "fertilize" lakes to increase fish yield. The increased spatial heterogeneity of the habitat can also give rise to an increase in species diversity. Excessive plant growth, however, has four main adverse consequences. The blanketing effect of macrophytes and filamentous algae can result in major faunal alterations owing to physical changes in the habitat. Respiration of dense plant growths can produce depressed dissolved oxygen levels not only at night when photosynthesis ceases but also during the day if the density of plant growth reduces light penetration. Some algal species, under the influence of elevated nutrient levels, "bloom"—that is they reproduce rapidly and dominate the flora. These algal blooms give rise to several problems, including tainting and discoloration of water (rendering it unsuitable for potable supply) and the production of toxins which are harmful to fish and invertebrates.

EUTROPHICATION

The phenomenon of eutrophication is particularly associated with lakes and slow flowing waters. It is widely, and erroneously, believed that pollution by plant nutrients and organic matter actually causes eutrophication. It is more accurate to say that pollution accelerates what is probably a natural process. To understand the causes and consequences of eutrophication requires some knowledge of the special characteristics of lakes.

In temperate latitudes, most lakes were formed by glaciation. Moving glaciers gouged out hollows in the earth, and when the ice retreated, these hollows became filled with water from the melting ice. Such lakes are not, therefore, geologically ancient phenomena. In modern times, substantial man-made lakes have become common in many parts of the world. A lake is a body of water which is very slow moving. Some lakes have rivers flowing into or out of them. Even those which do not, however, are not static; water moves slowly into or out of the lake via the ground. Because water moves only very slowly, some physical and chemical processes occur in lakes which do not occur in moving waters. Of particular importance are stratification and temporal variations in chemical quality of water.

Stratification occurs because the lake water is heated by the sun at the surface. Because warm water is less dense than cooler water, and water is a poor conductor of heat, during the warmer months of the year, an upper layer of warm water, the epilimnion, becomes established and sharply delineated from a lower layer of cooler, denser water, the hypolimnion. Between them is a very narrow zone, the thermocline, within which water temperature drops very sharply with only a slight increase in water depth. Little or no vertical mixing can take place, the lake being effectively divided horizontally into two distinct layers separated by the thermocline. Obviously, stratification cannot occur in very shallow lakes.

Photosynthesis can only occur in shallow water, where light can penetrate. At the lake margins, emergent plants and rooted aquatic macrophytes occur, but as the depth of water increases, primary production is possible only by phytoplankton in the surface waters, within the epilimnion. During the winter, phytoplankton growth is restricted by low temperatures and low light intensity. In spring and summer, increasing temperatures and light intensity stimulate phytoplankton growth, leading to an increase in population density and the depletion of nutrients in water of the epilimnion. Plant growth and reproduction slow down, and as the plant cells senesce and die, they sink into the hypolimnion and eventually to the bottom of the lake, where they begin to decompose. The inorganic nutrients which are the products of decomposition remain in the hypolimnion, however, as the stratification prevents vertical mixing of water and upward diffusion is slow. As the autumn approaches, reduced temperatures, light intensity, and limited nutrients accelerate the decline of the phytoplankton population. In the winter, the epilimnion cools and becomes denser. Its water sinks, displacing the hypolimnion which is now warmer and lighter than the epilimnion. The lake waters become thoroughly mixed, and nutrients from the hypolimnion are brought to the surface, bringing about conditions suitable for the start of the next annual cycle.

Underlying these annual cycles is a progressive change in the physical and chemical characteristics of the lake. At its formation, the lake contains few plant nutrients or dissolved minerals of any kind, and a negligible quantity of organic matter. With the passage of time, dissolved minerals including plant nutrients enter the lake from surface runoff and groundwater infiltration at a rate which depends largely upon the climate and the geology of the surrounding area. As the nutrient levels rise, a flora and fauna becomes established and develops, contributing an increased content of organic matter in the lake. Organic matter is also gradually accumulated from outside the lake, progressively building up a layer of sediment on the lake bottom. Airborne dust also falls into the lake, and the lake begins to fill up slowly. The rate at which this happens varies from the barely detectable up to a few millimeters per year. The gradual deposition of material on the floor of the lake basin causes the lake to shrink, new land being formed at its edges. This new land is colonized by terrestrial plants, and in some lakes, it is possible, by walking away from the lake's edge, to see clearly the various stages of development of the terrestrial flora, a classic example of ecological succession. In areas where these processes have occurred, for various reasons, at different rates in different lakes, it is possible to see contemporaneously all the stages of a lake's development from nutrient-poor, sparsely populated lakes of low productivity, through various stages of nutrient enrichment, to swamp or marsh and eventually dry land.

The term eutrophication is applied to the process whereby the nutrient levels of lakes increase from oligotrophic (nutrient poor) to eutrophic (nutrient rich). The transition from oligotrophic to eutrophic is accompanied by qualitative and quantitative changes in the biota. Since plant growth is commonly limited by nutrient levels, a gradual increase in nutrient levels would be expected to lead to successional changes in the plant community and corresponding changes in the animal community. Animals, in particular, are likely also to be affected by deoxygenation of the hypolimnion. In eutrophic lakes, the stratification which leads to nutrient depletion of the epilimnion also causes oxygen depletion of the hypolimnion. Oxygen demand due to aerobic decomposition of detritus is high in the hypolimnion, but the absence of either vertical mixing or photosynthesis in the hypolimnion prevents reoxygenation of the hypolimnion from the atmosphere.

THERMAL POLLUTION

Temperature is of such profound importance in chemical and biological processes that the effect of temperature alterations on aquatic biological communities is potentially large. Hot effluents from industrial processes and power generation can cause temperature increases in the receiving water of 10°C or more. Some effluents, such as water pumped from deep mines or regulating reservoirs, may be significantly colder than the receiving water, although the effects of cold effluents have received relatively little attention. Because the density of water alters with temperature, hot effluents often form a surface plume rather than mixing quickly with the receiving water. This can exacerbate some of the adverse effects, but may sometimes act to minimize the influence of the effluent on the benthic community, and fish can avoid the elevated temperature by remaining in deeper water. Elevated temperatures can influence aquatic organisms directly, as the organisms respond physiologically or behaviorally to the new conditions; or indirectly, as the changed temperature influences the chemical environment. For example, increased temperature reduces the solubility of oxygen in water. At the same time, it may increase BOD by stimulating more rapid breakdown of organic matter by microorganisms.

HEAVY METALS

"Heavy metals" is an imprecise term that is generally taken to include the metallic elements with an atomic weight greater than 40, but excluding the alkaline earth metals, alkali metals, lanthanides, and actinides. The most important heavy metals from the point of view of water pollution are zinc, copper, lead, cadmium, mercury, nickel, and chromium. Aluminum may be important in acid waters. Some of these metals (e.g., copper and zinc) are essential trace elements to living organisms, but become toxic at higher concentrations. Others, such as lead and cadmium, have no known biological function. Industrial processes, particularly those concerned with the mining and processing of metal ores, the finishing and plating of metals, and the manufacture of metal objects, are the main source of metal pollution. In addition, metallic compounds are widely used in other industries: as pigments in paint and dye manufacture; in the manufacture of leather, rubber, textiles, and paper; and many others.

Quite apart from industrial sources, domestic wastes contain substantial quantities of metals because water has been in prolonged contact with copper, zinc, or lead pipework or tanks. The prevalence of heavy metals in domestic formulations, such as cosmetic or cleansing agents, is frequently overlooked. Some forms of intensive agriculture give rise to severe metal pollution; copper, for example, is widely added to pig feed and is excreted in large quantities by the animals (Mance, 1987).

Heavy metals may be classed generally as toxic or very toxic to aquatic animals and to many plant species, though large interspecific differences in susceptibility occur even within closely related groups of organisms. Two features of heavy metal toxicity which should not be overlooked are their ability to form organometal complexes and their potential for bioaccumulation. There is some evidence that the presence of organic substances can reduce heavy metal toxicity considerably, at least as measured in conventional toxicity tests. Many metals, whether organically complexed or not, are known to accumulate in plant and animal tissues to very high levels, posing a potential toxic hazard to the organisms themselves, or organisms higher in the food chain including humans, which may consume them.

Ammonia, Cyanides, and Phenols

Ammonia and its compounds are ubiquitous constituents of industrial effluents because ammonia is a staple raw material in many branches of the chemical industry; it is, therefore, a common end product of industrial processes as well as an important byproduct of others, notably the production of coke and gas from coal, from power generation and from most processes involving the heating or combustion of fuel. It is also a natural product of the metabolism of organic wastes in treatment plants and receiving waters. The toxicity of ammonia to fish is well documented, and although less is known of its effect on invertebrates, it appears that the levels of ammonia which are tolerable to fish present little danger to most invertebrates (Alabaster and Lloyd, 1980). In aqueous solution, ammonia forms an equilibrium between unionized ammonia, ammonium ion, and hydroxide ions:

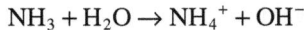

$$NH_3 + H_2O \rightarrow NH_4^+ + OH^-$$

Unionized ammonia is very toxic to most organisms, but ammonium ion is only moderately toxic. The toxicity of the solution therefore depends on the quantity of unionized ammonia. This in turn depends upon the pH and temperature of water; as pH and temperature rise, the proportion of unionized ammonia also rises. The effect of pH and temperature on ammonia toxicity is therefore considerable. In order to know whether a given level of total ammonia is likely to be toxic, it is necessary to use the pH and temperature values to calculate the corresponding level of free ammonia.

Cyanide is also a very common constituent of industrial effluents, being produced from processes involving coking and/or combustion such as steelworks, gas production, and power generation. Cyanides are also used in the hardening, plating, and cleaning of metals. The dissociation, and consequently the toxicity of cyanide, is pH-dependent, low pH favoring the formation of undissociated HCN which is highly toxic. Cyanide ions readily form complexes with heavy metal ions. The stability and toxicity of these complexes vary according to the metal and also with the pH. Thus, the toxicity of cyanides and effluents containing cyanides (which commonly contain substantial amounts of heavy metal) is, as with ammonia, greatly influenced by pH.

Phenolic substances include the monohydric phenols (phenol, cresols, and xylenols) and the dihydric phenols including catechols and resorcinols. They are found in a wide range of industrial effluents, and are particularly associated with gas and coke production; the refining of petroleum; power generation; many branches of the chemical industry; and the production of glass, rubber, textiles, and plastics. Phenolic substances rarely occur as pollutants except as components of complex effluents which contain a variety of other polluting substances. In practice, therefore, the major concern is to determine the extent to which phenols contribute to the overall toxicity of an effluent.

PESTICIDES

Pesticides are a diverse group of poisons of widely varying chemical affinities, ranging from simple inorganic substances to complex organic molecules. Of the latter, some are natural metabolites, particularly of plants, while others are synthetic derivatives of natural products or completely synthetic substances produced in chemical factories under conditions which do not exist in the natural world. They have in common only that each pesticide is highly toxic to some forms of life and of intermediate or negligible toxicity to others, and that they have been widely introduced into the natural environment. Pesticides are introduced into aquatic systems by various means: incidentally in the course of their manufacture and through discharge consequent upon their use. Surface water runoff from agricultural land and the side effects of aerial spraying are especially important, and many serious pollution incidents arise through the accidental or negligent discharge of concentrated pesticide solutions which have been used for agricultural purposes such as sheep dipping. Additionally, many pesticides are deliberately introduced into water bodies to kill undesirable organisms such as insect or molluscan vectors of human diseases, weeds, fish, and algae.

SUSPENDED SOLIDS

Virtually, all effluents contain suspended particulate matter, but especially those associated with mining and quarrying for coal, china clay, stone, and other mineral materials. Dredging, engineering works, and boat traffic commonly introduce particulate matter into suspension. Storm water drainage and surface water runoff also contribute substantial loads. Suspended matter may be organic or inert, and some forms are chemically reactive (for example, the ferric hydroxide precipitate associated with acid mine drainage). The effects of suspended matter on the receiving water biota are both direct and indirect. Direct effects include physical abrasion of body surfaces, and especially, of delicate structures such as gills. Physical damage of this kind interferes with respiration and renders the animals susceptible to infections. High levels of suspended particulates may interfere with the filter-feeding mechanisms of invertebrates and possibly with the feeding of fish which locate their food visually.

Indirect effects are mainly due to increased turbidity and the blanketing effect of the particulates when they eventually settle. Increased turbidity will reduce or prevent photosynthesis, leading to a reduction in primary productivity or the complete elimination of plants. Alternatively, certain forms of silting can, depending on the physical conditions, bring about major changes in the community by promoting the formation of stable weed beds. Insects which crawl upon the substratum may generally be disadvantaged in favor of species whose means of locomotion is better suited to a soft substratum, such as leeches, oligochaetes, and some molluscs. Thus, the input of even inert fine particulates can readily bring about major community changes.

EXTREME pH AND ACIDIFICATION

Many effluents, especially if untreated, are strongly acidic or alkaline. All natural waters have some buffering capacity, that is the ability to absorb acid or alkaline inputs without undergoing a change in pH. This buffering capacity is usually expressed in terms of the acidity (ability to neutralize alkalis) and alkalinity (ability to neutralize acids) of water, and it is determined by titration in the presence of a suitable indicator. The relationship between the pH, acidity, and alkalinity of water is not simple. Acid waters (pH < 7) can have measurable alkalinity, and alkaline waters (pH > 7) can have measurable acidity. Where the buffering capacity of water is exceeded by the input of an effluent, the pH of water will change. Unpolluted natural waters show a pH range from 3.0 to 11.0 or more; those lying between 5.0 and 9.0 generally support a diverse assemblage of species and this range may be considered broadly acceptable.

A very common form of pollution involving extreme pH is acid mine drainage. Coal mines are the most common source of acid mine drainage, but it can occur wherever mineral ores are mined. The effects of acid mine drainage are threefold. The low pH itself has adverse effects on the receiving water flora and fauna. It also promotes the solubilization of heavy metals, which exert their own toxic effects. Third, as the drainage water is diluted and the pH rises, ferric hydroxide precipitates and discolors water, producing the effects of suspended particles. As the hydroxide settles, it forms a gelatinous layer over and within the substratum, causing both direct and indirect effects on the receiving water community.

The consequences of acidification are undoubtedly serious. It is generally accepted that as pH decreases, both the diversity of species and the overall productivity of aquatic ecosystems decline. In addition, low pH values in theory could strongly and progressively reduce the rate of decomposition of organic detritus, presumably through the effect of low pH on the fungal and bacterial organisms responsible for this process. Many aquatic ecosystems depend on the decomposition of allochthonous detritus (i.e., organic material from outside the system, such as fallen leaves) as the main source of energy for the animals in the system. Primary (photosynthetic) production of organic matter in acid waters is naturally low, being normally limited by the low availability of nutrients; it may also be itself inhibited by the susceptibility of phytoplankton and macrophyte species to low pH. The alternative source of energy for animals, organic detritus, is only available in the presence of microbial decomposers, since animals cannot digest plant material unaided. Therefore, the overall productivity of the system will decline.

A third effect of acidification is to increase the threat of heavy metal toxicity. Aluminum, a metal which does not commonly cause serious problems of toxicity to aquatic life, has received particular attention. Within the pH range of approximately 5.5–7.0, aluminum is practically non-toxic and is certainly harmless at the concentrations found in most waters. The chemistry of aluminum and its compounds at low pH is poorly understood, but it appears that as pH falls aluminum compounds become more soluble and that the proportion of free aluminum rises. Below pH 4, the toxic effects of free hydrogen ion are so severe that the presence of aluminum is probably of little significance. However, between pH 4 and 5.5, the toxicity of aluminum is high, reaching a maximum at around pH 5. At this pH, the level of aluminum naturally present in water is acutely toxic to fish. High levels of calcium appear to offer some protection, although acid waters with high calcium levels are unusual. Aluminum toxicity is likely, therefore, to be a major contributor to the effects of acidification.

DETERGENTS

Synthetic detergents are an interesting group of pollutants because they were virtually unknown before 1945, yet within a few years became responsible for some spectacular water pollution problems which, unusually, came rapidly to the attention of the general public. The alkylbenzene sulfonate detergents rapidly replaced soap as domestic and industrial cleaning agents because of their cheapness and greater efficiency, and particularly because they did not cause precipitation of calcium salts in areas supplied with hard water. Unfortunately, they were not readily broken down by sewage treatment processes, giving rise to problems of toxicity to the receiving water biota and problems of foaming in watercourses and treatment works.

In areas where industrial usage of detergents was pronounced (for example, in textile-processing industries), whole towns were frequently covered in detergent foam; in waste treatment works, a number of serious accidents occurred through, for example, operatives falling into sedimentation tanks which were concealed under a thick layer of foam.

OIL AND PETROLEUM PRODUCTS

Oil pollution is commonly perceived as being a problem associated mainly with the marine environment. About one-quarter of all the oil released to the seas by human activity is estimated to enter via rivers. The sources of oil pollution are the usual ones—illegal, negligent or accidental discharges, plant failures, and so on—but to these must be added inputs from roads and railways, particularly in the case of traffic accidents, and discharges associated with oil extraction from inland oilfields and processing plants.

The effects of oil discharges vary enormously, because the characteristics of the receiving waters and the various types of oil are themselves very variable. Fortunately, many small discharges have relatively trivial effects, but substantial discharges can have severe impacts (Green and Trett, 1989). Generally, light oils are relatively volatile and disperse quickly, though they are extremely toxic and can cause severe local damage. Fast-flowing waters often recover rapidly, over a few weeks or months, but lentic waters are much more susceptible to long-term damage. Heavy oils are less toxic, but can have marked physical effects in the substratum and the banksides, which of course will be reflected in changes to the biological community. Heavy oils may remain in situ for long periods; they do, over time, decompose biologically and chemically, but in so doing impose a high BOD. Frequently, the effects of oil pollution are not dissimilar from those of heavy organic pollution, with varying levels of toxic effects adding to the overall impact. Oils, unlike most forms of water pollution, can also have damaging effects on terrestrial organisms, such as plants, birds, and mammals living at water's edge, since heavy oil contamination usually leads to deposition of oil on the banks of lakes or rivers.

Generally, the use of oil dispersants, which are commonly used in dealing with marine spills, is avoided in freshwater because of their high toxicity. Also, the physical conditions of freshwater habitats are often amenable to containment and recovery of oil by booms, skimmers, and/or the use of absorbent materials.

THE TOXICITY OF POLLUTANTS TO AQUATIC ORGANISMS

There are many circumstances in which the need to measure toxicity may arise. Many thousands of chemical substances are used for industrial, agricultural, and domestic purposes and their numbers increase annually. The toxicity of these chemicals and of their byproducts and degradation products to aquatic animals needs to be determined, since any compound manufactured and used in substantial quantities is likely to become a contaminant of watercourses. In the case of novel compounds or formulations, toxicity testing may precede large-scale manufacture and form part of the research into the feasibility of its commercial application. Toxicity tests may be incorporated into effluent monitoring schemes. Identification of the more toxic components of complex effluents may be a prerequisite for the development and improvement of effective treatment processes. The measurement of toxicity is essential in the formulation of quality standards for receiving waters. Finally, compliance with a toxicity standard may be a legal requirement for consent to discharge an effluent.

Toxic pollutants may exert their effects in several ways, depending upon the characteristics of the poison, of the receiving water, and of the biological community water sustains. In extreme cases, animals may be killed by the poison. In some circumstances, poisons—insecticides, herbicides, molluscides, and piscicides—are applied to water with the express purpose of killing some species and in the hope that others will be unaffected. Lower concentrations of poison may exert sublethal toxic effects. Some poisons appear to accumulate in the tissues of organisms during their lifetime and exert toxic effects after prolonged exposure to concentrations which are barely measurable by chemical means. It is widely suspected that some of these may pass from prey to predator organisms and achieve high concentrations in species at the top of a food web. Many poisons are known to be mutagenic, teratogenic, or carcinogenic, but the study of these phenomena in aquatic organisms is in its infancy.

From a biological point of view, any toxic effect is significant if it influences, or is likely to influence, the physiology or behavior of the organism in such a way as to alter its capacity for growth, reproduction or mortality, or its pattern of dispersal, since these are the major determinants of the distribution and abundance of species. Species which are not directly affected by a pollutant may nevertheless be indirectly influenced. For example, if a predator is deprived of its normal prey by the action of a pollutant on the prey, it may itself be numerically reduced. Alternatively, it may prey upon some other species, which itself may show a numerical response. Where two competing species are unequally affected by the pollutant, both may show a change in distribution and/or abundance. Thus, the effects of toxic pollutants can only be fully understood with some knowledge of trophic, competitive, and other interspecific relationships (Abel, 2002).

Since pollutants can exert such a variety of toxic effects at different levels of biological organization, an enormously wide range of investigative methods has been employed in their study. There is an enormous literature on the subject of pollutant toxicity to aquatic organisms, especially fish and invertebrates. Much of it is, unfortunately, of limited or even doubtful value, for reasons that will be made clear. For the purposes of this discussion, it is necessary at the outset to clearly define four basic terms which are widely misunderstood:

Lethal toxicity: toxic action resulting in the death of the organism.
Sublethal toxicity: toxic action resulting in adverse effects in the organism other than its death.
Acute toxicity: toxic action whose effects manifest themselves quickly (by convention, within a period of a few days).
Chronic toxicity: toxic action whose effects manifest themselves over a longer period (by convention, within periods measurable in weeks or months rather than days).

BIOACCUMULATION

Bioaccumulation is an aspect of sublethal toxicity which has received much attention, though many areas of uncertainty remain. Pollutants may, over long time periods, accumulate in tissues to levels which may be harmful to the organism. Since many aquatic species are utilized for human consumption, the public health significance of toxic substances accumulated in their tissues is obvious. Many national and international agencies set concentration limits for pollutants, particularly heavy metals, in tissues for human consumption, and promote research and monitoring programs. Study of the uptake, metabolism, and excretion of pollutants, and of their distribution in the various body organs and tissues, makes an important contribution to understanding their mechanisms of action. Levels of pollutants in the tissues of living organisms are widely used to indicate the degree of contamination of water in which they live, particularly when the pollutants are present only intermittently or in very low concentrations, making chemical analysis of water difficult. Finally, many poisons, particularly heavy metals and refractory organic compounds such as some pesticides, are widely believed to pass from the tissues of prey organisms into those of predators and to attain concentrations there which are several orders of magnitude higher than those in the tissues of the prey species. This phenomenon poses a specific threat to long-lived organisms at the higher trophic levels. Studies of bioaccumulation are carried out in the laboratory, in experimental ecosystems, and in the field. Laboratory investigations are usually concerned initially with determining the "bioconcentration factor," that is, the ratio between the concentration in the animals and the concentration in water, when the animals have been exposed for sufficiently long for an equilibrium or steady state to be achieved. This ratio is generally regarded as a valid indicator of the capacity of a pollutant to accumulate in animal tissues.

There are many models of bioaccumulation. In the simplest possible model, two compartments are considered: the organism and the environment. Pollutant will enter the organism at a certain rate, which is dependent upon the amount present in the environment. Pollutant will also be lost from the organism at a rate dependent upon the amount present in the organism. This simple model can be expressed mathematically and predicts that organisms exposed to a constant level of pollutant will eventually reach a "steady state"; that is, the concentration of pollutant will increase to a certain level and thereafter remain constant. Conversely, the model predicts that in contaminated organisms maintained in clean water, the concentration of pollutant in the organism will decline exponentially. These predictions are generally confirmed by experimental findings. Most organisms cannot be considered as a single compartment. Studies of the distribution of pollutants in animals invariably show that the pollutant is very unevenly distributed between the various body tissues. Different pollutants behave in different ways. Clearly, the animal will begin to suffer harm when the poison concentration in a particular organ reaches a critical level. Thus, the concentration of the pollutant in the whole body is not a good indicator of harmful effect. For this reason, models have been derived which treat the organism as a set of interacting compartments. These models treat discrete organs (e.g., liver, kidney, and brain) as interacting compartments connected via blood (itself considered a compartment) with each other and with the external environment. Further, in some models, the exchange of pollutants between the compartments is considered as a series of separate processes. For example, uptake of the pollutant from the environment may occur through the body surface, or by ingestion of food or non-food particulate material. Elimination of the pollutant may occur through outward diffusion, through renal or gastrointestinal excretion, or by metabolic breakdown.

The purpose of such models is twofold. First, if they can be validated by experimental findings (i.e., if the predictions of the models correspond with actual observations), they provide the means for making useful predictions based on comparatively simple experimental measurements. In other words, they can eventually become a substitute for actual experimentation, which may be time consuming and expensive. Second, they can provide valuable information about the mechanisms of bioaccumulation. If, for example, an experimental finding does not agree with a prediction of a model, it indicates that one or more of the assumptions in the model is wrong, and thus focuses attention on areas which require further investigation. Although it is by no means clear that any existing model is of general application, studies on bioaccumulation and its mechanisms are of great practical importance.

It is widely believed that many pollutants pass through succeeding trophic levels and accumulate in high concentrations in the tissues of long-lived predators. Although there are a small number of widely accepted examples of this, it is by no means well established that this is a general phenomenon, even for persistent pollutants like heavy metals and refractory organics. For example, comparing tissue levels of pollutants in field populations is very likely to produce biased results if, as is usually the case, mean levels of pollutants are compared. This is because of the differences in pollutant concentrations between individuals of the same species; frequently, mean values are biased by a very small number of individuals which have very high concentrations; that is, the frequency distribution of pollutant concentration values is highly skewed. Further, although under experimental conditions a steady-state concentration of the pollutant in the tissues generally is eventually achieved, it is not clear that this is the case in the field. Field populations are generally exposed to lower, and more widely fluctuating, pollutant concentrations in their environment. Clearly, a comparison between tissue pollutant concentrations in two different species is invalid if they both are not at their respective steady-state concentrations. It is also true that the interpretation of field observations, and the design of experimental investigations, often rests upon unverified, and unwarranted, assumptions about what animals actually eat.

Obviously in a simplified experimental system, predators will feed on prey which may not form their normal or natural diet in the field. A further example is the widespread assumption that large marine predatory fish such as tunas feed primarily or exclusively on smaller fish such as mackerel, herring, or sardine.

WATER POLLUTION AND PUBLIC HEALTH

It is widely but erroneously believed that the marked increase in life expectancy which has occurred among the populations of many countries since the end of the nineteenth century was due to advances in medical science. In fact, in the conditions which prevailed in the swelling and poorly sanitated cities of the nineteenth century, and which still prevail in much of the developing world today, the major hazards to life were epidemic diseases which exacted a heavy toll, particularly among the young. Certainly, medical scientists played a major role in establishing the nature and the means of transmission of these diseases. Nevertheless, the control of epidemic diseases was achieved largely through the development of proper methods for the treatment and disposal of wastes, and by attention to the provision of clean water supplies, together with education and legislation concerned with general hygiene. These developments preceded by some decades the availability of medical treatments, such as antibiotics, for the cure of waterborne infections. Today, those of us who live in the better developed parts of the world can be reasonably optimistic, should we catch a waterborne disease, of a successful outcome, since most of these diseases respond well to modern medical treatments. Nevertheless, we should not forget that the overwhelming majority of us will never become infected in the first place, and the reason for this is primarily the existence of adequate measures for the monitoring and control of pollution.

All this is of more than historical interest, since in many parts of the world today waterborne diseases remain a major hazard. They are endemic in those countries which have not yet established systems for the sanitary disposal of wastes. It is striking that in times of war or natural catastrophe, when sanitary systems cannot be maintained satisfactorily, waterborne diseases take very little time to spread through the human population. Even in countries where waterborne diseases are not considered endemic, the speed and frequency of international travel and the magnitude of international trade present a constant threat of the reintroduction and spread of infections. Further, increasing demands upon water resources even of developed countries present new public health problems.

Public health measures for the control of infectious and other diseases associated with water take various forms. Examples include good medical services for rapid diagnosis and treatment; a system of rapid reporting of infectious diseases so that epidemics and their sources can be quickly identified and eliminated; education and general public awareness of good hygienic practices; and controls over the processing and handling of food for human consumption. The monitoring and control of water pollution is a central element in the preservation of public health, because water is potentially the means by which many diseases can be spread.

WATER POLLUTION AND PATHOGENS

Pathogenic organisms which are spread by polluted water include bacteria, viruses, and parasites.

BACTERIAL PATHOGENS

Typhoid fever is a disease of the gastrointestinal tract which frequently gives rise to systemic infections. If diagnosed and treated with antibiotics, it is debilitating but rarely fatal. The disease is rare wherever public health and pollution control measures are adequate, but minor outbreaks are not uncommon, particularly among travelers recently returned from less-developed parts of the world. Typhoid infection is usually caused by ingestion of bacteria from fecally contaminated water or food. The causative organism is *Salmonella typhi*. If the bacteria survive passage through the acid of the stomach (a process which may be assisted if the bacteria are ingested with water), they colonize the intestine and enter the epithelial cells of the gut lining. Very few bacteria (between 100 and 1,000 cells) are required to establish a *S. typhi* infection; for other Salmonella infections, the minimum infective dose is rather higher, approximately 10^5 to 10^6 cells. Ulceration of the intestine occurs, and bacteria enter the bloodstream. At this stage, fever occurs and the patient becomes mentally and physically debilitated. Subsequently, the bacteria may become established in various parts of the body, especially in the lymph nodes, gall bladder, spleen, and skin—characteristic hemorrhagic spots on the skin may be seen in some patients. Infected individuals excrete large numbers of bacteria in the feces, providing a source of infection of further individuals. An interesting and dangerous feature of this disease is that in some cases the symptoms of infection can be mild and undiagnosed, although the patient becomes chronically infected with *Salmonella* without showing symptoms of the disease. These "carriers" are a persistent source of infection in the community; further, even if diagnosed, the disease in such patients does not readily respond to the normal treatment with antibiotics.

Bacterial dysentery (shigellosis) and its cause were first recognized in Japan at the end of the nineteenth century during an epidemic involving 90,000 cases. Shigellosis is caused by the consumption, in fecally contaminated water or food, of live bacteria of the genus *Shigella*; known pathogenic species include *Shigella dysenteriae*, *Shigella flexneri*, and *Shigella sonnei*. All of these appear to be associated exclusively with humans and some non-human primates, and transmission of the disease is usually by the fecal–oral route. *S. sonnei* outbreaks, although less serious than *S. dysenteriae*, are not uncommon in schools and similar establishments. If ingested, bacteria survive passage through the stomach, they invade the epithelial cells of the intestine and give rise to ulcerous lesions. It is believed that the bacteria also secrete a toxin which may have pathological effects. Infected patients excrete bacteria with the feces. Like typhoid, the disease typically takes a few days to manifest itself. The classical symptoms are the frequent passage, in small volume, of stools accompanied by blood and mucus, along with severe abdominal cramps; often, however, watery diarrhea is the only manifestation and many infected patients may be undiagnosed. Shigellosis is often a seriously debilitating disease which can give rise to chronic infection and periodic recurrence of symptoms. If untreated, death occurs in up to 15% of cases, most fatalities occurring among infants and elderly patients. Complications include hemolytic anemia and Reiter's syndrome, an arthritic condition. As with typhoid fever, chronically infected patients can act as carriers of the disease within the community without manifesting clinical symptoms of the disease.

Cholera is perhaps the most devastating of waterborne bacterial diseases, and well exemplifies the importance to public health of good sanitation and water pollution control. Cholera is caused by the bacterium *Vibrio cholerae*, and is transmitted by ingestion of live bacteria from polluted water or other material contaminated with fecal matter from an infected individual. The bacteria colonize the intestinal tract and produce a potent toxin which attacks the intestinal mucosa and interferes with the normal processes of salt and water balance which occur across the gut wall. The patient suffers severe diarrhea involving sudden and massive dehydration of the body and serious salt imbalance which impairs the normal function of many organs. Death, due to circulatory failure consequent upon dehydration and salt imbalance, can occur within hours of infection. In some cases, the bacterium can cause serious infections in other parts of the body.

Bacteria of the genus *Vibrio* have been studied mainly because of the medical importance of *V. cholerae*, which was first recognized about one hundred years ago (Colwell, 1984). Vibrios appear to be primarily free-living organisms of soil and water, particularly associated with saline environments such as estuaries and salt marshes. Apart from *V. cholerae*, several other *Vibrio* species are pathogenic to humans, and some are pathogenic in aquatic organisms including crustaceans and molluscs. These animals are primarily detritivores and filter feeders, and they frequently accumulate high concentrations of *Vibrio* bacteria in their bodies. This gives rise to a hazard of human infection, especially in areas where shellfish are cultured, harvested, processed, or consumed in significant quantities.

Viral Pathogens

The viruses of greatest interest in this context are the group known as enteric Viruses. Enteric viruses include agents from widely different taxonomic groups, but they have in common the fact that the intestine is their primary lodgment site. As they are commonly found also in water, they can potentially be transmitted by ingestion of water or other matter contaminated with fecal waste from an infected individual. However, the link water-patient-water-patient has so far been firmly established only for poliovirus and Hepatitis type A virus.

Poliovirus is the causative organism of poliomyelitis, probably the most familiar of waterborne viral diseases. In developed countries, public health policy generally dictates that the majority of the population is immunized in childhood against this disease. Poliovirus infection is primarily one of the alimentary tracts and may cause symptoms of fever, vomiting, and diarrhea. These symptoms may not be recognized as serious. In a minority of cases (up to approximately 1%), the virus enters the bloodstream of the infected individual and may penetrate the nervous system and give rise to paralysis; this may be permanent and, if vital organs are affected (such as respiratory muscles), fatal. Infected individuals, whether or not showing serious disease symptoms, excrete poliovirus in large quantities in the feces, giving rise to the risk of infection of further individuals.

Hepatitis A (infectious hepatitis) is clearly transmitted through contaminated water (Rao and Melnick, 1986). Viral hepatitis of the non-A, non-B form has also been shown to give rise to waterborne epidemics with fatality rates of up to 40% among susceptible groups of the population. There is strong circumstantial evidence of waterborne spread of rotavirus, Norwalk virus and of infections caused by consumption of shellfish from polluted water which contain high concentrations of parvovirus, rotavirus, and small round virus (SRV).

Parasitic Infections

Parasitic infections can also be spread through polluted water. Amoebic dysentery, caused by the protozoan *Entamoeba histolytica*, is a typical example. The most obvious symptom of acute amoebic dysentery is diarrhea, accompanied by blood and mucus. Chronic infection can occur giving rise to recurrent episodes of diarrhea alternating with constipation.

Other parasitic protozoans with relatively simple life cycles are known to spread in a similar manner. Examples include the flagellate *Giardia lamblia*, which causes serious gastrointestinal disorders (Zaman and Ah Keong, 1982).

Parasitic nematodes (roundworms) are important human parasites, and many have a simple life cycle involving direct transmission by the fecal–oral route. Others require passage through an intermediate host, but completion of the life cycle is facilitated by contamination of water, soil, or food with fecal wastes of infected individuals. In many species, the life cycle includes a stage which is specifically adapted for survival in the external environment, so that infective life stages may remain viable in water or soil for long periods. In some species, a period in the external environment is essential for completion of the life cycle. One example of a nematode with a simple life cycle which causes human infection is the whipworm *Trichuris trichiura*. Mild infections produce few serious symptoms, but provide a focus of infection from which more serious cases can arise. Heavy infection produces symptoms of dysentery, and leads to severe damage of the intestinal tract, anemia, and (especially in children) loss of weight. The feces of infected individuals contain eggs which require about 20 days in the external environment to complete their development; the infective stage can survive for several months in water or soil. Several other nematode species have similar life cycles.

Tapeworms (cestodes) include a number of species which infect humans and whose transmission is aided by contamination of water or soil with human fecal matter. The beef tapeworm, *Taenia saginata,* is a typical example; again, there are a number of other species whose life cycles are broadly similar. The adult *T. saginata* lives in the human intestine and sheds, in the feces of the host, proglottides which contain eggs. The eggs are ingested by cattle from polluted water or from pasture contaminated with human feces. In the cattle, the eggs develop through several stages, migrating through the intestinal wall and via the circulation to the muscles, where the cysticercus stage is formed. This encysted state remains in the muscle, and if ingested by a human in uncooked or partially cooked beef, develops into an adult tapeworm.

Trematode parasites which infect humans include a number of blood flukes (Schistosomes), liver flukes, intestinal flukes, and pulmonary flukes which cause serious disease. They are relatively rare in temperate zones and in countries where pollution control and public health practices are well established, but are major causes of ill health and economic loss in many parts of the world.

FURTHER READING

Abel PD. *Water Pollution Biology.* London: Taylor & Francis; 1996.

Abel PD. *Water Pollution Biology.* London: CRC Press; 2002.

Alabaster JS, Lloyd R (eds.). *Water Quality Criteria for Freshwater Fish.* Butterworths: Food and Agriculture Organization; 1980.

Bond RG, Straub CP (eds.). *Handbook of Environmental Control Vol. IV. Wastewater Treatment and Disposal.* Cleveland, OH: CRC Press; 1974.

Colwell RR (ed.). *Vibrios in the Environment.* New York: John Wiley & Sons; 1984.

Dugan PR. *Biochemical Ecology of Water Pollution.* New York: Plenum Press; 1972.

Green J, Trett MW. *The Effects and Fate of Oil in Freshwater.* London: Elsevier Applied Science; 1989.

Higgins IJ, Burns RG. *The Chemistry and Microbiology of Pollution.* London: Academic Press; 1975.

Mance G. *Pollution Threat of Heavy Metals in Aquatic Environments.* London: Elsevier Applied Science; 1987.

Moss B. *Ecology of Fresh Waters—Man and Medium,* 2nd ed. Oxford: Blackwell Scientific Publications; 1988.

RAO VC, Melnick JL. *Environmental Virology.* Wokingham: Van Nostrand Reinhold; 1986.

Zaman V, Ah Keong L. *Handbook of Medical Parasitology.* New York: ADIS Health Science Press; 1982.

3 Soil Pollution

Pollution of soil may arise from a wide range spectrum of sources. These might be discrete point sources or diffuse sources and the pollution process itself may be deliberate as in fertilization processes or following an accident as in the case of radio-nuclear accidents or oil spills.

POLLUTANTS OF AGROCHEMICAL SOURCES

Pollutants from agrochemical sources include fertilizers, manure, and pesticides. To these, we may add the accidental spills of hydrocarbons used as fuels for agricultural machines. As it was mentioned before, the main pollution effect caused by fertilizers and manure is the introduction of heavy metals and their compounds into the soil. Examples of these are the introduction of As, Cd, Mn, U, V, and Zn by some phosphate fertilizers or the soil contamination with Zn, As, and Cu when poultry or pig manure materials are used. Organic compounds used as pesticides, however, are of more far-reaching effects for the whole community depending on soil ecology.

Pesticides applied to plants or harmful organisms living on soil may (by successive adsorption and elution) move down the soil column, where they would be bound within the latticework of clay minerals or adsorbed on to soil organics. They may also join the soil water or the gas phase in the interstitial space, if the active ingredients are of suitable volatility.

The degree of penetration or sorption of pesticides into the tissues of their living targets whether animals or plants provides one of the basis for their classification. According to this, pesticides that remain as superficial deposits exerting only a local contact action are known as contact pesticides, while those with a local internal movement within the cuticles of leaves or the epidermis of animals are known as quasi-systemic. Pesticides that directly penetrate through the outer layers and are transported around the organisms of their targets are classified as systemic pesticides.

However, pesticides are generally classified into the following groups according to their mode of action and the specific organisms they are used to combat:

1. Insecticides are chemical compounds used to kill insects, whether specifically for a given type or generally for a variety of insects.
2. Herbicides are chemicals used to combat or suppress the growth of all or certain types of plants.
3. Fungicides are chemicals used to kill or suppress the growth of all kinds or of a certain type of fungus.

INSECTICIDES

The worldwide use of insecticides has been greatly increasing in agriculture and other fields since the end of the Second World War. Nowadays, there are a great

DOI: 10.1201/9781003272618-4

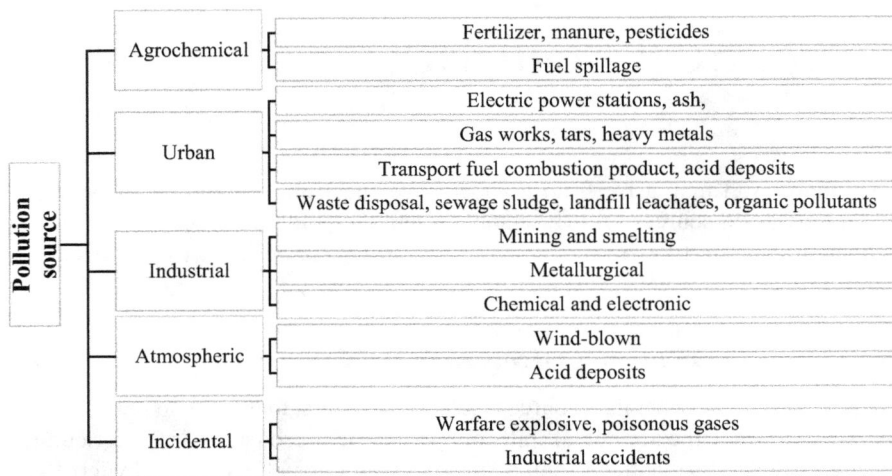

FIGURE 3.1 Sources of pollution (Mirsal, 2004).

number of commercial formulations for these products, yet they belong principally to four groups of organic compounds, providing a fundamental scheme for their classification. These are the organophosphorus compounds, the organochlorines, the carbamates, and the pyrethroids.

Organophosphorus compounds are technically nerve poisons, the basic technology of which was developed during the Second World War in Germany and Britain. They are used in many different ways in agriculture and animal hygiene. Some of them are used as fumigants; others are contact poisons, while still others are used as systemic pesticides (Figure 3.1).

Organochlorines

During World War II, a group of organochlorine compounds were found to be very effective in controlling pests responsible for diseases such as malaria and yellow fever. These compounds being cheap, easy to produce, and (at that time thought to be) safe to man and other warm-blooded animals were hailed as the best pesticides ever found by man. They belong to three chemical families: the DDT family, the BHC family, and the cyclodiene family.

Despite the fact that organochlorine compounds have been effectively used in the past in agriculture and hygiene, the later discovery (in the late fifties) of their persistence in the environment and their indiscriminate killing of beneficial as well as harmful insects has led to an emotional discussion about their use. This ended with a ban on their application in many developed countries. The ban decision is justified by the fact that the stability resulting from the inactive nature of the C–C, the C–H, and the C–Cl bonds forming these compounds makes them very persistent and hence dangerous for humans and animals. To this, we should also add the observation that due to their partition coefficients that favor the accumulation in biolipids, they tend to accumulate in body lipids of organisms exposed to their action. At present, organo-phosphorus and carbamate insecticides are largely replacing organochlorines.

Carbamates

These are derivatives of carbamic acid NH_2-COOH, of which about 40 commercial compounds, used as insecticides, molluscicides, or nematocides, are on sale. Their toxic effect, like that of the organophosphates, arises from their disruption of the nervous system by inhibiting cholinesterase. Carbamates are directly applied to the soil to control nematodes and snails or in order to be absorbed by root systems of weeds, where they operate as systemic pesticides after being translocated within the plant. Mild carbamate poisoning can affect behavioral patterns, reducing mental concentration and slowing the ability to learn (Hassall, 1982). Protein deficiency accentuates these symptoms. This renders their effect highly precarious; especially for poor farm workers and children in countries of the third world, where food shortage and protein lack is always a result of the bad economic conditions.

Natural and Synthetic Pyrethroids

Pyrethroids are originally quite effective natural pesticides, which were extracted from *Chrysanthemum cinerariaefolium.* Natural pyrethroids extracted from the dried pyrethrum flowers comprise four active ingredients known as pyrethrins I and II and cinerins I and II. The elucidation of the structure of natural pyrethroids made it possible to produce synthetic substances related to the pyrethroids and possessing similar or even higher insecticidal characters than the natural compounds. Some of these are preferred due to their lower toxicity, less persistence, and higher stability to light. Synthetic pyrethroids belong to four groups known as the allethrin, bioresmethrin, permethrin, and the fenvalerate groups.

HERBICIDES

The use of chemical weed control agents is a disputable problem among environmentalists since selectivity of these agents has never been completely achieved. After 1945, however, a considerable number of commercial organic compounds with some degrees of selectivity have replaced the older traditional herbicides such as copper sulfate solutions, dilute sulfuric acid, and petroleum oil. Main herbicides belong to one of the following groups:

Organochlorine Compounds

In this group, one encounters principally the derivatives of phenoxyacetic acid such as 2,4-dichlorophenoxyacetic acid, known as 2,4-D; 2,4,5 trichlorophenoxyacetic acid, known as 2,4,5 T; or 2-methyl-4, 6-dichlorophenoxyacetic acid, known as MCPA. Organochlorine derivatives of phenoxyacetic acid mimic natural growth hormones in weeds, leading to overproduction of RNA and death of the plants because their roots will not be able to deliver sufficient nutrition to support their abnormally induced growth. Besides derivatives of phenoxyacetic acids, derivatives of aniline are majorly high among organochlorine herbicides. Examples of these are propanil and alachlor.

Organophosphorus Herbicides

Organophosphorus herbicides, known as glyphosates, due to their effectiveness against weeds and their non-carcinogenic character, are widely used in agriculture. A glyphosate is a modified glycine. It mimics glycine and hence can be accepted by peptides, where it works as a synthesis inhibitor. It has a half-life in soil of about 60 days and is excreted by mammals unchanged.

Triazine Derivatives

Triazines are compounds in which three nitrogen atoms are incorporated into the benzene ring. Derivatives of these, like atrazine and simazine, are used as systematic weed control agents of relatively low toxicity for mammals. Water solubility of both atrazine and simazine is enhanced by enzymatic action of soil organisms leading to replacement of the chloro-substituent by a hydroxyl group. The same was also found to occur through dealkylation of these compounds by UV radiation. Accordingly, and after discovering that the use of triazine base herbicides polluted water supplies in the Thames Valley, the UK- government has banned use of both compounds. Some EU countries have also done the same.

Pyridine Derivatives

In pyridine, one nitrogen atom is incorporated into a benzene ring. Bipyridyl known under the name Diquat is used as systemic herbicide.

Aliphatic Compounds

There are few aliphatic compounds, used as herbicides. Of these, the product known under the commercial name Dalapon was found useful in controlling the couch grass. It is not persistent because of being readily hydrolyzed to pyruvic acid.

FUNGICIDES

Fungicides comprise a group of chemicals ranging from inorganic to organic compounds of comparable structures as the foregoing pesticides. Of these, the following are examples:

Inorganic and Organic Compounds of Heavy Metals

Examples are mixtures of copper-bearing inorganic compounds (e.g., Bordeaux mixture) or organometallic compounds such as organotins, which may be represented by tributyltin acetate or triphenyltin acetate.

Derivatives of Phthalic Acid

Example here is given by phthalimide, which is a compound produced by the reaction of phthalic acid with ammonia. This is marketed under several commercial names (e.g., Captan, Captafol).

Benzimidazole

Benzimidazole, a compound related to histamine, which is known for its blood pressure reducing character, is used as a systemic fungicide. The pentagonal ring in histamine is known as imidazole ring. Its fusion with a benzene nucleus gives the benzimidazole.

Derivatives of Barbituric Acid

Barbituric acid on treatment with phosphorus oxychloride followed by reduction with hydroiodic acid gives a group of compounds known as pyrimidines. These are used as fungicides.

FUEL SPILLS IN FARMS

Fuels through accidents or careless handling in farms may pollute soils. Fuels used in agriculture machines are mostly petroleum products that may contain organic contaminants like benzene; heptane, hexane; and isobutane, toluene, phenol, tetraethyl, and tetramethyl lead and zinc (anti-knocking compounds).

SOIL POLLUTANTS OF URBAN SOURCES

Soil pollution by materials of urban sources is a problem as old as urbanization itself. Archaeological studies show that, through construction and demolition of domestic concentrations and public centers of human activities (temples, sport arenas, etc.), a great deal of polluting substances was always dumped or disposed of on soils, resulting in their physical or chemical degradation. The damage of soil in those ancient days was of limited scale; yet since the beginning of the industrial revolution, it has taken dimensions that are hardly controllable in modern times. The main sources of urban soil pollution, however, are power generation emissions, releases from transport means and waste disposal (Bridges, 1991).

POWER GENERATION EMISSIONS

Emissions from power generation plants include COx, NOx, SOx, UOx and polycyclic aromatic hydrocarbons (PAHs) from coal-fired power stations and radionuclides from nuclear power plants. These may be introduced into the soil either directly as fall-out (dry deposition) or in a wet form after being dissolved in precipitation.

A number of organic and inorganic soil pollutants including tars, CN, spent iron oxides, Cd, As, Pb, Cu, sulfates, and sulfides may be released in sites of abandoned gas stations. The most abundant radionuclides found in soils originating from nuclear power generation are ^{137}Cs and ^{134}Cs.

In soils with a high cation exchange capacity (CEC) and pH values near 7.0, these radionuclides are normally absorbed onto clays and humic materials. Electric power generation in coal-fired power plants contributes not only to the addition of inorganic and organic pollutants to the soil through air born fly ash, but it also adds to the radioactive nuclides content of the soil. In the USA, many studies have been done on the concentration of uranium in fly ash, showing that uranium in fly ash may reach concentration of between 1 and 10 ppm. Despite the fact that these concentrations may not represent severe danger to individuals and life in general, chemical conditions under which uranium may be leached from fly ash and be concentrated in soil are still not completely understood.

Soil Pollution through Transport Activities

Transport activities in and near urban centers constitute one of the main sources of soil pollution not only because of the emissions from internal combustion engines and petrol spills but rather from these activities and their accompanying changes as a whole. To explain this, we should consider the breathtaking increase in highway construction projects all over the world. One also should not ignore the secondary or satellite land use activities attracted to the sites of newly constructed highways such as gas stations, moles, shopping centers, and all other services offered to car owners and commuters.

In fact, the impact of highways on the hydrogeologic environment may cause considerable transformations on the terrain leading to physical and/or chemical degradation of soil. These may be summarized in the following (Richard, 1973):

a. Water quality changes due to sediment damage to surface and groundwater supplies.
b. Pollution due to highway activities such as accumulations of oils, chemicals, and hazardous substances through accidental spills.
c. Pollution resulting from maintenance activities requiring the use of chemicals such as weed and insect control compounds as well as salts used to control the formation of ice in winter.
d. During highway road construction, cuts may expose pyrite-bearing strata that in turn would produce acid and other chemically polluted waters.
e. Enhanced new economic activities attracted to the highway site may result in producing huge amounts of roadside litter and debris.

The principle contribution of transport activities to soil pollution is caused by emissions from vehicles and airplanes, especially supersonic ones. Emissions from transportation means driven by internal combustion engines include oxides of carbon, nitrogen, and sulfur as well as some heavy metals. These pollutants may be transported to the soil by deposition of particulate matter or by being washed from the atmosphere.

Sulfur and nitrogen oxides on oxidation by photochemical reactions in the atmosphere react with water droplets in the air to produce strong acids such as HNO_3 and H_2SO_4. These acids are produced by reaction with bases (existing in the atmosphere mainly as particulate matter), a mixture of basic and acid radicals that dissolve in the rain, forming what has been known as the phenomenon of acid rain, causing great devastation in soils and plants.

As the concentration of these radicals together with carbon dioxide in surface and pore water approaches equilibrium, a great deal of change in the chemical environment of soil takes place, leading to drop in pH and increase in acidity of the soil. As a result, increasing intensity of weathering combined with the release of toxic, Al-ions from clay minerals as well as leaching of nutrients from the upper soil takes place.

SOIL POLLUTION BY WASTE AND SEWAGE SLUDGE

Of all urban sources contributing to soil pollution, waste and sewage sludge disposal occupies a central role in this environmental problem. In highly developed OECD countries despite retreating rates of population growth, the production of waste is still increasing, especially in the industrial sector. In developing and underdeveloped countries, high rates of population growth and increasing waste and sludge production, combined with lack of municipal services, create a dangerous situation. Waste produced by households is known collectively as municipal waste to differentiate it from waste originating from industrial processes. It includes various types of materials that may contribute to changing the environment of soil.

Municipal waste disposal by landfills and incineration may in both cases lead to concentration of heavy metals such as Cd, Cu, Pb, Sn, and Zn either directly from landfill leachates that may be polluting soil and underground waters or by ash fallout from incinerating plants. To this, we may add the effect of landfill gases that may pass to neighboring soils, causing a change in their soil air environment. The disposal of sludge produced by sewage treatment poses a great problem as well since in almost all developed countries the disposal of this sludge by dumping at sea is being phased out and the principal method of disposal is now shifting to land use. In fact, the mere use of sludge to amend soils is an advantageous process in itself. It adds essential organic matter as well as useful nutritive elements like phosphorus and nitrogen to the soil. Yet, pollutants such as heavy metals which are normally concentrated in the sludge may accumulate within the soil and eventually would be taken up by food crops as leafy vegetables, which are known to preferentially take up cadmium—one of the heavy metals that are normally abundant in sewage sludge.

Besides heavy metals, sewage sludge may include various organic micropollutants such as PAHs, PCDDs (polychlorodibenzo-p-dioxins), and PCDFs (polychlorodibenzofurans).

PCDDs or simply the dioxins are represented by over twenty isomers of a basic chlorodioxin structure and can be differentiated from each other through the number and positions of the chlorine atoms in a molecule. The most common form of dioxins is the 2,3,7,8-tetrachlorodibenzo-p-dioxin. Dioxins are considered the most toxic man-made chemicals. PCDFs such as 2,3,7,8-tetrachlorodibenzofuran are compared to 2,3,7,8- tetrachlorodibenzodioxin for toxicity and are considered as examples of the most lethal synthetic chemicals.

The above-mentioned substances are synthetic chemicals and none of them has been found to form as a result of any natural process. Their main sources are the following activities:

a. Municipal waste incineration
b. Chemical industry
c. Coal combusting power plants
d. Iron and steel industry
e. Car traffic
f. Hospital ovens
g. Forest industry

POLLUTION MECHANISMS AND SOIL–POLLUTANTS INTERACTION

Pollutants behavior and interaction with soil comprise various physical, chemical, and biological processes that take place in all three (solid, gas, and liquid) components of the soil medium. They generally include three main groups of processes:

1. Retention on and within the soil body
2. Infiltration, diffusion, and transport by soil solutions
3. Alteration, transformation, and initiation of chemical changes within the soil.

While the first two groups include mainly physical processes by which pollutants are transported and distributed in the soil, the third group comprises only chemical and biological processes by which pollutants are transformed or stored as residues in the interstitial space.

Physical processes of soil/pollutant interactions are those processes including transport and retention. They depend mainly on the physical parameters of the medium (temperature, grain size, electric charges, etc.), while chemical processes depend largely on the type of pollutants and their chemical nature. Both groups of processes are further classified according to the mechanisms involved. As for biological or biologically controlled soil pollution processes, we may include all processes of biotransformation and biodegradation, each depending on the microbial ecology, the depth, and the oxygen availability at the site of pollution (Figure 3.2).

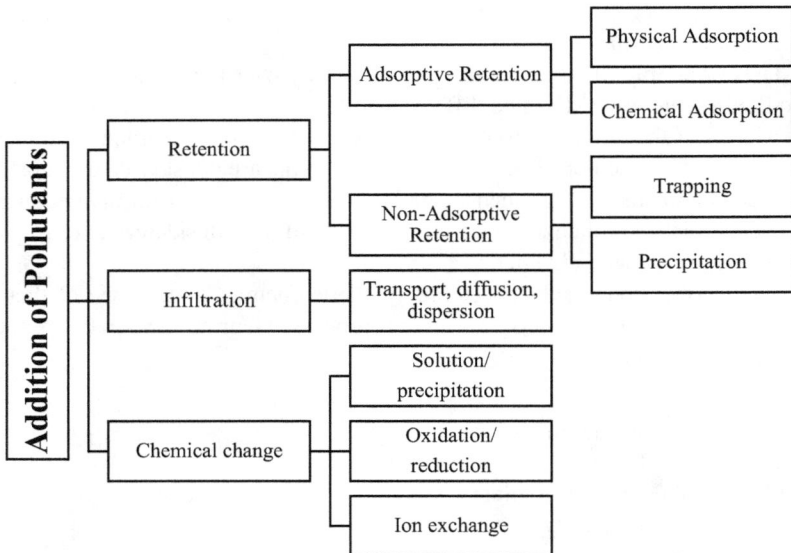

FIGURE 3.2 Mechanisms involved in soil pollution (Mirsal, 2004).

PHYSICAL PROCESSES AND MECHANISMS OF POLLUTION

Pollutants on encountering soil grains will either be retained by adsorption on the surface of theses grains or be accumulated in their intergranular space, where they may form concentrations retaining their original chemical composition or substances that have been altered by various chemical reactions. Pollutants retained thus on the soil surface or in its interstitial space may be organic, inorganic, or a mixture or complexes of both. They reach the soil in various physical conditions as solutes, water-immiscible liquids, or suspended particles. The mechanisms of their interaction with the soil will thus depend upon physical parameters prevailing in the soil medium such as temperature, moisture content, or salinity of the soil water, as well as upon their own physical and chemical properties. Adsorption and its accompanying phenomena are considered as the most important physicochemical mechanisms of pollutants retention on the surface of soil grains. In the following, these phenomena will be treated in some detail.

ADSORPTIVE RETENTION

Molecules of pollutants can be retained on the surfaces of soil grains in two ways. In physical adsorption, which is also known as physisorption, molecules of pollutants will be attached to the surfaces of soil grains by Van der Waals forces, which are known to stand for a long range, yet weak interaction. The amount of energy involved in such attachment is normally of low magnitude and is not sufficient for bond breaking. Thus, pollutant molecules sticking to the soil surface will retain their chemical identities, although they might be stretched or bent on account of their proximity to the surface.

   In chemisorption or chemical adsorption, the pollutants attach themselves to the grain surfaces as a result of the formation of a chemical (usually covalent) bond. In this case, the energy of attachment is very much greater than in physical adsorption. Thus, a molecule undergoing chemisorption can be torn to satisfy valency considerations arising from bond formation with the surface atoms. Although it is very difficult to differentiate between physical and chemical adsorption, one can generally say that the amount of physically adsorbed material decreases with increasing temperature, while this relation for chemically adsorbed material is reversed.

   Normally, various adsorbents exist in soil medium. Some examples of these are given by clay minerals, zeolites, iron and manganese hydrated oxides, aluminum hydroxide, humic substances, bacterial mucous substances, and plant debris. Many rock forming minerals such as micas, feldspars, some pyroxenes, and some amphiboles are also considered as good adsorbents of pollutant molecules. The capability of clay minerals and colloids in general to adsorb foreign molecules on their surfaces is attributed partially to their high surface energy and partially to the existence of a net surface charge ($\sigma5$) which may be caused by some functional groups (e.g., M-OH). These normally possess charges that are dependent on the sign and magnitude of the composition of the ambient liquid phase, as well as on the nature of the surface they are bound to. Such net charges attract ionic pollutants to the surfaces of the adsorbent material. However, non-ionic pollutants can also be adsorbed by soil grains. This occurs principally through electrostatic forces.

CHEMICAL ADSORPTION OR CHEMISORPTION

As it was stated before, formation of covalent bonds during chemical adsorption makes the energy of attachment very much greater than that in the case of physical adsorption. A molecule undergoing chemisorption can be torn to satisfy valency considerations arising from bond formation with the surface atoms.

Generally, the amount of adsorbed material in physical adsorption is inversely proportional to temperature. This relation is reversed in chemical adsorption. In chemisorption, molecules undergoing this process normally lose their identities as the atoms are rearranged.

FACTORS AFFECTING ADSORPTION

The intensity of adsorption depends upon several factors including physical and chemical properties of the pollutants themselves as well as the soil matrix, composition, and surface properties. It is generally possible to summarize all these factors as follows:

- Mineralogical composition of the soil
- grain size distribution in the soil
- the content and distribution of humic substances in soil
- chemical and physical properties of the soil solution
- CEC of organic and mineral components
- the pollutants, their nature, and chemical constitution
- to the above, we may add external conditions such as climatic conditions and agricultural practices.

MINERALOGICAL COMPOSITION OF THE SOIL

As mentioned before, clay minerals are the most important adsorbents in the soil environment followed by some silicates and organic components. Accordingly, the intensity of adsorbance in soils will largely depend on the clay content of the soil as well as on the share of other silicates in the mineralogical composition. The negative framework of the clays consists essentially of sheet structures of aluminum silicates in which the exchangeable cations occupy interlayer positions or are located adjacent to the particle surfaces.

SOIL MATRIX (GRAIN SIZE DISTRIBUTION)

It has generally been observed that the rate of adsorption is higher on finer sediments than on coarser ones (Kennedy and Brown, 1965; Malcolm and Kennedy, 1970).

Despite the fact that concentration of certain cations in finer sediment fractions is also known from other sediments (e.g., carbonates), the following possibilities should be taken into consideration when interpreting this phenomenon for silicates. In the case of sandy sediments, when Ca and Na are concerned, relative hardness of certain silicate minerals (e.g., feldspars) and their lower resistance to abrasion, relative to harder silica compounds with low Na and Ca content (e.g., Quartz), may control the distribution of such cations between different grain size fractions in the same sandy layer.

HUMIC SUBSTANCES AND THEIR DISTRIBUTION IN THE SOIL

Humic substances containing carboxyl and phenolic hydroxyl functional groups increase the CEC of the soil. In general, the presence of active functional groups (e.g., carboxyl, hydroxyl, carbonyl, methoxy, and amino groups) is thought to be of positive influence on the CEC of a soil.

CHEMICAL AND PHYSICAL PROPERTIES OF THE SOIL SOLUTION

A major part of pollutants, however, passes in solution or in particulate form to the vadose or even saturated groundwater zone. In the presence of clays, water molecules are adsorbed on their surfaces to form hydration shells; these provide adsorption sites for pollutant molecules. Water adsorbed on clay molecules generally has higher rates of dissociation providing surfaces of acidic character that may increase the exchange capacity of the soil. Mechanisms by which pollutants are transported to deeper horizons of the soil are collectively called infiltration. Descending contaminated fluids that might end in joining the vadose or saturated zones of groundwater are generally known as leachates. In the vadose zone, leachates spread horizontally in the direction of groundwater flow.

THE POLLUTANTS, THEIR NATURE, AND CHEMICAL CONSTITUTION

The composition and nature of contaminants control to a considerable extent not only solution and diffusion processes but also adsorption on the soil grains. Such control may be explained by the fact that ion exchange and hydrolysis reactions are particularly sensitive to the parameters (pH, Eh) of the chemical environment created by the contaminants in their direct vicinity. An example of this may be provided by the adsorption of organophosphorus pesticides on clay surfaces.

NON-ADSORPTIVE RETENTION

TRAPPING

Entrapment of solid particles and large dissolved molecules in the pore space of the soil forms one of the major mechanisms of retention of pollutants in the soil (Boulding, 1995).

CAKING

This may occur physically when the pollutant particles are larger than the soil pores. In this case, the entrapped particles form a layer (cake) on the surface where the pore sizes become too small. Caking may also result from biological activities through which particles cluster in bigger lumps that clog the soil pores.

STRAINING

Straining occurs when pollutant particles are about the size of the soil pores. They move down the pores until they are entrapped at the entrance to a pore which is too small.

PHYSICAL–CHEMICAL TRAPPING

Limitation of flow through clogging of pore space may occur because of physical or chemical transformation such as the production—by chemical reactions—of new products having molecular sizes that exceed that of the soil pores. An example is the flocculation of colloidal material resulting from the precipitation of iron and manganese oxides.

PRECIPITATION

Retention of contaminants in soil may often occur through passing of contaminants from a dissolved form to an insoluble form in the course of geochemical reactions taking place within the soil pores. Precipitation reactions are controlled by acid-base equilibria and redox conditions. They are reversible and may lead to dissolution of formerly precipitated compounds if conditions are changed.

INFILTRATION

This is perhaps the most common mechanism of contamination of soil solutions in the vadose zone as well as deeper regions of the saturated zones of groundwater. As fluids move downward under the influence of gravity, they dissolve materials to form leachates that contain inorganic and organic constituents. As they reach the saturated zone of groundwater, the contaminants spread horizontally and vertically by joining the main cycles of geochemical flows.

FURTHER READING

Boulding R. *Practical Handbook of Soil, Vadose Zone and Ground Water Contamination: Assessment, Prevention and Remediation.* Boca Raton, FL: Lewis Publishers; 1995.

Bridges EM. Waste materials in urban soils. In Bullock P, Gregory PJ (eds.) *Soils in the Urban Environment.* Oxford: Blackwell; 1991:28–46.

Duarte A, Cachada A, Rocha-Santo, T. In: *Soil Pollution. From Monitoring to Remediation.* 1st ed. Academic Press; 2018.

Finkl CW. Soil mineralogy. In: *Mineralogy. Encyclopedia of Earth Science.* Boston, MA: Springer; 1981.

Hassall KA. *The Chemistry of Pesticides: Their Metabolism, Mode of Action and Uses in Crop Protection*. Verlag Chemie; 1982.

Kennedy VC, Brown TC. Experiments with a sodium ion electrode as a means of studying cation exchange rates. *Clays Clay Min*. 1965;13:351–352.

Malcolm RL, Kennedy VC. Variation of cation exchange capacity and rate with particle size in stream sediments. *J. Water Pollut. Control. Fed*. 1970;42:153.

Mirsal IA. *Soil Pollution: Origin, Monitoring & Remediation*. Germany: Springer Berlin Heidelberg, Germany; 2004.

Mirsal IA. *Soil Pollution: Origin, Monitoring & Remediation*. Germany: Springer Berlin Heidelberg, Germany; 2008.

Richard RP. Impact of high ways on the hydrogeologic environment. In: Coates DR (ed.) *Environmental Geomorphology and Landscape Conservation. Vol III, Benchmark Papers in Geology*; 1973.

4 Air Pollution

Earth's atmosphere formed from physical and biological processes that took place over billions of years. Although the atmosphere is a continuous cloak of gas molecules and other particles, this cloak can be divided into layers or strata. These strata are not based on the composition of the atmosphere itself, because the types of gases and their abundance relative to each other do not vary much (Desonie, 2007).

Primarily, the atmosphere is divided into layers based on how the air temperature decreases with height in the lower atmosphere and increases with height in the upper atmosphere. The density (mass per unit volume) of gases decreases with the increase in height above mean sea level, as well. Nearly the entire atmosphere of the Earth—99%—lies within 19 miles (30 km) of sea level. At around 435 miles (700 km) out, the atmosphere is almost completely devoid of matter.

THE EVOLUTION OF THE ATMOSPHERE

The Earth formed from a swirling cloud of dust and gas about 4.5 billion years ago. Gases entered the atmosphere from two main sources: inner space and outer space. Gases that were trapped in the Earth's interior were part of its inner space. Just as they do today, gases—about 80% water (H_2O), 10% carbon dioxide (CO_2), and a few percent nitrogen (N)—erupted from volcanoes and steam vents. From outer space came comets that were composed of ices and trapped gases—primarily H_2O, CO, CO_2, methane, and ammonia, along with some organic compounds. These gases were liberated as the comets struck the planet.

These gases, from both inner and outer space, formed the planet's early atmosphere. Water vapor condensed, formed clouds, and precipitated, leading to the creation of rivers, lakes, and oceans on the barren surface. The early ocean absorbed CO_2, yet the gas was still many times more abundant in the atmosphere than it is today. Thus, the greenhouse effect was much stronger at that time. The planet was not hotter than it is now because the Sun did not burn as brightly. Under these early Earth conditions, surface water was stable as a liquid, a necessity for the origin of life.

As of now, nitrogen was the most abundant gas, but the second most abundant gas today, oxygen (O_2), was barely present. This is because there were no plants, and without them, there was no photosynthesis, which is the source of nearly all of the planet's oxygen. Without oxygen, there was no protective ozone layer, and life could not evolve beyond very simple, single-celled organisms. Once single-celled plants evolved and colonized the planet, they supplied O_2 to the atmosphere.

For hundreds of millions of years, O_2 was taken up by many elements and compounds that were waiting to combine with oxygen, a process called oxidation. (Iron will readily oxidize to rust when oxygen is available, for example.) Only after the exposed minerals were oxidized could O_2 build up in the atmosphere and only then could animals arise. Once multicellular plants were able to survive, between 1 billion

DOI: 10.1201/9781003272618-5

and 543 million years ago, they added oxygen to the atmosphere rapidly. At about 450 million years ago, the planet reached its present oxygen level of 21%. With the proliferation of complex life forms, the composition of the atmosphere became fairly stable. Quantities of nitrogen and oxygen and most other gases have remained about the same for hundreds of millions of years. Carbon dioxide levels have always been variable, depending on input from volcanic eruptions and input or output into the oceans and plant materials. Water vapor is variable depending on local conditions.

There is a synergy between the Earth's atmosphere and its organisms. It is no accident that plants and animals live in an atmosphere that is mutually favorable to them. Organisms evolve to live in the environment that is available. What is important in biological evolution is adapting to, and staying adapted to, the environment. But, in very important ways, the composition of the Earth's atmosphere has also been shaped by the organisms that live in it. With their intertwined cycles of photosynthesis and respiration, plants and animals have kept O_2 and CO_2 at favorable levels for themselves for hundreds of millions of years. Not only have they adapted to their environment but they also have worked together to create it.

By adding CO_2 and other greenhouse gases to the atmosphere, modern human society has disturbed the long-term balance of the atmospheric gases. Chemicals that have never existed in nature, such as chlorofluorocarbons (CFCs), have been introduced into the atmosphere.

Ozone is being added to the lower atmosphere, where it does harm by helping to create smog, and depleted from the upper atmosphere, where it does good by protecting the Earth from the dangerous forms of UV radiation.

ATMOSPHERIC PRESSURE AND TEMPERATURE

Air is made of gas molecules that can move freely. Like all matter, air molecules are attracted to the Earth by its gravity, which draws objects to the center of the planet. Gravity is strongest at lower altitude (the height above sea level) and air molecules are packed closest together at sea level. Air is also compressed by the weight of all the air above it. The weight of a column of air from the top of the atmosphere onto a person's shoulders at sea level is more than one ton. But people and animals are not crushed because billions of molecules inside our bodies are pushing outward to compensate. The force of the air weighing down over a unit of area is known as its atmospheric pressure or air pressure.

Atmospheric pressure decreases with increasing altitude because there is less gravity and less air to weigh down from above. Air density, then, is greatest at the Earth's surface and decreases with altitude. Each 3.7-mile (6-km) increase in altitude reduces the weight of the atmosphere above it by half so that at 18,000 feet (5,500 m) above sea level, air pressure is only half of what it is at sea level. People feel changes in atmospheric pressure while on an airplane or driving through the mountains when their ears "pop" as they go up or down in altitude. This occurs because the air molecules inside their ears maintain the density of the previous altitude until they have had a chance to equilibrate with the air pressure at the new altitude. The dense packing of air molecules near the Earth's surface restricts the ability of each molecule to move. At sea level, a molecule can travel an average distance of less than one millionth of a centimeter before it collides with another molecule. Each collision between molecules releases heat, so the air at sea level is relatively warm. At higher altitudes, where the molecules are not packed in so tightly, they are less likely to collide, so the air becomes cooler.

The transfer of heat is important for driving the motions of the atmosphere. Heat transfer occurs in two different ways. Conduction is the transfer of heat through a substance that has different temperatures in different parts. Because warm atoms and molecules move more vigorously than cold ones, the particles in the warmer region strike their neighbors, transferring heat until it is evenly distributed. Convection transfers heat by the movement of currents. Think of a room with a floor heater. As the air near the heater becomes warmer, the air's density decreases and it rises. The air near the ceiling is pushed sideways by the rising air. The sideways movement of air is called advection. Because the air pushed along the ceiling is now far from the heater, it is relatively cool. When it becomes denser than the air beneath it, it sinks. The air then moves by advection along the ground until it again is near the heater. The circuit described above is a convection cell. Warm air rises to make a low pressure zone, cool air sinks to make a high pressure zone, and air moves between the two. Something else happens to air as it rises or sinks. Warm air can hold more moisture than cold air, so as warm air rises and cools, it is able to hold less moisture, which may result in precipitation. More water may result in evaporation of water from the Earth's surface.

THE LAYERS OF THE ATMOSPHERE

The composition of atmospheric gases is about the same at different altitudes, with the important exception of the ozone layer. Despite being similar in its composition throughout, the atmosphere is divided into layers that are defined primarily by temperature gradient, which is the change of temperature that occurs with distance (or, in this case, altitude).

The layer nearest to the Earth's surface, rising from sea level to about 6 miles (11 km), is the troposphere. The primary heat source for the troposphere is infrared energy (heat) that radiates from Earth's surface. This layer measures a decrease in temperature of about 3.6°F per 1,000 feet (6.5°C per 1,000 m) of altitude. The value fluctuates with the day, the location, and the season. Sometimes, a portion of the troposphere has a temperature inversion and the situation is reversed: Air temperature increases with height. Almost all of the weather found at the planet's surface is due to the vertical movement of air in the troposphere.

The stratosphere rises from the top of the troposphere to about 30 miles (45 km). Since this layer is heated by the Sun's UV rays, it gets warmer with increasing proximity to the Sun. The warm air of the upper stratosphere "floats" on the cooler air of the lower stratosphere, since it is less dense. With warmer air above cooler air, the stratosphere is very stable vertically. This layer usually experiences very little turbulence, which is why commercial airliners fly at this level. Not only does air within the stratosphere not mix but also there is almost no mixing between the stratosphere and the troposphere beneath it. Ash and gases shot into the stratosphere by a volcanic eruption may remain there for many years.

The stratosphere contains the ozone layer, which lies between 9 and 19 miles (15 and 30 km) from the Earth's surface. Even here, the ozone concentration is quite small, measuring only about 12 ozone molecules for every 1 million air molecules. As small as it is, the ozone concentration is one reason that the stratosphere warms with altitude. The ozone molecules absorb the high-energy UV as they break apart into molecular oxygen (O_2) and atomic oxygen (O).

Air density decreases in the layers of the atmosphere that lie beyond the stratosphere. In each of these layers, the air molecules are very far apart and the air is very cold. Beyond the atmosphere is the solar wind, which is made up of high-speed particles traveling rapidly outward from the Sun.

AIR POLLUTION

An atmosphere is the gases and particles that surround a planet or a moon. Some of the Earth's atmospheric gases, like oxygen and carbon dioxide, are essential for plants and animals as they carry out life processes. A few gases trap heat and keep the planet's temperatures moderate. One particular gas protects Earth's life from the Sun's harmful radiation (Table 4.1).

GASES NEEDED FOR BIOLOGICAL PROCESSES

Nitrogen (N_2) and oxygen (O_2) make up 99% of the gases found in the atmosphere. Although other components comprise only the remaining 1%, some are extremely important. The table above is a list of atmospheric gases and their concentrations. Many gases are in balance in the atmosphere; that is, the amount that enters, the

TABLE 4.1
Concentrations of Atmospheric Gases (Desonie, 2007)

Gases	Concentration (%)
Nitrogen	78.08
Oxygen	20.95
Water vapor	0–4
Argon	0.93
Carbon dioxide	0.036
Neon	0.0018
Methane	0.0017
Helium	0.0005
Hydrogen	0.00006
Nitrous oxide	0.00003
Xenon	0.000009
Ozone	0.000004
Particles	0.000001
Chlorofluorocarbon	0.00000002

input, equals the amount that leaves, the output. Nitrogen, the most abundant atmospheric gas, is in balance. This gas is input by the decay of plants and animals and is output by the activities of bacteria in the soil. Argon, neon, helium, and xenon—the noble gases—are also in balance. These gases are colorless, odorless, tasteless, and chemically inert; they do not undergo chemical reactions with other elements or compounds. Carbon dioxide (CO_2) and oxygen (O_2) are the most important gases for living organisms. CO_2 is a tiny component of the atmosphere, accounting for only 36 out of every 100,000 gas molecules, yet it is fundamental to nearly all life on Earth. The gas is essential for photosynthesis, the process by which plants take raw materials that are abundant in the environment and turn them into food.

Carbon dioxide is absorbed from the atmosphere in several ways. Most importantly, CO_2 is used by plants during photosynthesis. CO_2 is also stored in plant tissue and in soil, where it goes when plants decay. In the oceans, CO_2 moves freely between the sea surface and the atmosphere, but when it sinks into the deep sea, the gas is sequestered from the atmosphere. CO_2 gas also enters the atmosphere by respiration, during volcanic eruptions, and when plant materials decay or burn. The CO_2 content of the atmosphere is currently not in balance. Although the inputs and outputs from respiration and photosynthesis are equal, humans are increasing the input of CO_2 by burning plants for agricultural purposes and by burning fossil fuels, which are made from ancient plants that the Earth's processes have transformed into oil, gas, or coal for use by humans as fuel.

Ozone for Protection from Ultraviolet Radiation

Ozone is a molecule composed of three oxygen atoms (O_3). In the layer of the upper atmosphere known as the stratosphere, ozone filters out the Sun's harmful high-energy ultraviolet (UV) radiation. The ozone molecule forms in the stratosphere when UV energy breaks down some O_2 molecules to make single O-atoms. These O-atoms then bond with other O_2 molecules to form O_3. The reverse process also takes place in the stratosphere as UV energy breaks apart O_3 molecules to make O_2 and O. The breaking down of O_3 into O_2 and O absorbs the most dangerous UV radiation, UVC, which comes in from the Sun. By filtering out UVC, stratospheric ozone protects living things at or near the Earth's surface. Stratospheric ozone is concentrated in what is called the ozone layer, where the high-energy UV is broken down. Normally, inputs and outputs of O_3 into the ozone layer are equal, but in recent decades, human activities have brought about a decrease in stratospheric ozone.

A Reservoir for Water

To keep water moving between the atmosphere and the Earth's surface, the gaseous form of water, or water vapor, must pass through the atmosphere. While air is never dry, the amount of water vapor it contains varies from place to place and from time to time. Humidity is the concentration of water vapor in the air. Up to 4% of the volume of the air can be water vapor. This vapor is created in the atmosphere when liquid water at the Earth's surface changes from a liquid to a gas, a process called evaporation. Although water vapor is invisible, it may convert into tiny liquid droplets, in a process known as condensation, to form clouds. The droplets can come together to create precipitation—rain, sleet, hail, snow, frost, or dew.

GREENHOUSE GASES FOR INSULATION

Carbon dioxide and several other gases, both natural and man-made, are greenhouse gases. The presence of greenhouse gases makes complex life on the Earth possible. Although UVC and some UVB are filtered out by the ozone layer, most of the Sun's radiation (lower energy UV and visible light) passes through the atmosphere unimpeded. When this radiation hits the planet's surface, the energy is absorbed by soil, rock, concrete, or water and then is reemitted as heat.

Greenhouse gases in the atmosphere trap some of this heat and cause the atmosphere to warm, a property known as insulation. The warming of the atmosphere is called the greenhouse effect because it works somewhat like the glass on a greenhouse. Without the greenhouse effect, the Earth's average atmospheric temperature would be a very low 0°F (−18°C). Temperatures also would be extremely variable, scorching in the daytime and frigid at night, such as those on the Moon and the planets that have no atmosphere.

Most greenhouse gases are present naturally in the environment. For example, carbon dioxide is input by respiration, volcanic eruptions, and the burning of plant material. Another greenhouse gas, methane, is a hydrocarbon gas (an organic compound composed of hydrogen and carbon) with a variety of inputs including the breakdown of plant material by bacteria in rice paddies and the biochemical reactions that occur in cow stomachs (i.e., by cows passing gas). The nitrous oxides, NO and NO_2—referred to together as NO_x—are produced naturally by bacteria. Ozone is found naturally in the lower atmosphere in small amounts. Some man-made gases, such as CFCs, are not present naturally and are greenhouse gases in the lower atmosphere.

PARTICULATES FOR CONDENSATION

Solid and liquid particles in the atmosphere, called particulates or aerosols, are necessary for the development of clouds and precipitation. Particulates provide a nucleus for water vapor to condense on to form clouds and precipitates such as raindrops and snowflakes. Particulates are solid particles that are light enough to stay suspended in the air and include windblown dust and soil, fecal matter, metal beads, saltwater droplets, smoke from fires, and volcanic ash. Some particulates are the result of human activities, such as fossil fuel burning.

The atmosphere provides a place for the gaseous waste products of modern human society to go. For this reason, it contains gases that were never before in the atmosphere, such as CFCs, or that are present in unnatural locations or quantities. Air pollution has a variety of effects, from raising global temperature, to destroying natural atmospheric processes, to simply dirtying the air.

Fossil fuel burning releases enormous quantities of pollutants, including nitrogen dioxide (NO_2), sulfur dioxide (SO_2), carbon monoxide (CO), and hydrocarbons. Some pollutants do not come directly from fossil fuel emissions but are the result of secondary chemical reactions. For example, the action of sunlight on nitrogen oxide and hydrocarbon pollutants in the lower atmosphere forms ozone (O_3). While ozone in the stratosphere is beneficial, in the lower atmosphere—the troposphere—this gas is a pollutant and is the primary component of photochemical smog, which is air pollution that results from a chemical reaction involving pollutants and sunlight. Also known as "bad" ozone, it can be extremely harmful to animals, plants, and humans. It is also a greenhouse gas.

While atmospheric greenhouse gases are inarguably good for the planet and its life, increased levels of greenhouse gases are not. Additional greenhouse gases amplify the insulating properties of the atmosphere and result in a boost in global temperatures. Increased atmospheric greenhouse gases are responsible for at least some of the current warming of the planet. Of the greenhouse gases that are increasing, carbon dioxide is not the most potent, but it is the one that is most on the rise. Plants store CO_2 in their tissues, and plant materials that have been converted to fossil fuels emit CO_2 when burned. As a result, atmospheric CO_2 levels have been rising sharply since the Industrial Revolution began about 150 years ago. Methane levels have been going up for the past century due to the expanded agricultural production necessitated by the swelling human population. Water vapor levels are increasing because warm air holds more water vapor than cool air.

CFCs are greenhouse gases that are released into the lower atmosphere and destroy the ozone layer when they reach the upper atmosphere. These man-made chemicals were once widely used. These compounds are extremely stable and they continue to be present in the atmosphere.

Atmospheric gases support life by assisting with the production (photosynthesis) and utilization (respiration) of food energy, by protecting life on Earth from harmful solar rays (the ozone layer), by keeping global air temperatures moderate (the greenhouse effect), and by providing a reservoir for water by forming clouds and precipitation. Besides providing life support, the atmosphere gives humans a sink for gaseous pollutants.

AIR POLLUTANTS AND AIR POLLUTION

POLLUTANTS FROM FOSSIL FUELS

Air pollution today is largely caused by the burning of fossil fuels. Primary pollutants enter the atmosphere directly from a smokestack or tailpipe. Secondary pollutants form from a chemical reaction between a primary pollutant and some other component of air, such as water vapor or another pollutant. Ozone is the major secondary pollutant.

Fossil fuels come from decayed and transformed ancient organisms. Plants store CO_2 in their bodies, so plant materials that have been converted to fossil fuels emit CO_2 when burned. The two major types of fossil fuels are coal and petroleum. Coal forms in swamps, where plants grow and die in rapid succession. The plant bodies accumulate so quickly that little oxygen can get to them and they do not decay effectively. If this material is buried deeply enough, it is compressed and transformed into coal.

Petroleum forms in ocean regions, primarily along the margins of the continents, where plankton—tiny plants (phytoplankton) and animals (zooplankton)—flourish.

When these organisms die, they fall to the sea floor and are buried by sediment. If they are buried deep enough, high temperatures and pressures convert them to oil and natural gas. Petroleum is a mixture of different hydrocarbons, which are compounds composed of hydrogen and carbon. When burned completely, these compounds produce CO_2 and water vapor. The energy that comes from burning fossil fuels can be thought of as ancient solar energy that was taken in by long-dead plants and animals.

The CO_2 and water vapor produced by the complete burning of pure coal and petroleum are both greenhouse gases. While greenhouse gases contribute to the warming of the Earth, they do not pose any direct threat to human health. But fossil fuels are rarely pure and often release other pollutants such as carbon monoxide (CO), nitrogen dioxide (NO_2), sulfur dioxide (SO_2), and hydrocarbons.

Carbon monoxide is a colorless, odorless gas that is lethal in high quantities. CO kills by substituting for O_2 in the blood, which starves the brain and other body parts of the much-needed O_2. In the United States, motor vehicle exhaust accounts for about 60% of all CO emissions and as much as 95% of CO emissions in cities. CO does not build without limit in the atmosphere, because it is removed by microorganisms in the soil. Nonetheless, this toxic gas causes problems in poorly ventilated tunnels or parking garages; because people cannot detect it by smell, it kills without warning. CO is a greenhouse gas.

The nitrous oxides (NO_x) are also greenhouse gases. Nitrifying bacteria in the soil and in plant roots naturally produce nitrous oxides during decomposition. These gases also are produced by human activities. When coal and petroleum are burned, they emit nitrogen, which reacts with O_2 to form NO_2 and NO. NO_2 is a noxious, reddish brown gas that contributes to the mucky russet color and odor of the air in polluted cities. The concentration of NO_x in urban areas is 10–100 times greater than in rural areas. In wet air, NO_x reacts with H_2O to form nitric acid (HNO_3), a component of acid rain, which is rain that is considerably more acidic than normal rainwater. Another major component of acid rain is sulfuric acid (H_2SO_4), which forms when sulfur dioxide (SO_2) mixes with water vapor (H_2O). SO_2, a common emission of low-grade coal and petroleum combustion, is a main ingredient of industrial pollution. When sulfur from burning impure coal reaches the air, it combines with oxygen to form sulfur oxides, mainly sulfur dioxide (SO_2) and sulfur trioxide (SO_3).

Mercury is a heavy metal that is released by burning coal, municipal and medical wastes, and by volcanic processes. According to the EPA, mercury is the most hazardous material emitted by power plants. Mercury does not do its damage as a gas. Although it is released into the air by combustion, it turns into aerosol droplets as it cools. These droplets can travel hundreds of miles through the atmosphere before falling to the ground or into the water, where they are deposited in sediments. Bacteria in the sediments then convert the droplets to organic mercury, usually the dangerous compound methyl mercury. When an organism takes in methyl mercury, the substance is stored in fat and muscle instead of being excreted. This means that animals accumulate all the methyl mercury that is contained in all the organisms that they eat. During their lifetimes, small organisms take in only a small amount of methyl mercury, but large, predatory animals, such as tuna and polar bears, eat more and store more of the substance. This process is called bioaccumulation. Because of it, predatory fish that live at the top of the food chain, such as tuna, may have extremely high methyl mercury concentrations. This is a great risk to the birds, mammals, and humans that eat food that is located high on the food chain.

Lead is the most common toxic material found in humans and enormous amounts of it are produced each year. Tetraethyl lead (a compound in which lead is combined with carbon and hydrogen) was first included as an ingredient in gasoline in the 1920s to increase fuel efficiency and to counter engine knock; it was also an ingredient in paints and other materials. Even when it was first used, tetraethyl lead was known to be toxic, but safer compounds were more expensive. In industrialized nations, lead has been banned from many uses, including gasoline, although it is still used in diesel fuels.

POLLUTANTS FROM SECONDARY CHEMICAL REACTIONS

Some pollutants, most importantly ozone (O_3), do not come directly from fossil fuel combustion but are the result of a two-step process. First, hydrocarbons from incompletely burned gasoline react with nitrogen oxides and atmospheric oxygen to form ozone. Second, the ozone reacts with automobile exhaust to form photochemical smog. This noxious, bad-smelling gas damages the lungs and is extremely harmful to animals and plants. Ozone is also a greenhouse gas. As a secondary pollutant, ozone production can be decreased only if both nitrogen oxides and hydrocarbons are reduced. Therefore, it is a difficult problem to combat. Ozone is not restricted to cities; it also is produced in the countryside, where hydrocarbons are given off by vegetation and nitrogen oxide pollution drifts out from urban areas to add to it.

POLLUTANTS FROM BIOMASS BURNING

Burning plant and animal material also produces pollutants. Biomass is the amount of living material found in an environment; in this case, material primarily from plants. Slash-and-burn agriculture, the preferred method for farming in tropical regions, is an enormous source of pollutants. In this type of agriculture, rain forests are chopped down and burned, and the land is then farmed. But rain forest soil is infertile, and in a few years, the farmer needs to slash and burn another patch of forest. Other biomass that is commonly burned includes savanna, fuel wood, leftover material from crops, and dried dung, which is burned for fuel. The pollutants that are emitted from biomass burning are similar to those emitted by fossil fuels, because fossil fuels are ancient, transformed plant materials. Of primary concern are CO_2, CO, methane, particulates, nitrogen oxide, hydrocarbons, and organic and elemental carbon. The first three are greenhouse gases. Burning forests increases atmospheric CO_2 levels in two ways: by releasing the CO_2 that had been stored in the plants into the air and by stopping the forest from sequestering more CO_2.

Particulates are the byproducts of fossil fuel and biomass burning. They also can enter the atmosphere from natural sources such as volcanic ash or windblown dust. Most particulates are non-toxic, but they may cause health problems if they penetrate the lungs.

VOLATILE ORGANIC COMPOUNDS

Volatile organic compounds (VOCs) are mostly hydrocarbons that enter the atmosphere primarily by evaporation. Some VOCs form naturally, but most are synthetic (man-made). VOCs are found in paint thinners, dry cleaning solvents, petroleum fuels, and wood preservatives, among other materials. There are tens of thousands of different VOCs in the air: Some are harmless, others are poisonous, and the effects of many are not fully known. Between 50 and 100 airborne VOCs are typically monitored. Methane (CH_4) is the most common VOC; it poses no immediate threat to human health, but it is a greenhouse gas. Although methane is produced naturally, it now enters the atmosphere in unnatural amounts due to an increase in certain agricultural practices. Some VOCs contribute to ozone production in the lower atmosphere.

CFCs are synthetic VOCs. Although they were once widely used, they are now being phased out. These compounds are extremely stable. Long after their manufacture, they continue to rise into the upper atmosphere, where they break down stratospheric ozone. CFCs will eventually break apart and will no longer be able to destroy the ozone layer. CFCs are also greenhouse gases.

Some pollutants are compounds that are not manufactured or used; they are simply a byproduct of another process. The most important example is dioxin, which serves no useful purpose but is widespread in the environment. Dioxin is a byproduct of the production of herbicides, disinfectants, and other chemicals. This chemical also forms when specific compounds are burned, such as the plastic polyvinyl chloride. In this instance, some of the chlorine reacts with organic compounds to form dioxin. No living creature, no matter how remote its home, is completely untainted by dioxin. The harmful effects of dioxin are not yet fully understood, but it may restrict the body's ability to manufacture proteins and decrease immune system function. Some studies suggest that dioxin may cause cancer—the family of illnesses characterized by uncontrolled cell growth—and other serious illnesses.

AIR POLLUTION AND THE ENVIRONMENT

Air quality not only depends on the types and amounts of pollutants that are in the air but also on external factors such as wind speed, atmospheric stability, and landscape. Stagnant air, such as the air beneath a temperature inversion or in a windless location, will collect more pollutants because clean air is not coming in and pollutants are not being removed. Pollutants have an adverse effect on the environment. Particulates reduce visibility and obscure sunlight. Ozone alters the plant species found in an ecosystem and reduces crop yields. Even in areas that have been set aside for protection, such as national parks, air pollution can impact the natural system and the visitor experience.

EXTERNAL FACTORS THAT AFFECT AIR QUALITY

The amount and type of pollutants entering the air varies by time and location. Bad air days can occur in the winter, when there is a buildup of wood smoke or a temperature inversion, or in the summer, when photochemical smog is at its worst. The air quality of a location also depends on external features, such as winds, temperature inversions, and the local topography (the ups and downs of the landscape). Winds move polluted air away from a region or bring in fresh air to dilute it. On days when winds are strong, the air is rapidly cleansed. When there is little or no wind, the air stagnates, with little input of clean air and little chance for output of pollutants.

THE EFFECTS OF PARTICULATES ON THE ENVIRONMENT

Particulates primarily affect the environment by impairing visibility, reducing solar radiation, and altering climate. Particulates impair visibility in large cities, small towns where there is a lot of industry, and even in the national parks. Particulates also reduce the quantity and quality of radiation received by plants, sometimes obstructing photosynthesis. Particulates can alter climate by changing temperature and precipitation; increase nutrients, such as nitrogen and iron; and cause acid rain. Some of these effects may inhibit plant growth and some may increase it. Deviation from natural conditions may favor non-native species so that an ecosystem will be overrun with species that do not belong within it.

THE EFFECTS OF OZONE ON THE ENVIRONMENT

Ozone affects trees and other natural plants and food crops. Commonly, ozone slows plant growth rates. Food crops vary in their sensitivity to the gas; some are affected a great deal and some not at all. Spinach, for example, develops spotted leaves, which make it unmarketable; soybeans suffer reduced crop yields. Different amounts of ozone damage occur under different environmental conditions. Ozone enters plants through small openings in their leaves known as the stoma. Conditions that cause the stoma to open more, such as strong light or high temperature and humidity, maximize ozone damage to the plant. Conditions that cause the stoma to close, such as a shortage of water or the presence of some pollutants, lessen the damage.

Ozone may also alter natural ecosystems. Forests are especially vulnerable since trees live a long time and ozone effects accumulate over many years. Some trees are more sensitive to ozone than others, and if a forest's native trees become sick or die, they may be replaced by more ozone-tolerant species. Tree species that are sensitive to ozone include American sycamore (*Platanus occidentalis*), flowering dogwood (*Cornus florida*), Jeffrey pine (*Pinus jeffreyi*), paper birch (*Betula papyrifera*), quaking aspen (*Populus tremuloides*), red maple (*Acer rubrum*), and white pine (*Pinus strobus*). If ozone damage becomes too great, these important species may be lost and non-native species may move into forest areas. With different trees forming the forest framework, entire ecosystems can be altered.

Other pollutants besides particulates and ozone cause damage to the environment. Nitrogen oxides and sulfur oxides in the atmosphere create acids that fall as acid rain. Toxic pollutants, such as metals and VOCs, can bioaccumulate, causing harm to animals—including people—who are high on the food chain.

AIR POLLUTION AND HUMAN HEALTH

People live in areas where the air is unhealthy have the serious effects amount to nothing more than itchy eyes or the need to curtail outdoor activities. But for many, particularly children, the elderly, and people with other health conditions, air pollution can cause debilitating illnesses or hasten death.

Not all people are equally at risk for health problems from air pollution. Children are vulnerable because their lungs are growing and developing; they also have a larger lung-surface area per unit body weight, so they take in 50% more air per unit body weight than adults.

For example, infants have up to eight times the amount of ozone entering the deep portions of their lungs compared to adults. Children are also less able to fight off illnesses because of their developing immune systems. In addition, young people spend more time outdoors than adults, particularly in the summer when ozone levels are high. When children exercise or play, they take in more air, which increases their exposure to pollutants. If lung growth and function are harmed, the child may develop respiratory tract illnesses or even long-term lung damage.

The elderly are also at increased risk for health problems from air pollution, which may exacerbate the health problems they already have. People of any age who have chronic health problems, such as asthma and heart and lung disease, are also more susceptible to the effects of air pollution. For example, people with type 2 diabetes are at higher risk for cardiovascular troubles if they ingest airborne particles (Table 4.2).

TABLE 4.2
Health Consequences of the Six Major Air Pollutants

Pollutant	Effects
Ozone	Irritates respiratory system, reduces lung function, aggravates asthma, damages the cells lining the lungs, aggravates chronic lung disease, causes permanent lung damage in children
Particulates	Aggravate asthma and other respiratory illnesses, may cause chronic bronchitis, decrease lung function, bring about premature death
Sulfur oxides	Intensify asthma and other respiratory disease, cause respiratory illnesses, bring about difficulty in breathing and premature death
Nitrogen oxides	Irritate respiratory system, aggravate respiratory conditions such as asthma and chronic bronchitis
Carbon monoxide	Reduces oxygen delivery to organs and tissues; exacerbates cardiovascular disease; brings about visual impairment, reduced work capacity, reduced manual dexterity, poor learning ability, difficulty performing tasks; can be poisonous at high concentrations
Lead	Brings about gastrointestinal pain, nervous system damage, and encephalitis; chronic exposure can lead to brain, kidney, nervous system, or red blood cell damage; children can experience lowered intelligence and visual motor problems

Metal Poisoning

Toxic metals cause serious problems in humans, especially in children. One example is mercury. Fossil fuel burning releases the metal into the environment, and people ingest it when they eat large predatory fish, such as tuna, in which it has bioaccumulated. Methyl mercury enters the body's cells, causing nervous system and brain damage, including loss of motor control, limb numbness, blindness, and loss of ability to speak. Mercury poisoning is irreversible. Elevated mercury levels that are not high enough to cause problems in an adult may cause problems in a developing fetus or nursing infant since growing bodies are far more easily damaged than adult ones. In children, mercury poisoning shows up as brain or developmental damage, learning disabilities or reduced cognitive ability, and problems with motor skills such as walking, talking, or hand-eye coordination.

Asthma

Asthma, a chronic inflammatory respiratory disease characterized by periodic attacks of wheezing, shortness of breath, and a tight feeling in the chest, is the most visible health impact of dirty air. As with other diseases, developing asthma requires two factors: having genes that make a person susceptible to it and being exposed to something critical in the environment. Children who are exposed to secondhand smoke or polluted outdoor or indoor air are especially vulnerable. For example, children living within 650 feet (200 m) of a busy street are more likely to develop asthma. Children who live near truck traffic, which emits large amounts of particulates, are also more vulnerable. The biggest culprits for triggering asthma attacks and worsening the condition are O_3, SO_2, and particulates. Diesel engine exhaust particles inflict more damage than other particulates.

Lung Cancer

Lung cancer is the main respiratory cancer and its primary cause is smoking. Studies have also found a strong link between the disease and long-term exposure to air pollution, primarily from fine particulates.

Premature Deaths

Premature deaths, from a variety of causes, can be attributed to air pollution. The World Health Organization estimates that bad air kills 600,000 people worldwide each year. Air pollution has also been linked to spontaneous abortions (miscarriages) and increased infant mortality.

ACID RAIN

Acid rain is different from the other types of air pollution. For one thing, tailpipes and smokestacks do not emit acids; the acid rain forms from a combination of pollutants and water vapor in the atmosphere. Acid rain does not do its major damage in the atmosphere, but in lakes, ponds, and streams. Besides making surface waters more acidic, acid rain strips the soil of its nutrients, thus damaging trees by depleting their nutrient supply as well as by harming their leaves and needles. Acid rain leaches metals from soil; the metals are then deposited near the top of the soil, where they harm trees, or are carried into ponds, where they can hurt freshwater.

Acid rain is rain that is more acidic than normal; by definition, it has a pH of less than 5.0. The pH of a substance is a measure of its acidity or alkalinity. Natural rainfall is slightly acidic, with a pH of about 5.6. The acidity of natural rain is due to the small amount of CO_2 that dissolves in rainwater and forms mild carbonic acid. There are several steps to the creation of acid rain. Sulfur dioxide (SO_2) and the nitrogen oxides (NO_x) are released during the combustion of coal and petroleum, or from the refining of metal ores. In the United States, about two-thirds of all SO_2 and one-quarter of all NO_x come from electric power plants, which mostly burn coal for energy. The gases then react with water vapor in the air to produce sulfuric acid and nitric and nitrous acid. These acids dissolve in water droplets, which fall as rain—in this case, acid rain. Acid rain is not the only form of precipitation that has a lower pH; any type of precipitation can be acidic and is called acid precipitation. For example, acid fog can do more damage than acid rain, since rain falls on the upper surface of an object, but fog surrounds an object. Acid fog, with an average pH of 3.4, is also more acidic than acid rain. Part of the reason for the higher acidity of fog is that the small droplets have more surface area for pollutants to dissolve into. Acid particles may act as condensation nuclei, and the small droplets of acid fog will contain many more condensation nuclei than the larger droplets of acid rain. Acid snow, another form of acid precipitation, may lie on the ground for months, bringing a sudden rush of acid into lakes and streams during the spring thaw.

THE EFFECTS OF ACID RAIN

Acid rain has little effect on some regions yet is very harmful in others. Besides the pH of the water, the local environment plays a role in how much damage acid precipitation can cause. Some materials, including some rock and soil, can neutralize an acidic or an alkaline solution, a quality called buffering capacity. Rocks and soils that contain calcium carbonate (sometimes called lime) are the best acid buffers. These include limestone, marble, and their soils. The pH of ponds and streams situated in those rock types will be nearly normal. Even in these locations, though, if the buffering capacity of a soil or rock is exceeded, the region will be vulnerable to acid damage. Most rock and soil have little buffering capacity, so acidic waters can cause a great deal of damage. Acid rain causes damage as it filters through soil. Hydrogen ions from the acid invade minerals in the soil and replace elements that are good plant nutrients, such as calcium, magnesium, and potassium. The freed nutrients are then washed away so the soil is no longer able to nourish plants. Acidic waters can also leach metals, such as aluminum, from the soil and transport them to freshwater lakes, ponds, and streams where they accumulate and may become toxic to fish.

ACID RAIN AND FRESHWATER ECOSYSTEMS

Water with low pH diminishes the quantity and variety of life in lakes and streams. Most aquatic plants grow best in water with a pH of 7.0–9.2. As pH decreases, populations of submerged aquatic plants decline, reducing food for some water birds. Numbers of freshwater shrimp, crayfish, clams, and some fish start to dwindle. At pH 5.5, the bacteria that decompose leaf litter and other debris begin to die, cutting off the supply of organic material for plankton. Aluminum leached from the soil by acidic water enters the lake in great quantities, putting fish populations under even more stress. Young fish hatching into acidic, metal-rich waters do not survive into adulthood, or they may be deformed or stunted in their growth. Females under stress will not spawn, and fish eggs will not hatch if the pH is less than 5. Animals that live in harsh environments are more vulnerable to other problems, such as disease. With pH below 5, adult fish die; at less than 4.5, lakes are entirely devoid of fish. Organic material lies undecayed on the bottom, and the sides of the lakes are covered with moss. Although most acid enters lakes and ponds continuously, melting snow or heavy downpours can bring in excess runoff and temporarily raise the acid content of streams and lakes sharply. Temporary acidification can completely upset an ecosystem and result in massive fish kills. Some organisms tolerate or even thrive in an acidic environment; these include some plants and mosses and black fly larvae. Frogs can tolerate lower acidity than fish, but they cannot live in a lake without food. Birds and mammals that depend on the lake for fish or plants also experience a decline in population. Therefore, whole freshwater ecosystems can be turned around, with organisms living in lakes that are not supposed to be there and other organisms absent that should be present.

THE EFFECTS OF ACID RAIN ON FORESTS AND AGRICULTURE

Sulfuric and nitric acids are detrimental to all plant life. Even if the soil is well-buffered, forests can be damaged by acid fog. Acid deposition of all sorts ruins the waxy coatings of leaves, harming the tree's ability to exchange water and gases with the atmosphere. Trees weakened by acid experience slower growth or injury and are more vulnerable to stresses such as pests or drought. Acid-damaged plants are easily identified: the leaves of leafy plants turn yellow and damaged pine needles become reddish orange at the tips before they die. Acid rain also leaches soil nutrients, which stunts tree growth. When trees are deficient in calcium, they are less able to withstand freezing. Acid rain can destroy forest ecosystems. Acidic soil wipes out snail populations, which lowers the calcium intake of songbirds and causes them to produce eggs with thin shells. Birds and mammals that eat calcium-deficient plants may produce young that have weak or stunted bones; mammals may produce less milk.

ACID DESTRUCTION OF CULTURAL MATERIALS

Acid rain takes a toll on stone buildings and other structures including those that are culturally significant. Just as limestone and marble buffer acidic water, acid rain dissolves buildings and statues made of these materials. The decreased pH of rain and fog is taking its toll on cultural objects, a phenomenon that has long been recognized. In limestone buildings, the calcite mineral that makes up the stone reacts with sulfur dioxide pollutants and moisture in the air to form the mineral gypsum. This mineral grows into a network of thin crystals that traps particles of dirt. A dark crust forms on its surface, turning the building black. Since the gypsum crust dissolves in water, it accumulates in sheltered areas protected from rainfall. The result is that much of the detail carved into many old limestone buildings appears black and dirty.

REDUCING ACID RAIN DAMAGE

The best way to reduce damage from acid rain is to lessen the emissions of SO_2 and NO_x into the atmosphere so that acid precipitation does not form. Methods for reducing acid-producing emissions also reduce other air pollutants. The alternative (and far less effective) approach is to mitigate the damage from acid precipitation. Just as rocks containing calcium carbonate buffer acidic water in nature, lime can be added to acidic lakes or ponds. The downside to this is that the lime only neutralizes the water and does nothing to change soil chemistry or improve forest health. This approach is expensive and must be done repeatedly if acid rain continues to fall.

FURTHER READING

Akimoto H. Global air quality and pollution. *Science*. 2003;302:1716–1719.

Desonie D. *Atmosphere: Air Pollution and Its Effects*. Chelsea House Publishers; 2007.

Dockery DW, Pope CA. Acute respiratory effects of particulate air pollution. *Ann. Rev. Public Health*. 1994;15:107–132.

Fuzzi S, Baltensperger U, Carslaw K, et al. Particulate matter, air quality and climate: lessons learned and future needs. *Atm. Chem. Phys.* 2015;15:8217–8299.

Graedel TE, Hawkins DT, Claxton LD. *Atmospheric Chemical Compounds: Sources, Occurrence, and Bioassay*. Orlando, FL: Academic Press; 1986.

Houghton J. *Global Warming: The Complete Briefing*. Cambridge: Cambridge University Press, 2004.

Kidd JS, Kidd RA. *Into Thin Air: The Problem of Air Pollution*. New York: Facts On File; 1998.

Landsberg HE. *The Urban Climate*. New York: Academic Press; 1981.

Middlebrook AM, Tolbert MA. *Stratospheric Ozone Depletion*. New York: University Science Books; 2000.

Seinfeld JH. *Atmospheric Chemistry and Physics of Air Pollution*. New York: Wiley; 1986.

Seinfeld JH, Pandis SN. *Atmospheric Chemistry and Physics: From Air Pollution to Climate Change*. 2nd ed. New York: Wiley; 2006.

Stern AC (ed.) *Air Pollution*, 3rd ed., Vol. IV. New York: Academic Press; 1977.

Vallero DA. *Environmental Contaminants: Assessment and Control*. Amsterdam: Elsevier Academic Press; 2004.

Vallero DA. *Fundamentals of Air Pollution*. Academic Press; 2008.

Weart, S. *The Discovery of Global Warming*. Cambridge, MA: Harvard University Press; 2003.

Part II

Bioremediation

The choice of method and the determination of the final remediation standard will always be chiefly governed by site-specific factors including intended use, local conditions and sensitivities, potential risk, and available timeframe. For this reason, it is appropriate to take a brief overview of the available technologies at this point, to set the backdrop for the discussions of the specifically biotechnological methods to come.

REMEDIATION METHODS

The currently available processes for soil remediation can be divided into five generalized categories:

- biological;
- chemical;
- physical;
- solidification/vitrification;
- thermal.

BIOLOGICAL

Biological methods involve the transformation or mineralization of contaminants to less toxic, more mobile, or more toxic but less mobile, forms. The main advantages of these methods are their ability to destroy a wide range of organic compounds, their potential benefit to soil structure and fertility, and their generally non-toxic, "green" image. On the other hand, the process end-point can be uncertain and difficult to

DOI: 10.1201/9781003272618-6

gauge, the treatment itself may be slow, and not all contaminants are conducive to treatment by biological means.

CHEMICAL

Toxic compounds are destroyed, fixed, or neutralized by chemical reaction. The principal advantages are that under this approach, the destruction of biologically recalcitrant chemicals is possible and toxic substances can be chemically converted to either more or less biologically available ones, whichever is required. On the downside, it is possible for contaminants to be incompletely treated, the reagents necessary may themselves cause damage to the soil, and often there is a need for some form of additional secondary treatment.

PHYSICAL

This involves the physical removal of contaminated materials, often by concentration and excavation, for further treatment or disposal. As such, it is not truly remediation, though the net result is still effectively a cleanup of the affected site. Landfill tax and escalating costs of special waste disposal have made remediation an increasingly cost-effective option, reversing earlier trends which tended to favor this method. The fact that it is purely physical with no reagent addition may be viewed as an advantage for some applications and the concentration of contaminants significantly reduces the risk of secondary contamination. However, the contaminants are not destroyed, the concentration achieved inevitably requires containment measures, and further treatment of some kind is typically required.

SOLIDIFICATION/VITRIFICATION

Solidification is the encapsulation of contaminants within a monolithic solid of high structural integrity, with or without associated chemical fixation, when it is then termed "stabilization." Vitrification uses high temperatures to fuse contaminated materials. One major advantage is that toxic elements and/or compounds which cannot be destroyed are rendered unavailable to the environment. As a secondary benefit, solidified soils can stabilize sites for future construction work. Nevertheless, the contaminants are not actually destroyed and the soil structure is irrevocably damaged. Moreover, significant amounts of reagents are required and it is generally not suitable for organic contaminants.

THERMAL

Contaminants are destroyed by a heat treatment using incineration, gasification, pyrolysis, or vitalization processes. Clearly, the principal advantage of this approach is that the contaminants are most effectively destroyed. On the negative side, however, this is achieved at typically very high energy cost, and the approach is unsuitable for most toxic elements, not least because of the strong potential for the generation of new pollutants. In addition, soil organic matter, and, thus, at least some of the soil structure itself are destroyed.

A common way in which all forms of remediation are often characterized is as in situ or ex situ approaches. These represent largely artificial classes, based on no more than where the treatment takes place—on the site or off it—but since the techniques within each do share certain fundamental operational similarities, the classification has some merit.

In Situ

The major benefit of approaches which leave the soil where it is for treatment is the low site disturbance that this represents, which enables existing buildings and features to remain undisturbed, in many cases. They also avoid many of the potential delays with methods requiring excavation and removal, while additionally reducing the risk of spreading contamination and the likelihood of exposing workers to volatiles. Generally speaking, in situ methods are suited to instances where the contamination is widespread throughout, and often at some depth within, a site, and of low to medium concentration. Additionally, since they are relatively slow to act, they are of most use when the available time for treatment is not restricted.

These methods are not, however, without their disadvantages and chief among them is the stringent requirement for thorough site investigation and survey, almost invariably demanding a high level of resources by way of both desktop and intrusive methods. In addition, since reaction conditions are not readily controlled, the supposed process "optimization" may, in practice, be less than optimum and the true end-point may be difficult to determine. Finally, it is inescapable that all site monitoring has an in-built time lag and is heavily protocol dependent.

Ex Situ

The main characteristic of ex situ methods is that the soil is removed from where it originally lay for treatment. Strictly speaking, this description applies whether the material is taken to another venue for cleanup or simply to another part of the same site. The main benefits are that the conditions are more readily optimized, process control is easier to maintain, and monitoring is more accurate and simpler to achieve. In addition, the introduction of specialist organisms, on those occasions when they may be required, is easier and/or safer and generally these approaches tend to be faster than the corresponding in situ techniques. They are best suited to instances of relatively localized pollution within a site, typically in "hot-spots" of medium to relatively high concentration which are fairly near to the surface.

Among the main disadvantages are the additional transport costs and the inevitably increased likelihood of spillage, or potential secondary pollution, represented by such movement. Obviously, these approaches require a supplementary area of land for treatment, and hence, they are typically more expensive options.

APPROACHES TO SOIL AND AQUIFER BIOREMEDIATION

Numerous industries have viewed soil as a waste disposal site, dumping vast quantities of hazardous materials—oil and oily sludges from refineries; solvents and a variety of chlorinated hydrocarbons from the cleaning, printing, and many other industries; radionuclides from nuclear power plants and nuclear weapon facilities;

coal tars from town gasification plants; creosote from wood treatment facilities; and pesticides from agriculture. Large quantities of hazardous substances carelessly disposed of in the environment are creating enormous pollution problems in soils and waters around the world. Some are so heavily contaminated that they are designated Superfund sites.

The contamination that has resulted from the intentional spillage and burying of hazardous materials, accidental spills, and migration of hazardous substances from spillages that have occurred elsewhere has rendered many soils and waters dangerous for human health and harmful to the ecology of the area. Elevated rates of human cancer, dead birds, and landscapes devoid of plants are the unfortunate results of uncontrolled human activities. The environmental consequences and the human health threats of industrial pollution led to public demands for environmental cleanup and restoration and a cadre of government regulations and legal actions. The result has been a search for remediation technologies that can be actively applied to soil and aquifer cleanup. Most of the contaminants that commonly cause concern originate above ground or in surface soils where pollutants are spilled or buried. Soil contaminants become mixed with the naturally occurring soil. Many of the contaminants in the soil become physically or chemically attached to soil particles or become trapped in the small spaces between soil particles. Over time, many of these contaminants, however, begin to migrate into nearby waterways and into the underlying groundwater.

Soil permeability and pores and fractures in the rock are critical factors controlling the movement of water from the surface into the underlying layers. If water can move rapidly through the rock material pores and fractures, an aquifer forms. The water that accumulates in an aquifer can be contaminated when surface water, which recharges an aquifer, is polluted or when hazardous substances soak through the soil into the groundwater. Groundwater begins to accumulate within the unsaturated zone of soil, a layer that contains air and water filling the pores. Below this layer lies the saturated zone in which all the pores and rock fractures are filled with water. The top of the saturated zone is referred to as the water table. Soil overlying the water table provides the primary protection against groundwater pollution. Some potential pollutants nevertheless reach the groundwater.

The potential vulnerability of an aquifer to groundwater contamination is in large part a function of the susceptibility of its recharge area to infiltration. Areas that are replenished at a high rate are generally more vulnerable to pollution than those replenished at a lower rate. Unconfined aquifers that do not have a cover of dense material are susceptible to contamination. Bedrock areas with large fractures are also susceptible by providing pathways for the contaminants. Confined, deep aquifers tend to be better protected with a dense layer of clay material. There are three general approaches to cleaning up contaminated soil: (1) soil can be excavated from the ground and be either treated or disposed of (ex situ treatment), (2) soil can be left in the ground and treated in place (in situ treatment), or (3) soil can be left in the ground and contained to prevent the contamination from becoming more widespread and reaching plants, animals, or humans (containment and intrinsic remediation). Containment of soil in place is often done by placing a large plastic cover or concrete barrier over the contaminated soil to prevent direct contact and keep rainwater from seeping into the soil and spreading the contamination. In situ and ex situ treatment approaches can include flushing contaminants out of the soil by using water, chemical solvents, or air; destroying the contaminants by incineration; encouraging natural organisms in the soil to break them down; or adding material to the soil to encapsulate the contaminants and to prevent them from spreading.

Bioremediation is one of the technologies that can be applied by each of these general approaches. In some cases, one relies upon the intrinsic biodegradative capabilities of the indigenous microbial communities and monitors the movement and progressive slow decline in contaminant concentrations. This monitored natural attenuation approach is valuable when there are no acute threats to human health and where the impact is not spreading rapidly. In other cases, active remediation is needed to curtail the impact. Depending upon the nature of the problem, it may be necessary to excavate the contaminated soil and move it to a site for its safe disposal or treatment. This ex situ approach to bioremediation is analogous to what is done for traditional sewage and solid waste treatment. Bioremediation may also be conducted in situ, for example, by bioventing, in which air is used to move the contaminants from the groundwater into a phase where evaporation and biodegradation can occur simultaneously. Depending upon the nature of the pollutant, there may already be sufficient populations of microorganisms to degrade the contaminant—in which case stimulation of those microbial populations by environmental medications (e.g., addition of fertilizer or oxygen) may be all that is required. In other situations, it may be beneficial to consider augmenting the indigenous microbial populations by seeding with specialized microorganisms, including possibly with genetically modified microorganisms.

Because groundwater is a hidden resource, it is too easy to forget that its misuse is a hidden problem. For the purpose of context, the global importance of groundwater needs to be emphasized. As many as 2 billion people rely directly on aquifers for drinking water, and 40% of the world's food is produced by irrigated agriculture that relies largely on groundwater.

Despite this importance, the number of instances of groundwater contamination due to accidental spills or unsatisfactory disposal is beyond counting. Up to a certain point, natural processes, especially biodegradation, can attenuate contamination. In this regard, the biological active zone is the vadose (unsaturated) zone, where attenuation rates are highest. Contaminant removal continues in the saturated zone but usually at much lower rates, and migration of contaminants to the saturated zone can have the effect of dispersion of the contaminants.

While bringing about dilution, the latter process often cannot be relied upon for complete decontamination. The easy availability of groundwater and its vast supply (95% of the freshwater on the planet, apart from the locked water of the polar ice caps, is groundwater) have been its undoing. A great deal of it lurks fairly close to the surface, but intrusive disturbance of the subsurface has very high potential for destroying its flow and distribution. Therefore, techniques for remediating contaminated groundwater should operate by minimal disturbance. The sheer scale of the problem dictates that the remediation technologies should be as inexpensive as possible, and it is for that reason that bioremediation is often considered.

5 Ex Situ Bioremediation Technologies

In many cases, it is necessary to move the contaminated soil or groundwater to a site where a suitable treatment system can be engineered. Contaminated soil may be excavated and moved to landfills; to thermal treatment systems, e.g., incinerators; or to a variety of bioremediation systems including biopiles, windrows for composting, landfarms, and soil slurry reactors. All have their merits. The choice of which technology to use often is driven by the required performance criteria, i.e., the nature of the contamination and the levels of cleanliness that must be achieved and the cost of remediation, including the cost of transporting the contaminated soil or water from the site of contamination to the site of treatment. Bioremediation technologies often can be performed at or very near the site of contamination, reducing the cost of transporting contaminated materials and thereby making the economics of bioremediation more favorable than the other physical disposal and treatment technologies. If the economic and technical analysis favors bioremediation and if a risk-based remedial design has concluded that ex situ treatment is the optimal approach, there are a variety of technologies available which can be considered.

Two technologies—biopiles and windrow composting—currently dominate the ex situ bioremediation market for treatment of contaminated soils. Both are aerobic processes in which the soil is excavated and heaped into a defined space for treatment. In composting, an organic material is added so that microorganisms generate heat through their metabolism, often causing the temperature to rise to at least 60°C (Jorgensen et al., 2000). Except for the addition of organic material to support heat generation, biopiles and composting are essentially identical processes.

Both technologies involve preparation of the contaminated soil to favor aerobic microbial metabolism of the contaminants; this may involve the addition of bulking agents, fertilizers, and water. Composting windrows for contaminated land bioremediation are aerated by periodic turning of the windrows with a modified windrow turner. In a biopile, aeration is accomplished through a network of slotted plastic pipes, either passively or by forced aeration. If space at a site is a constraint, then biopiles would be favored since a compost system requires sufficient space between windrows for access for the turning equipment. Biopiles can be formed with much larger volumes of soil than can be achieved for windrows, since the height of the windrow is limited by the size of the machinery used to turn it. Biopiles and windrows have been used successfully at pilot scale and full scale for the bioremediation of a wide range of contaminants. While the majority of applications have been at petroleum hydrocarbon-contaminated sites (Namkoong et al., 2002; Van Gestel et al., 2003), they have also been used for manufactured gas plant sites (Sasek et al., 2003), pharmaceutical wastes (Guerin, 2001), chlorophenols (Laine and Jorgensen, 1997), creosote (containing high concentrations of polynuclear aromatic hydrocarbons [PAHs])

DOI: 10.1201/9781003272618-7

(Atagana et al., 2003; Civilini, 1994), pesticides (Michel et al., 1994), polychlorinated biphenyls (PCBs) (Michel et al., 2001), and nitroaromatics (Boving et al., 2000).

An emerging market for ex situ bioremediation by biopiles and windrows is cold-climate cleanup of fuel and oil spills. Crude oil is generally more persistent in Arctic tundra than in other regions, and a number of oil fields are situated in cold regions, e.g., in Alaska and Siberia. The feasibility of Arctic tundra ex situ bioremediation has been demonstrated, and the need for temperature, nutrient, and moisture control makes ex situ bioremediation the likely technology of choice in the Antarctic. These technologies appear to be applicable to arid desert contaminated sands also.

The third generic choice of ex situ technology is landfarming, which is more of a niche choice that has been used in the oil industry, from such diverse climates as Canada to Bolivia (Zhu et al., 2004) to Saudi Arabia (Hejazi and Husain, 2004), for decades for the treatment of refinery residues. It is a shallow treatment that is land intensive, which makes it unpopular for most applications, particularly inner-city brownfield sites, typified by former gas stations and sites with leaking underground storage tanks (USTs). The other generic technology is the use of slurry bioreactors, which offer quite different advantages over the other technologies, principally in the level of process control that is possible in a bioreactor and flexibility of operation.

BIOPILES

Biopiles range in size from sub-cubic meters for pilot-scale investigations to $10,000\,m^3$ at full scale (Kodres, 1999). Design is critical for achieving optimal performance, and remediation contractors have their own proprietary designs and engineering specifications for biopiles. The design and engineering considerations are similar but differ significantly with respect to the construction of the base upon which the contaminated soil is piled. Permanent installations usually have a concrete base, whereas temporary biopiles make use of different types of plastic liners. Effective biopiles have been constructed in a large variety of shapes and sizes. There are no guidelines or limits for height or width of biopiles, but it is wise to build them such that the maximum reach of the front-end loader being used to form the piles is not exceeded. If this guidance is not heeded, then the frontend loader will inevitably run over the previous lift when adding the subsequent lift, thus compacting the soil and undoing the careful work previously done in soil preparation. In the early stages of construction, this might also destroy the piping runs at the lower levels in the biopile. It is important that the shape and size of the pile should be considered as a means to creating conditions of even aeration throughout the pile. Sides that do not have a high slope might lead to over-aeration of peripheral soil relative to soil in the core. This can also lead to "wasted" aeration, in that air is lost from the sides, which might have the added consequence of removing volatile contaminants with it, creating unwanted volatile organic compound (VOC) emissions and aerosols, and cause preferential drying of the peripheral soil if the pile is not covered. In practical terms, shallow sloping sides also waste space and complicate maneuvering with a front-end loader.

Pile height is again governed by aeration. The very simplest biopiles have no forced aeration and rely largely on temperature gradient driven convection to create airflow through the slotted pipes. This limits the size of the piles considerably. Forced aeration, either by blower or vacuum pump, relieves this restriction. Experience has shown that a single piping layer close to the base of the pile is sufficient for piles up to 3 m in height. Electing to build pipes higher than this would necessitate adding a second layer of pipes, which greatly complicates the construction.

The width and length of the biopile are determined by the total volume of soil to be treated (after amendments) and the amount of space available on-site. For large sites, it is common practice to build multiple biopiles. They may all be identical or may be treating soils that have been identified as more or less contaminated.

Biopiles have a variety of space requirements so that the site must be considerably larger than the biopile itself. A soil storage space is required to stockpile soil before processing. The size depends on the total volume of soil to be processed and the coordination with soil processing. It is quite feasible on a large site that a relatively small stockpiling area can be used since stockpiling, soil processing, and biopile construction can proceed at the same time. The majority of the soil processing space is required for stone removal, soil sieving, and shredding. A tank and manifold system may be required for leachate treatment. Another tank for water and/or nutrient supply may be needed. A secure container can be used to house blowers, with diesel generators kept close by outside. The container can also house spare parts, sampling equipment, and other sundry equipment. All of these have to be arranged logically on a site. At an early stage of site design, the turning circle of mobile plant, such as front-end loaders, tractors and trailers, and telescoping fork trucks, should be considered, especially in the critical areas of soil storage and processing. These should be located close to the biopile areas to maximize the efficiency of use of soil processing equipment. It is desirable to have the access road as short as possible and close to the soil processing area.

Soil sieving and stone removal equipment can be brought in on hire for short periods; the awkward size and shape of this equipment necessitate that the access make it easy to maneuver. All of the space at a biopile treatment facility needs to be secure from vandalism. Access to the site for workers should be through a single-entry point, where clothes can be changed. The idea of a "clean" and "dirty" side at the entry point reinforces the idea that a biological facility is being entered. The housing for this can also be used to store other materials deemed necessary for health and safety and emergencies, e.g., respirators, disinfecting solutions, and trays. The site office should be located on the clean side, and its size and facilities depend on the size of the project and the number of staff assigned to it on a full-time basis.

BIOPILE COMPONENTS

The biopile design includes piping for aeration and optional other design components, which may include an irrigation system and cover.

BIOPILE BASE

The biopile base should be built on a relatively solid surface. At its most engineered, the base consists of a soil or clay foundation (up to 25 cm), an impermeable liner, and a bund to contain leachate. The biopile base should have a slight slope of 1″ or

2″ to allow drainage of leachate to an appropriately sited leachate collection sump located at a corner of the biopile. The impermeable liner, usually clay or a synthetic material, is then placed over the base. Clay liners are not recommended for highly soluble contaminants such as phenol. Synthetic liners of a high-density polymer are recommended, with thicknesses from 1 to 2 mm. High-density polyethylene (HDPE) with heat welded seams is ideal for this purpose. Thinner liners should be capable of taking the weight of heavy, even-tracked plant without tearing. Clean soil can be compacted on top of the liner to further protect it.

AERATION SYSTEM

Once a biopile base has been constructed, the aeration system is installed on top of the liner. Slotted plastic pipes of various thicknesses and materials are available as the main element of the aeration system within the pile. Polyvinyl chloride (PVC) slotted pipe (2–4 in. diameter) is a common choice. The pipes are embedded within a highly permeable matrix, such as new wood chips or gravel, to act as an aeration manifold, whether operated with a blower or a vacuum pump. The depth of this layer is variable but obviously the pipes must be sufficiently covered to minimize short-circuiting from them. The length of the biopile determines the length of each aeration pipe run. At the manifold header side, each aeration pipe run starts with a solid, not slotted, length of pipe of the same diameter as the slotted pipe. This is typically of the order of 3 m long, the actual length dictated by the distance to the aeration manifold. It should proceed about 3 m into the pile before it is joined to slotted pipe to ensure that short-circuiting does not occur right at the start of the pile. Next, the solid pipe is connected to the slotted pipe with a rubber or plastic connector. The length of the slotted pipe is dictated by the length of the biopile and, as at the start of the pipe run, should terminate some 3 m short of the far end of the biopile to prevent short-circuiting through the sloping edge. The slotted pipe is terminated with an end cap.

The start of the pipe run is connected to the header manifold by a gate valve of appropriate diameter. When several pipe runs are used to aerate a wide biopile, the gate valves are used to equalize the flow of air through each pipe run. The zone of influence of each pipe run is influenced by rate of airflow and also by soil porosity. Too rapid airflow causes drying of the pile and may drive off VOCs creating an environmental concern in the vicinity of the pile. It also uses excessive electricity. By using relatively low-power blowers, this problem can be circumvented. To ensure even aeration within the pile, then, it is necessary to add several pipe runs parallel to each other. Once the biopile is built, it is very difficult to influence the porosity of the soil. As a general rule, parallel pipe runs are spaced about 2.5–3 m apart. Air velocity can be easily measured in each pipe run.

Connection of the header manifold to the blower then completes the system. A low-power centrifugal blower is sufficient for aerating large volumes of soil, and it is wise to install several small blowers of various powers (1–5 hp). Varying the rate of aeration over the duration of a project makes sense, as the need is greater at the start when there is a high level of contamination and microbial activity must be stimulated. Variable speed blowers are therefore advantageous. It is not advisable to control aeration by on–off cycling, especially in the early stages, as even short periods with the blower off can lead to anaerobiosis in regions of the pile, and this is very difficult to monitor. As an average figure, the blower should be capable of delivering about $0.14\,m^3$ of airflow per pipe run per min.

Depending upon the location of the project, operating in blower mode might have a high potential for stripping VOCs and creating an odor and health risk on and even off the site. Local legislation may then require containment measures that are rather expensive. In such circumstances, it is better to operate aeration by vacuum, as the captured air can be passed through a VOC removal system, typically granular activated carbon (GAC), and thus ameliorate the odor and risk. There has to be good justification for this, however, as operating in the extractive mode complicates the aeration setup. Pulling air through the pile will entrain condensate, and even leachate may be pulled through, so the vacuum pump must be preceded by water knockout and collection tanks. GAC treatment of off-gases considerably increases the cost. It has been shown under laboratory conditions that VOC volatilization can be suppressed by the addition of activated carbon as a soil amendment. On a full-scale biopile, however, this would represent a large cost. If the extractive mode is deemed necessary, the pump should be capable of removing at least 15 pore volumes per day.

Passively aerated biopiles have also been used in which a similar engineering system is used but without mechanical aeration. Airflow is created by differences in temperature between the soil and the outside air driving convective currents, facilitated by the slotted pipes, and by wind-induced pressure gradients. Naturally, this can achieve more limited aeration and might lead to uneven aeration as a result of a limited radius of influence as the air leaves the pipes and enters the soil mass. The central region of a pile would be particularly prone to local oxygen deficit.

COVERS

Many bioremediation projects have been done without covering the biopiles, but covers offer some advantages. A primary advantage is that a cover prevents leachate formation and simplifies the biopile design accordingly. Another is that covered biopiles lose very little water. A typical ex situ bioremediation contract for petroleum hydrocarbon cleanup might last 3 or 4 months, during which time a covered biopile might lose only 1%–2% of its initial water content. Thus, it would be sufficient to amend the water content during the construction phase only. If the alternative is to install an automated sprinkler or drip-type system to maintain water content within defined tolerances, then a cover is a much simpler engineering option. The moisture content, nevertheless, must be monitored since forced aeration, by either blower or suction, tends to remove moisture as the air entering the pile usually does so at less than 100% humidity.

The cover is often waterproof plastic sheeting, Visqueen, or a thin grade of HDPE liner, sufficiently thick to prevent tearing while being manually removed and refitted. If extractive aeration is being performed, completely impermeable liners may not allow sufficient air circulation. Framing systems have been tried but complicate the installation and are prone to water ponding and collapse. Alternatively, the pile can be covered with wood chips, which retain heat but allow air to flow.

Recently, fleece liners have been adopted from the composting of green wastes, Fleece liners are made from blown polypropylene. They have properties similar to fleece garments. They resist moderate rainfall but are not completely impermeable to water and are also gas permeable. They allow some water to pass through to a biopile during operation, which will replace water lost through aeration. Gas permeability overcomes the problem with completely impermeable plastic covers mentioned above.

IRRIGATION SYSTEMS

If water has to be added, it is preferable to do it at the biopile formation stage if the batching procedures described above are being used. If an alternative water addition system needs to be employed, the preference is for a dripline irrigation system. Water flow from dripline irrigation system can easily be monitored as the low rate of application prevents runoff and it can be set up to evenly irrigate a biopile without supervision. However, it is inevitably a further engineering complication that is best done without if possible.

BIOPILE FORMATION

The standard way to form a biopile is to add all necessary amendments to the graded, sieved soil and then start lifting the soil onto the base and aeration system with a front-end loader. An alternative is to form the biopile with graded soil and any bulking agents and then add liquids at a later stage. The latter strategy carries risks in that it is difficult to achieve uniform distribution of materials. Usually, nitrogen and phosphorus are applied together. If a liquid source is sprayed onto the top of a biopile, the mobility of the nitrogen source will allow it to travel through the pile. However, phosphorus interacts with metals and other soil components and will be immobilized within a meter of the top of the biopile. Also, if the occasion dictates the use of inocula, then very quickly the microorganisms used will be immobilized and achieving uniform mixing by application after biopile formation is all but impossible. Many laboratory and field trials have shown that bacteria do not move appreciably through soil as a result of physical filtration by small soil pores and adsorption to soil particles.

Therefore, the safest way to guarantee the maximum level of homogeneity of materials in a biopile is to mix everything just prior to, or even during, biopile formation. The standard practice is to hire in soil grading (screening) equipment to remove stones and grade the soil down to an average particle size, usually 30–50 mm. Parallel bar screens remove large stones, and then trommel or vibrating screens are used to grade the soil down to a smaller size. Large amounts of clay will require soil shredding as the next step in the process.

Knowing the capacity of the loader being used to feed the screening equipment and the speed of the conveyor belt, it should be possible to spray liquid amendments at the screening stage with some accuracy. It should also be possible to add bulking agents to a known percentage (usually by volume). If the soil requires shredding, then the effort will have been wasted. After screening and shredding, various equipments are available for soil mixing, and this is the optimum stage for adding amendments. Contaminated soil may be batched with the other liquid and solid amendments, and the mixing action allows efficient aeration. A large, slowly rotating reel drives materials forward to two blending augers. The combined action of the augers mixes, aerates, and discharges the materials, and the discharge can be done directly to a forming biopile or compost windrow. This equipment can be truck mounted or tractor mounted and can be driven electrically or by self-contained diesel engine. As soil is batched, it is easier to control the volumes and thus the final concentrations of any amendments.

Conventionally, a biopile is formed from back to front, the back being the end at which the aeration header is located. Common sense precautions to be taken are to make sure that previous lifts are not driven over, since this compacts the soil; driving over the aeration, system will destroy it.

WINDROW COMPOSTING

The main difference in materials used in windrow bioremediation and biopiles is that some organic, heat-generating material is added to the windrows. The objective is to increase the metabolic rate of indigenous hydrocarbon oxidizers. Most bacteria grow over a range of approximately 40°C, whatever their optimum temperature for growth. Within the normal temperature range for growth of a bacterium, growth rate obeys the Arrhenius relationship between reaction rate and temperature. Increasing the temperature of a contaminated soil windrow from 20°C to 30°C would be expected to more or less double the growth rate of hydrocarbon-oxidizing bacteria, and therefore, speed up the bioremediation process. Another effect of increasing temperature might be to increase the bioavailability of poorly water-soluble contaminants since increasing the temperature should increase solubility, as well as decreasing viscosity of oily contaminants. During composting of contaminated soil, the thermophilic stage is not reached, and the temperature does not exceed 45°C (Semple et al., 2001).

In this regard, bioremediation of contaminated land by windrow treatment is a very different operation from composting of organic wastes or sewage sludge. The main difference between contaminated soil composting and traditional composting is that the former lacks the concentration of organic materials of the latter. A direct consequence is that heat generation is normally lower in contaminated land composting. For traditional composting of waste materials, the high temperature generated in the procedure is essential: several ordinances stipulate that every part of the compost must reach a temperature of 65°C at a water content of at least 40% for three consecutive days (Illmer, 2002) for the purpose of pathogen kill. For contaminated soils, the likelihood of the material containing large pathogen loads is much less, and also high temperatures are not suited to the metabolism of most hydrocarbon-oxidizing bacteria. In addition, such high temperatures would involve water loss at a level that would complicate windrowing operations. One of the purposes of turning windrows is to dissipate heat, and the increased porosity achieved by adding bulking agents aids heat dissipation.

WINDROW COMPONENTS

A large variety of organic amendments has been used in composting bioremediation. Many are based on the application of manure, from either cows, pigs, or chickens.

Sewage sludge is abundantly available globally, and it has been successfully used as an amendment in composting bioremediation. Virtually any putrescible material available in bulk can be used, such as vegetable wastes, spent mushroom compost (SMC), and even garden waste. The use of composting approaches to bioremediation of organic pollutants generally, and specifically, the use of composting to treat PAHs have been reviewed. The use of SMC is an interesting case. SMC is the residual compost waste generated by the mushroom production industry. It is readily available, as mushroom production is the largest solid-state fermentation industry in the world; in the United Kingdom alone, there are some 400,000–500,000 tons of waste mushroom compost produced per annum. It consists of a combination of wheat straw, dried blood, horse or chicken manure, and ground chalk composted together. It is a good source of general nutrients (0.7% N, 0.3% P, 0.3% K plus a full range of trace elements), as well as a useful soil conditioner. A fascinating feature of SMC is that it may contain a relative abundance of extracellular ligninolytic fungal enzymes, which are relatively nonspecific in their substrate preference. Hence, they may assist in the biodegradation of aromatic molecules such as PAHs, giving SMC an additional role in composting bioremediation.

The construction of more robust turning equipment, based on compost windrow turners but strong enough to turn large soil windrows, has allowed full-scale windrow bioremediation of contaminated land. Much of the discussion is similar to that for biopiling. For example, soil preparation is identical in the need for water, nutrients, and bulking agents, although the initial care in mixing need not be so rigorous since the whole point is that aeration is brought about by pile turning. The turning naturally mixes the contents of the windrow: the more often it is turned, the greater the homogeneity of materials, including moisture. Windrows are inherently simpler in design; therefore, this section will focus more on the differences between windrow composting and biopiling.

The natural shape of a windrow is more like a triangular prism than the characteristic trapezoidal, or incomplete pyramid, shape of the biopile. Windrows are suited to long, narrow sites, as they can be constructed to any length required. Typically, windrows for bioremediation of contaminated soil would be of the order of 1.5–2 m high and perhaps 3–4 m wide. With the arrival of large, self-propelled windrow turners, windrows can now be constructed to greater heights and widths than before. Doing so necessitates more regular turning to prevent anaerobiosis, though. Given that compost windrows can be much longer than they are wide, the small volume of material at each end can be ignored in sizing. That is why the volume of a triangular prism is a realistic choice.

WINDROW FORMING AND TURNING

Comments similar to those for bed construction pertain to windrow composting and biopile construction, except that for windrows there is no added complication of installing piping runs for aeration. Mixing can be done during windrow formation. For example, if wood chips are to be used for bulking and the soil-to-chips volume ratio has been calculated, then the wood chips can be laid out on the windrow bed along with, say, pellet NPK fertilizer, then the moist soil, and any additional heat-generating organic material, and the windrow turner will mix them together while forming the windrow.

For medium-scale operations, a custom made turner driven from a tractor power takeoff is suitable for windrows around 1.25 m high and 4.35 m wide. Large-scale operations require large, self-propelled windrow turners, which can turn windrows around 2.5 m high and 5.5 m wide. Paradoxically, space is saved by using large self-propelled vehicles: the space between windrows is of the order of 1.25–3 m, whereas using a tractor-driven turner requires a space between windrows of 3 and −4.5 m, and self-propelled vehicles build higher windrows.

IN-VESSEL COMPOSTING

At first sight, in-vessel composting, i.e., treatment within a bioreactor, offers some advantages, mostly relating to the higher degree of process control that can be applied, e.g., better temperature control, control of odors, and improved mechanical mixing. However, it is not a popular full-scale practice for at least two reasons.

Very often, the volumes of soil are too large to make this a cost-effective approach. Mobile in-vessel plant would be too small for mass application, and transporting large volumes of contaminated soils from the field to large fixed facilities raises a number of concerns, including safety, and not least, the extra treatment cost associated with transportation. Bioreactors, however, can be used in some cases for treatment of contaminated groundwater.

LANDFARMING

The basis of landfarming is the controlled application of waste on a soil surface to allow the indigenous microorganisms to biodegrade the contaminants aerobically (Martins, 2001). Because it is a shallow treatment with a large exposed surface area, the relative contributions of biodegradation and volatilization have always been debated. Volatilization of crude oil in temperate climates is minimal, whereas in hot climates it might approach 40%. Landfarms differ from biopiles and windrow composting technologies in that landfarms are fixed facilities to which the materials for remediation are transported. The oil industry has a long history of using landfarming; landfarms around the world are used to treat various upstream (drilling wastes) and downstream (refinery wastes) materials, including oily sludges (Morgan and Watkinson, 1989).

Landfarms also have been proven effective in reducing concentrations of all components of fuels found in USTs including petrol, diesel, and kerosene, as well as primarily non-volatile oils such as heating and lubricating oils. For the oil industry, it has been seen as a relatively cost-effective and simple technique for dealing with refinery wastes, but there are environmental concerns over landfarming operations.

LANDFARM DESIGN

The preparation of the site for a landfarm bears similarity to the methods for preparation of the base for biopiling, but of course the area to be prepared has to be much larger. The amount of land required depends on the volume of materials to be treated and the depth of the landfarm soil. The depth of landfarm soils is usually between 30 and 45 cm, although very powerful soil tillers can still down to about 60 cm. An advisable upper sludge loading is 150 g of sludge per kg of soil. Waste is typically applied in layers of no more than 20 cm. Time should be invested to maximize the accuracy of the land requirements, as a key to economic landfarm operation is to size the installation properly and then run it close to capacity.

LANDFARM BASE

Landfarm surface areas are highly variable but may be of the order of 4 ha (40,000 m^2). The total land area is divided up into treatment cells, generally square and of variable size. By so dividing the land, wastes can be sequenced in time so that continuous operation can be maintained. Given the large area and the shallowness of a landfarm, the base has to be properly designed and built to accommodate local climate. Leachate control is essential for wet and temperate climates. To this end, a high-integrity liner is required. A compacted clay layer of 0.6 m with a hydraulic conductivity of 10^{-7} cm/s is suitable. A 1-mm-thick HDPE liner is a suitable alternative, and at permanent sites, both may be used.

The leachate collection system to sit on top of the liner consists of slotted-pipe laterals embedded in a granular drainage layer. The grade of a treatment cell should be between 0.5% and 2%, depending on local rainfall. The granular drainage layer is formed from compacted, well-sorted gravels and coarse sand, with a minimum compacted hydraulic conductivity of 10^{-2} cm/s. Gravels greater than 13 mm in diameter may damage the HDPE liner, and it would be advisable in such a case to have a protective layer of sand or geotextile between the granular drainage material and the liner.

PERIMETER DIKE

The perimeter dike is designed to prevent runoff and storm water run-on based on a 25-year flood. A minimum freeboard of 0.3 m between the top of the dike and the surface of the treatment cells, and also from the top of the dike to the exterior surface of the cell, is recommended.

Stockpile Area

Stockpiling is necessary for the storage of contaminated material for treatment, already treated material awaiting haulage out of the farm, and oversize material. Stockpiling areas should be lined with a chemically resistant impermeable geomembrane to a minimal thickness of 1 mm. Rain is prevented from entering the stockpile with a 0.25-mm-thick impermeable geomembrane.

LANDFARM OPERATION

The main activities of landfarm operation are placing of contaminated material, aeration, watering, fertilization, and removal of treated material. The most important of these is aeration. The soil amendments are as for other ex situ bioremediation technologies. The critical issues, as always, are water and oxygen (Figure 5.1).

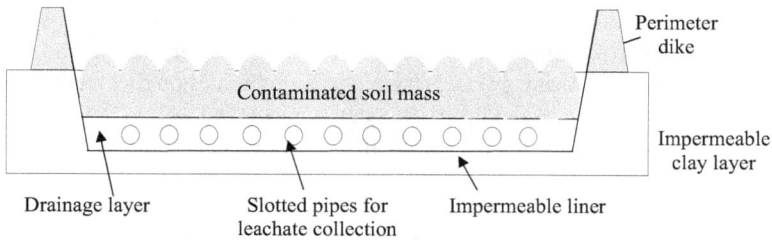

FIGURE 5.1 Landfarm schematic (Kuyukina et al., 2003).

IRRIGATION

As in all bioremediation processes, a fine balance has to be sought, between having enough water for essential microbiological needs and not so much that it would fill the soil pores and inhibit oxygen diffusion. A sprinkler-type irrigation system is the norm, with overlap between adjacent sprinkler patterns. The water delivery rate is variable, but the recommended lower rate is 0.7 L/s (1,000 m^{-2}). The key issue is that the system should be able to deliver water evenly over the whole surface to prevent short-circuiting. As with other technologies, the moisture should be kept to between 40% and 80% of field capacity (water holding capacity). Unlike for the other technologies, greater effort has to be expended in monitoring the moisture because of the large surface area involved.

NUTRIENT ADDITION

Slow-release fertilizers are preferred for landfarming. After a new batch of contaminated soil or sludge is applied on a landfarm, standard practice is to measure ammonia, nitrate, and ortho-phosphate at 2-week intervals, especially during the first few weeks of treatment, and thereafter every 6 weeks for the duration of the treatment. A simple rule of thumb is to reapply fertilizer when the level of available nitrogen falls below 50 mg kg/soil, and the application rate should be adjusted so that the level does not exceed 100 mg/kg. The trigger level for ortho-phosphate is 5 mg/kg.

AERATION

The technique for aeration is simple and well established. The soil is tilled with agricultural equipment. The timing of tilling is the most important variable. If done too soon after rainfall with a soil containing a significant amount of clay, this can form clay clods in which the subsequent aeration will be very poor.

SOIL SLURRY REACTORS

Slurry-phase ex situ bioremediation has several apparent advantages. The addition of water makes the materials easier to move, by pumping, and the ability to set up liquid-phase bioreactors means that a greater degree of process control is possible.

A great deal of knowledge of the functioning of fermentors and chemostats has been accrued, and the interrogation of kinetics in slurry-phase systems should accordingly be simplified. Another advantage of slurry-phase bioremediation that has been championed is the great degree of flexibility that can be achieved.

Slurry reactor contains several reactors in series means that the operating conditions can be easily and rapidly modified. Indeed, even the metabolism can be altered drastically. For example, attempting to treat, say, a waste containing a highly chlorinated phenol aerobically may be folly. Since the molecule is already highly oxidized due to the presence of electronegative chlorine atoms, further electrophilic microbial attack may prove futile. Under such circumstances, the first reactor might be better run anaerobically to bring about at least partial reductive dichlorination (Kodres, 1999). Removal of chlorines and substitution with hydrogen would make the molecule more amenable to oxidative microbial hydroxylation and subsequent ring cleavage, which could be carried out in a second bioreactor stage. Subsequent stages would involve slurry thickening and then filtration to form a treated cake. Such an arrangement might speed reaction not only by optimizing growth conditions but also by increasing the apparent concentrations of contaminants in solution by encouraging desorption from soil to the aqueous phase. Indeed, some studies do convince that this approach is efficacious.

The reasons that soil slurry reactors are not popular for contaminated land bioremediation are due to practical consideration. Large amounts of water would be required, and the engineering is complicated by the need for water recovery and recycling. Water itself is an expensive commodity. Another reason is that the large volumes of soil at many sites mean that the bioreactors would be impractically large. This would virtually dictate that such facilities were fixed and so would require the transportation of large volumes of contaminated soil, most probably by road. Soil slurry bioreactors would have to have a good record of full-scale success before the large investments necessary would become available.

IN SITU BIOREMEDIATION TECHNOLOGIES

The very term bioremediation should tell us that this technology is highly dependent upon external conditions, which is key to determining whether bioremediation can be performed in situ. Are the in situ environmental conditions suitable for the microbial activities necessary for successful bioremediation to occur? Can the environment be modified to create conditions that favor microbial activities and biomediative removal of the pollutants? The conditions of greatest importance in this consideration are the physicochemical and chemical conditions that exist in the contaminated soil and water. Conditions that have to be optimal or near-optimal to allow bioremediation to proceed at a reasonable rate include the following: dissolved oxygen for aerobic processes; electron acceptors for anaerobic processes; pH; temperature; nutrient availability, especially with regard to nitrogen and phosphorus; water content (for soil); soil composition; alkalinity; salinity; metal concentrations; and concentration bioremediable contaminants (biodegradability versus toxicity). This list is not exhaustive, but already it is obvious that microbial activity in a contaminated site is affected by a wide variety of site conditions. Often, having to deal with microorganisms and having to consider this complex list of environmental parameters are off-putting to the remediation engineer. However, previous experience can simplify matters. In most cases, it is possible to simplify this list to the most important conditions, which are (in no particular order of importance):

- water content or moisture,
- pH,
- temperature,
- nutrient status, and
- electron acceptor status (usually oxygen).

All of these factors are more or less influenced by a fundamental property of soils that requires some discussion before proceeding further. That property is soil porosity, which is the ratio of the volume of voids to the total volume of a soil and is expressed as a percentage. Above all, soil porosity has a defining influence upon the transport process that will make or break the applicability of bioremediation. Water availability, contaminant availability, gradients of oxygen and biodegradation of waste products, pH, and even heat transfer are all strongly influenced by porosity. Porosity also influences the movement of contaminants into underlying groundwater and away from the site of contamination. While the porosity gives an indication of the applicability of bioremediation to a contaminated site alone, it is not a sufficiently detailed measure. Soil is a three-phase system composed of solids, liquids, and gases. The solid components determine the porosity, and the pores are more or less filled by the liquids and gases.

Hydraulic conductivity is a velocity term and is often expressed as centimeters per second. The lower the porosity of a soil, the lower the hydraulic conductivity, and at low values, there is a serious inhibition for the application of bioremediation. In particular, the very low hydraulic conductivity associated with clay makes clay soils difficult for bioremediation, where soil modification is difficult or impossible.

In situ bioremediation is best performed on sandy soils, with a lower permeability limit of about 10^{-5} cm/s. The hydraulic conductivity of dry, unconfined bentonite is 10^{-6} cm/s; when the bentonite is saturated, the conductivity drops to less than 10^{-9} cm/s. While the porosity problem can often be overcome with ex situ bioremediation technologies by adding bulking agents or shredding the soil, even then low clay or silt content is required, with a minimum void volume of 25% recommended. Compared to ex situ bioremediation systems, there is very little to see at an in situ site; the bulk of the activity happens underground. Most of the quoted advantages of in situ compared to ex situ bioremediation relate to this fact. At in situ bioremediation sites, there is very little excavation done, so site disturbance is much less. This should make in situ bioremediation cost-competitive since the bulk of the transportation and excavation cost has been nullified. As there are no biopiles or windrows to be constructed, the construction engineering is less involved, and in situ treatment is highly suited to small, inner-city sites, such as old gas stations. Naturally, it allows treatment of the deep subsurface, which is not feasible with ex situ methods unless the cost and space implications of deep excavation can be borne. A related issue is that in situ bioremediation allows treatment around and under buildings, which means that working sites can be treated as well as abandoned ones. Because there is limited excavation, there is much less exposure of the workforce and others in the vicinity of the site to volatiles.

This lack of excavation also accounts for most of the quoted disadvantages of in situ systems. Because the contractor is effectively working blind in the subsurface, detailed site assessment is required, and the hydrogeology and contaminant distribution must be known as accurately as possible. Gathering the necessary borehole information and laboratory analyses can make the site assessment phase very expensive. Working at depth in very inhomogeneous matrices makes process control difficult: whereas altering the temperature or pH of a biopile is straightforward, this is not the case with in situ technologies. It is also difficult to accurately predict end points, and careful monitoring of the process, while necessary, is difficult, as the subsurface cannot be made homogeneous. All things considered, as more experience is being gained with in situ treatments, it is becoming more popular, whereas in the early development of bioremediation ex situ treatments were much more common.

BIOVENTING

Bioventing bears great similarity to the physical extraction technique of soil vapor extraction (SVE). The primary engineering objective of both SVE and bioventing is stimulation of airflow in the vadose zone. However, whereas SVE is designed to maximize contaminant volatilization, bioventing is operated at much lower airflow rates to optimize oxygen transfer and utilization by microorganisms. SVE aims to extract volatile compounds from groundwater; it does not involve transformation of the compounds. Bioventing aims to achieve transformation of the contaminant through microbial attack. The two technologies use the same equipment, especially for aeration.

Bioventing differs markedly from SVE in that nutrients and moisture are often added to the subsurface to stimulate biodegradation during bioventing. The two functions, however, are not mutually exclusive: SVE always entails a variable component of biodegradation, and likewise bioventing can involve an element of volatilization.

CONTAMINANTS BIODEGRADED BY BIOVENTING

Bioventing has found its niche in the in situ treatment of fuel spills. Whereas fresh gasoline may be too volatile to be treated by true bioventing, bioventing has been used many times for the full-scale bioremediation of diesel and kerosene spills. It has also been used successfully on PAHs and a mixture of acetone, toluene, and naphthalene. Bioventing is not considered appropriate for the treatment of PCBs and other chlorinated hydrocarbons. However, deliberate encouragement of co-metabolism has proven successful in the biodegradation of trichloroethylene (TCE) by injecting oxygen and phenol, and the advent of anaerobic bioventing offers the potential for treatment of chlorinated compounds, such as DDT and energetic compounds (Figure 5.2).

BIOVENTING DESIGN

As with ex situ bioremediation, bioventing relies crucially on effective aeration. Unlike ex situ treatment, at a potential bioventing site, it is geology that dictates

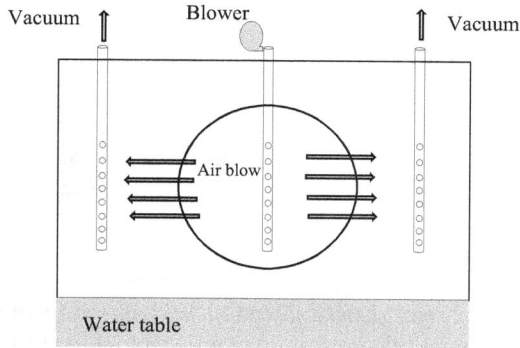

FIGURE 5.2 Bioventing schematic (Cookson, 1995).

whether it is applicable (De Paoli, 1996). Soil gas permeability is once again the governing issue, which determines the relationship between applied pressure, or vacuum, and gas flow rate. To achieve the optimum aeration regimen whereby biodegradation is maximized and volatilization is minimized is the ideal one, and inducing uniform airflow is the best way to achieve this, although in practical terms this may not be possible.

The basic steps involved in designing a bioventing system are as follows (USEPA, 1995):

1. Determine the required airflow system.
2. Determine the required airflow rates.
3. Determine the working radius of influence.
4. Determine well spacing.
5. Determine vent well requirements.
6. Provide detailed design of blower, vent wells, and piping.
7. Determine monitoring point requirements.

There are three primary physical characteristics that are used in bioventing design: soil gas permeability, contaminant distribution, and the radius of influence of oxygen.

PERMEABILITY: GAS AND VAPOR TRANSFER

Flux of gas through a soil matrix is described by Darcy's law, which shows that flux of gas is proportional to permeability. Generally, a significant flux decrease occurs when the permeability is less than 10^{-4} cm/s. At this point, the flux drops exponentially with decreasing permeability. While bioventing is certainly possible in relatively low-permeability soils, when the soil gas permeability falls below cm/s, gas flow will be through secondary porosity, such as fractures, or through more permeable strata present. In such cases, the feasibility of bioventing is site specific. In higher permeability soils, the most important bioventing limitation becomes soil moisture. A high level of soil moisture replaces soil gas with water, which drastically reduces air diffusion.

CONTAMINANT DISTRIBUTION

With consideration to partitioning phenomena, a petroleum hydrocarbon spill onto soil will allow contaminants to be present in any or all of the phases: sorbed to soils in the vadose zone; in the vapor phase of the vadose zone; free floating on the water table as residual saturation of the vadose zone; or in the aqueous phase, either dissolved in the pore water of the vadose zone or dissolved in the groundwater. Due to their higher density, dense non-aqueous phase liquids (DNAPLs) partition to the vadose and saturated zones, whereas light non-aqueous phase liquids (LNAPLs) distribute primarily to the vadose zone. As bioventing is really a vadose zone treatment, it is the LNAPLs that are the focus of attention. LNAPLs are more likely to migrate to the capillary fringe relatively uniformly in a sizeable spill. Then, they will spread laterally along the surface of the saturated zone. The fluctuating water table allows migration of LNAPLs below the water table, but they cannot permeate the saturated zone unless a critical capillary pressure is exceeded.

LNAPLs can remain as free product, partition to the vapor phase or the aqueous phase (pore water), or sorb to solids. The equilibrium concentration of most hydrocarbons in the aqueous or vapor phase is determined by the immiscible phase, if present, or the sorbed phase, if an immiscible phase is absent. Only limited oxygenation of the capillary fringe by bioventing is possible due to the limitation of oxygen diffusion into water, and the pore space is water saturated in the capillary fringe. However, when bioventing is operated in air injection mode, the positive pressure depresses the water table. This dewaters the capillary fringe, allowing for more effective treatment.

AIRFLOW SYSTEMS

Air injection is easier to operate and maintain and also less expensive. In the presence of surface buildings or basements within the radius of influence of a bioventing system, air extraction may be preferred to prevent the accumulation of gases within these buildings. Air extraction, however, does not move air outward to create an extended bioreactor zone, and therefore, there is a relatively greater contribution from volatilization than biodegradation. Another consequence of the negative pressure created is that with extraction systems, it is possible to cause the water table to rise. The effect on biodegradation at the capillary fringe would be the opposite of the situation with air injection: water saturation decreases oxygen diffusion, leading to a drop in biodegradation rate. For air extraction, an explosion-proof blower is required when working with petroleum hydrocarbons, a knockout drum and storage tank are needed upstream of the blower to remove condensates, and the off-gas may require treatment, which will significantly increase the cost of the treatment. Air injection bioventing may cause soil desiccation to the point that microbial activity would be limited.

BIOSPARGING

One way to remove contaminants from groundwater is to pump and capture the contaminants or otherwise separate the contaminants from the water, which is returned to the aquifer. In theory, prolonged pumping could eventually flush out all the contaminants, but the solubility properties of many contaminants of aquifers make reliance on physical flushing alone prohibitively slow and expensive. The widely used pump-and-treat method for the decontamination of groundwater may require long periods of recirculation of groundwater from the aquifer to the surface—to clean an aquifer by simple water flushing may take 15–20 years and several thousand times the volume of the contaminated portion of the aquifer.

In situ air sparging has emerged as a popular alternative. This method consists of injecting a gas, usually air, into the saturated subsurface, below the lowest point of contamination. This promotes the partition of volatile and semi-volatile contaminants from the dissolved and free phases into the vapor phase. Inevitably, as the dissolved oxygen concentration of the groundwater is increased, microbiological activity is stimulated. To turn this into a deliberate bioremediation technology rather than a volatilization, one requires process modification, and the technique that is emerging has been called biosparging. A schematic of biosparging is shown in Figure 5.3.

In a similar fashion to bioventing, the equipment required is rather simple and mostly inexpensive, and time and money have to be spent to understand the local subsurface in sufficient detail to guarantee end points. Unlike bioventing, of course, biosparging is a technique of the saturated zone, which creates the need for engineered differences from bioventing. The effectiveness of biosparging is governed by two overriding factors: soil permeability, which determines the rate of transfer of oxygen from the gas phase to the aqueous phase and eventually to the microorganisms, and contaminant biodegradability.

CONTAMINANTS TREATED BY BIOSPARGING

In a manner analogous to the comparison between SVE and bioventing, air sparging and biosparging are suited to the treatment of different contaminants, although there is inevitable overlap. Air sparging utilizes higher airflow velocities and the treatment is therefore dominated by volatilization with some incidental biodegradation. The lower flow velocities of biosparging favor biodegradation over volatilization.

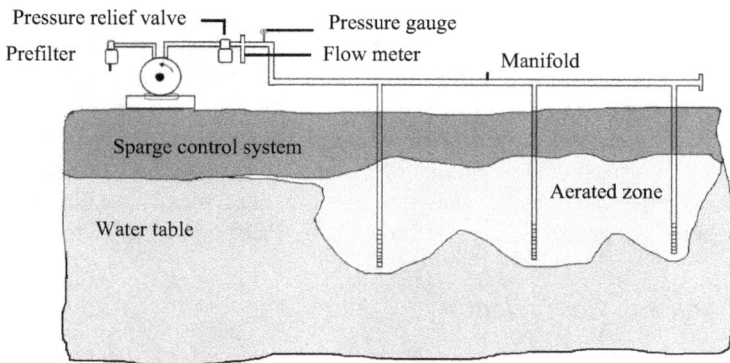

FIGURE 5.3 Biosparging schematic (Cookson, 1995).

Biosparging is most often used at sites with groundwater contamination by middle distillate fuels, such as diesel and kerosene. The more volatile components of gasoline are also more toxic to microorganisms, and fresh gasoline spills in groundwater would be better treated by air sparging. Longer chain hydrocarbons, such as those present in lubricating and heating oils, are intrinsically biodegradable but at lower rates than, say, medium-chain length n-alkanes. However, these components are still amenable to biosparging, but the process necessarily takes longer. These larger hydrocarbons are also less volatile and therefore less amenable to air sparging.

BIOSPARGING DESIGN

Biosparging is much less established as a treatment technology than bioventing. Specifically, biosparging should not be used if free product is present in significant quantities; if basements or other confined utilities are located underground at the site, unless another technology such as SVE is being used for vapor control; or if the contaminants are in a confined aquifer, since the sparged air will have no escape path. Setting such exclusion conditions limits the number of suitable sites. Many of the criteria for suitability of bioventing also apply to biosparging.

PERMEABILITY

The intrinsic permeability of soil is the single most important factor that determines the suitability of biosparging at a particular site. A hydraulic conductivity of greater than 10^{-4} cm/s at 20°C is suitable for biosparging, but at less than 10^{-5} cm/s it is unlikely to be.

AIRFLOW SYSTEMS

An oil-free compressor of suitable size for the flow rate (middle of the operational range) is required. As demand should change through the duration of a biosparging project, the compressor should be sized according to maximum demand and should be rated for continuous duty at this demand. The compressor should be equipped with a particulate filter to prevent downstream contamination. Flow rate and pressure should be measurable. Typical airflow rates for biosparging are low compared to those for air sparging and would be in the range of 85–700 L/min per injection well. A pulsed air supply has been suggested as a means of improving mixing and distribution of air in the saturated zone. This has been suggested as a possible reason for observed improvements in remediation rate. The increase in hydrostatic pressure with depth of injection must be taken into account, but typically the air pressure will be in the range of approximately 69,000–103,000 Pa (roughly 10–15 lb/in^2).

WELL DESIGN AND CONSTRUCTION

The choice of horizontal or vertical wells is governed by the same reasoning as for bioventing. Horizontal wells are best used at shallow sites, if ten or more sparge points are required, or if the area affected is under a building or some other

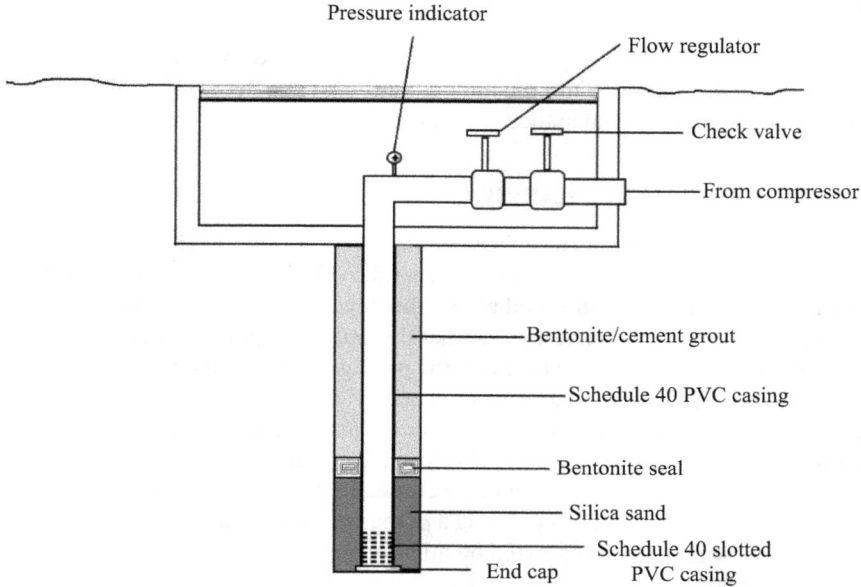

FIGURE 5.4 Biosparging well design (USEPA, 1995).

surface structure. Vertical wells are required for deep contamination (greater than about 8 m) and where only a few wells are required. Construction is quite standard (Figure 5.4) and similar to that of bioventing wells but with some significant differences. The wells are usually fabricated from 1- to 5-in diameter PVC or steel pipe. The slotted, or perforated, screened interval is usually about 0.3–1 m long and is set about 1.5–4.5 m below the lowest point of contamination. Proper capping is essential, especially because of the elevated pressures, to prevent air from short-circuiting back to the surface. To enable even distribution of air (or indeed, to divert more air to where it is needed), each well should be fitted with a pressure gauge and flow regulator. The piping manifold can be buried in a shallow trench depending on the site. Metal pipe should be connected directly to the compressor because of the elevated temperature of the exit air. At the pressures in the system, PVC pipe can be used otherwise.

Number and Spacing of Wells

The required number and spacing of wells are defined by the bubble radius. The bubble radius is the greatest distance from a sparging well at which a sufficient sparge pressure and airflow can be induced to enhance the biodegradation of contaminants. It is determined mainly by the hydraulic conductivity of the aquifer that is being sparged and is generally in the region of 1.5 m for fine-grained soils to 30 m for coarse-grained soils. Closer well spacing is appropriate in zones of high-level contamination to improve oxygen delivery.

NUTRIENT DELIVERY

Laboratory trials are used to determine if nutrients need to be added. In particular, the addition of ammonium ions might be tightly regulated or even specifically banned, depending on local authority regulations.

PERMEABLE REACTIVE BARRIERS

The permeable reactive barrier (PRB) is an interception technology for the remediation of contaminated groundwater (Figure 5.5). PRBs are also known as passive treatment walls. They are installed across the flow path of the ground water and are constructed from porous materials, so that the water can pass through the wall. Yet the wall contains materials that prevent the passage of the pollutants. The pollutants are either degraded within the wall or retained in a concentrated form. The wall materials are various and often consist of zero-valent metals [mostly zero-valent iron, $Fe(O)$], chelators, sorbents, or compost. It is clear that this is not a bioremediation technology, then. However, an inevitable consequence of flowing water containing a low concentration of pollutants entering a porous material of high surface area is that the materials will become colonized by microorganisms. Thus, with time, it is likely that a contribution to the treatment is biological.

In fact, as most PRBs are not bioreactors, the growth of microorganisms within the barrier material has been perceived as a possible detrimental effect due to the potential for biofouling to decrease the permeability of the barrier. It is necessary to keep the reactive zone permeability greater than or equal to the permeability of the aquifer to avoid diversion of the flowing water around the barrier. Most PRBs are based on $Fe(0)$ as the reactive material (about 80 laboratory, pilot, and full-scale installations worldwide), mostly due to its ability to reductively dechlorinate troublesome chloroaliphatics such as TCE. They are also able to remove some metals. The working lifetime of these PRBs may be limited by precipitation of secondary minerals due to reaction with groundwater. This could be exacerbated by microbial growth. It is generally believed that the high pH within the treatment zone of

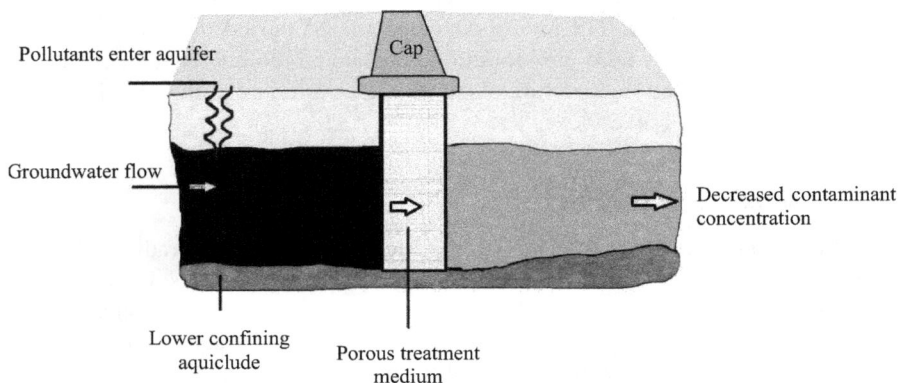

FIGURE 5.5   PRB schematic (Birke et al., 2003).

Fe(0) (around ten) would discourage microbial growth, and some studies show the impact of microorganisms on the performance of Fe(0) PRBs to be minimal. Data that show that diverse microbial communities can establish in this high pH, highly reducing environment are emerging, but how this will affect the long-term performance of Fe(0) barriers is not yet known. The role of microorganisms in PRBs is likely to be enhancement, rather than a standalone biotechnology on its own. In this context, bio-enhanced PRBs are being researched for removal of metals from acid mine drainage, explosives, chlorinated solvents, and inorganic pollutants. The metal removal is usually mediated by metal sulfide precipitation due to the action of the dissimilatory sulfate-reducing bacteria. Metal sulfides have extremely low-solubility products. Flow through columns has shown the feasibility of microbial colonization of Fe(0) to enhance the removal of TCE, sulfate, nitrate, and Cr(VI). Other biological materials, although non-living, are being used, primarily for their sorptive properties, e.g., bone meal apatite and bone charcoal for the removal of uranium; peat for the removal of benzene, toluene, ethylbenzene, and xylene (BTEX); and chitosan-coated sand for the removal of copper.

An attractive feature of PRBs combined with biological treatment is the possibility for sequenced treatment, with, for example, anaerobic, reductive chemical and biological treatment followed by aerobic biodegradation of dechlorinated hydrocarbons and other, non-halogenated pollutants that might be present in a pollutant cocktail. Large numbers of mucoid bacteria are pumped into injection wells. A growth substrate is added to stimulate growth. Bacterial growth and extracellular polymer production substantially reduce the hydraulic conductivity of the formation.

PRB DESIGN

As the PRB is not primarily a biotechnology, there are few design details for bio-enhanced PRBs. One of the great advantages of PRBs is cost savings, because pumping, large-scale excavation, and off-site disposal are eliminated. The conventional construction technique is trench and fill, which is a relatively hazardous and costly way to build a PRB. Reactive Fe(0) barriers can be installed by using biodegradable guar gum slurry without significantly decreasing the reactivity or long-term treatment characteristics of the iron. Biopolymers produce high viscosity for suspending the iron for injection or as a liquid shoring for trenching.

BIOSLURPING

Bioslurping evolved as a combination of several technologies to deal with one of the big challenges to in situ remediation: the presence of free product, specifically LNAPLs. Such is the evolution of the technology that it has at least one book dedicated to it. Bioslurping combines vacuum-assisted LNAPL recovery with bioventing and SVE. Thus, bioslurper systems simultaneously recover free product and remediate the vadose, capillary, and saturated zones. They use an aboveground vacuum pump to create enough vacuum airlift to draw LNAPLs from the subsurface, along with soil gases and small amounts of groundwater. Bioslurping can greatly enhance LNAPL recovery compared to conventional skimming and pumping technologies.

Most of the equipment involved is similar to that already described in previous sections. The main components of a bioslurper system are as follows:

- recovery well,
- vacuum pump capable of extracting liquids and vapors (usually a liquid ring pump),
- liquid–vapor and oil–water separator units, and
- water and vapor treatment systems, if required.

The heart of a bioslurping system is the in-well dual drop tube assembly (Figure 5.6). The use of the dual drop tube assembly greatly reduces the risk of emulsification of the free product and groundwater, along with solid entrainment, by allowing the separation of the free product and groundwater in the well before vacuum extraction. An aboveground vacuum pump enhances subsurface migration of LNAPL to the extraction well, where it is withdrawn from the upper drop tube (the secondary drop tube), or "slurp tube," in the dual assembly. The primary drop tube, usually a 1-in. (2.54-cm)-diameter PVC pipe, is shielded in the lower end with a larger diameter, open-ended pipe that extends both above and below the end of the secondary tube. This shield is usually made of 2-inch PVC pipe and is termed the fuel isolation sleeve. It is conventionally 4 ft (about 1.2 m) long, extending about 2 ft above and below the end of the slurp tube.

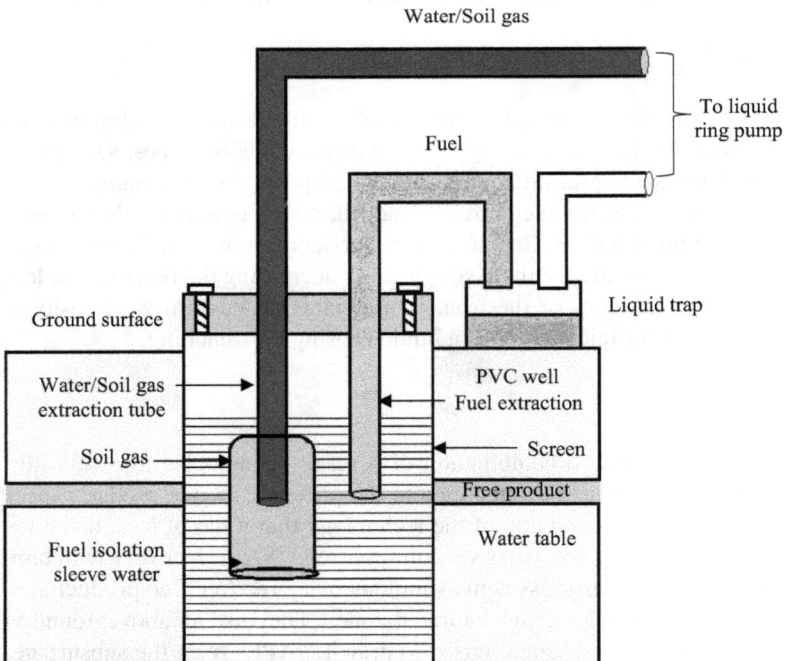

FIGURE 5.6   Bioslurping dual drop tube (Place et al., 2001).

When the vacuum pump is switched on, the negative pressure induced in the well promotes LNAPL flow toward the well and draws LNAPL trapped in pore space above the water table. In response to pumping, the LNAPL level decreases to the point where the slurp tube draws in and extracts vapor in a manner akin to SVE. This vapor extraction promotes airflow in the unsaturated zone, which stimulates bioremediation in a manner akin to bioventing. When the vacuum causes the water table to raise slightly, the system reverts back to extraction of LNAPL and ground-water. By avoiding high levels of groundwater rebound, large water table fluctuations are prevented, and this greatly reduces the chance of LNAPL being trapped below that water table, the "smearing" problem often encountered with more conventional free-product recovery systems.

The drop tube position within the well is altered by using a section of flexible tubing connecting the extraction manifold to the drop tube. Like other airflow-based technologies, bioslurping is ineffective in low-permeability soils but is reported to also be cost-effective, over and above the advantages cited above. The primary cost saving is due to the reduction in the amount of extracted groundwater, which mini-mizes storage, treatment, and disposal costs.

BIOAUGMENTATION

Bioaugmentation in the context of bioremediation can be considered as the inocula-tion of contaminated soil or water with specific strains or consortia of microorgan-isms to improve the biodegradation capacity of the system for a specific pollutant organic compound(s).

Bioaugmentation often is considered for the bioremediation of compounds that appear to be recalcitrant, i.e., contaminants that persist in the environment and appear to be resistant to microbial biodegradation. In this context, bioaugmentation may have value; i.e., if indigenous microorganisms lack the capability to degrade the contaminant and if organisms that will be active can be introduced, bioremediation can become useful in that situation. Many still look to modern biotechnology, i.e., the application of recombinant DNA technology, as the panacea to problems of pollution with highly resistant contaminants.

There are three fundamental approaches to bioaugmentation of a contaminated site, although little attention is paid to these during operations. The first is to increase the genetic diversity by inoculation with allochthonous microorganisms. By increas-ing the genetic diversity of the soil or water, it is assumed that this increases the catabolic potential and thereby the rate of removal of the contaminant(s) by biodeg-radation will increase.

The second is to take samples from the site and use them as initial inocula for serial enrichments with the contaminant(s) in question as the sole source of carbon (Figure 5.7). Typically, the selected strains are a subset of the fast-growing micro-organisms. During this procedure, each subsequent enrichment should increase the proportion of the biodegradative population compared to the remainder. This inocu-lum is then returned to the site in large numbers in order to increase the rate of biodegradation. This approach, then, does not rely on increasing genetic diversity. Rather, it solely increases the catabolic potential of the specific strains capable of

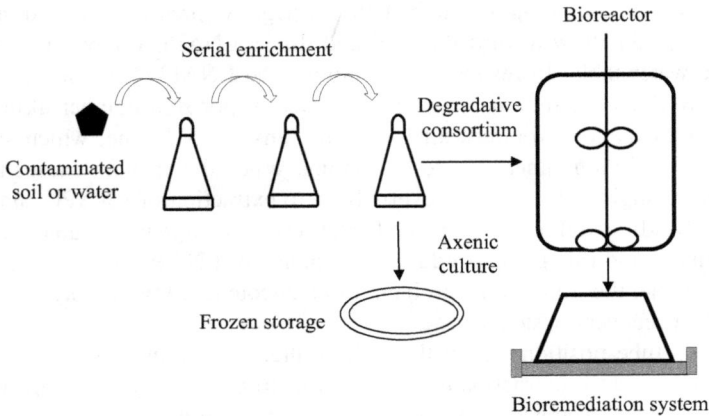

FIGURE 5.7 Typical serial enrichment procedure for bioaugmentation (Rittmann and Whiteman, 1994).

degrading contaminating chemicals. The strains used as inocula in these cases typically demonstrate exceptional degradative capacities and rapid growth rates; however, upon bioaugmentation, many, if not most, strains provide only a brief burst of pollutant biodegradation, only to decline by several log orders into the background of the indigenous microbial community due to competition.

The third approach is often performed in practice but is not valued in the scientific community. This involves the addition of uncharacterized consortia present in materials such as sewage sludge and compost. These are easy sources of inocula for companies to sell to engineers, who faced with a contaminated site, often feel compelled to do something and who do not fully evaluate the microbial value of bioaugmentation. Even materials such as garden waste provide extra microbial communities, even though that is not the primary function in the bioremediation, which is normally to provide heat-generating materials during composting.

From the plethora of studies of bioaugmentation, a trend that should be taken account of is that, despite selection of strains with great catabolic potential, the organisms have often failed to survive in the highly competitive environments of water and soil. Perhaps, it is time for a shift of focus to survival rather than catabolic potential in strain selection. Yet the ability to survive is much more difficult to select for than a simple catabolic trait. Molecular biology is starting to furnish more methodical approaches to strain selection in this context.

SELECTION OF ADDITIONAL TRAITS

A large number of desirable traits could be imagined. Where oil or solvent contamination levels are very high, it is likely that catabolic strains will also possess solvent tolerance mechanisms. The hydrophobicity of many pollutants adds to their toxicity to microorganisms by the nonspecific effects on membranes. The ability to combat this toxicity would give a catabolic strain a competitive edge over others without the ability. Multiple responses to solvent action are known in bacteria, among which two of the best known are cis-trans isomerization and efflux pumps. The cis-trans modification is a rapid response to exposure to solvents that takes place within a minute of exposure, and this is common within the genus *Pseudomonas*.

PLANT-ASSOCIATED STRAIN SELECTION AND DELIVERY

The pairing of catabolically relevant microorganisms that can colonize plant roots with a plant circumvents some of the problems faced by traditional bioaugmentation. It was identified that a strain indigenous to a host plant's rhizosphere that exhibited preferential growth on plant root exudates and enhanced pollutant-degrading ability and rhizosphere competence. The approach is aided not only by the abundance of pollutant-degrading microorganisms typically associated with the rhizosphere but also by the fact that the highly competent pollutant-degrading inoculum is sustained on a carbon source that is constantly being replenished by the plant. This would aid survival of strains when the pollutant is present in low available concentrations.

ADAPTATION OF STRAINS: VALUE OF PRIMING OR ACTIVATION

Priming is generally described as predisposing an isolate or population of microorganisms to future conditions in which they are designed to perform a function. Priming for bioremediation might consist of enriching clean soil for particular pollutant-degrading microorganisms by repeated exposure to the relevant pollutant(s). In other words, this bears similarity to the traditional serial enrichment procedure, but it is done in soil, not liquid culture. The resulting soil, with its highly competent degrader microbial community, is then used as the inoculum for the target soil. This approach has several advantages: the consortium of indigenous microorganisms within the primed soil are all utilized in the bioaugmentation, not just a laboratory-grown isolate; the primed soil consortium is maintained within its native soil, minimizing the competitive elements (i.e., microorganisms, predation, or chemicals) of the target soil; and otherwise less culturable, yet potentially highly competent, pollutant degraders are included obligatorily. This approach, while logical and easier to set up than the traditional shake flask enrichment, is certainly among the most understudied bioaugmentation strategies.

BIOAUGMENTATION WITH GMOS

Genetically engineered microorganisms (GEMs), which are also referred to as genetically modified microorganisms or genetically modified organisms (GMOs), have shown great potential for bioremediation applications in soil, groundwater, and activated sludge environments, exhibiting enhanced degradative capabilities encompassing a wide range of chemical contaminants that are found at numerous soil and groundwater sites. Various investigators have used modern biotechnology to create recombinant bacteria with potential for environmental applications (Sayler and Ripp, 2000).

The first living organism ever patented was a hydrocarbon-degrading pseudomonad that was genetically engineered by Ananda Chakrabarty of General Electric

118

Environmental Biotechnology

(GE); the bacterium had been designed to degrade components of crude oil for potential applications to oil spills. Chakrabarty was able to fuse plasmids to genetically engineer a strain of *Pseudomonas* having the single-cell capability for multiple separate degradative pathways, including camphor, salicylate, and naphthalene degradative pathways, which gave the organism-specific beneficial oil-degrading capabilities. The initial attempt to obtain a patent for the GEM was denied by the U.S. Patent Office on the basis of the argument that living bacteria are not patentable because they are products of nature. General Electric appealed the decision, which was eventually decided upon by the U.S. Supreme Court in 1980 in a 5 to 4 decision in the case of Diamond v. Chakrabarty (Sydney Diamond was the Commissioner of Patents who sought to prevent the granting of a patent). This landmark decision said that life could be patented, opening the door for major investments in biotechnology and the possibility of using genetic engineering for creating organisms that could be used for bioremediation. Since the granting of the patent to Chakrabarty, many researchers and start-up bioremediation companies have considered genetically engineering microorganisms for bioremediation applications.

Besides strains engineered to have specific metabolic capabilities, GMOs that are able to function in hostile environments, such as those containing high concentrations of heavy metals, have been created. *Deinococcus geothermalis*, a radiation-resistant bacterium for the treatment of mixed radioactive wastes containing ionic mercury, was engineered. Another example is the construction of multiple heavy metal-resistant phenol-degrading pseudomonads. A strain of *Mesorhizobium huakuii* was engineered to express tetrameric human metallothionein. When this bacterial symbiont forms nodules, the host plant accumulates high concentrations of the heavy metal, e.g., cadmium. The plant can then be harvested, thereby removing the heavy metal from the soil.

Although various GEMs that may be useful for bioremediation have been developed, the only field trials conducted have involved the use of reporter genes. Reporter genes, such as the lux genes that confer bioluminescence and the green fluorescent protein gene, allow tracking of the environmental fate of GEMs. For example, a bioluminescent reporter strain of *Ralstonia eutropha* was constructed for the detection of PCBs in the environment. Similarly, *Pseudomonas fluorescens* strain that exhibited bioluminescence when degrading naphthalene was genetically engineered (Layton et al., 1998).

The lack of field trials to test the efficacy of biodegradation by GEMs in soils and aquifers is a serious impediment to the development of GEM-based bioremediation applications, since the only manner in which to fully address the competence of GEMs in bioremediation efforts is through long-term field release studies. Therefore, it is essential that field studies be performed to acquire the requisite information for determining the overall effectiveness and risks associated with GEM introduction into natural ecosystems. One way to eliminate the regulatory hurdles of releasing GMOs into the environment is to use enzymes, i.e., to eliminate the living organism. Site-directed mutagenesis was done to develop an enzyme that is capable of attacking tetrachlorinated dibenzo-p-dioxins that could be applicable to the bioremediation of soils contaminated with dioxin. Mixed cultures capable of oxidizing and hydrolyzing endosulfan may be a good source of enzymes for use in enzymatic bioremediation of endosulfan residues. Enzymes derived from organophosphate-resistant sheep blowfly have also been found to detoxify orthophosphate insecticides. Microorganisms have been engineered to degrade organophosphates and to produce enzymes that can detoxify organophosphate-contaminated water. These enzymes have been shown to work under field conditions and have the potential to degrade parathion, malathion, and monocrotophos; carbaryl; many synthetic pyrethroids; and endosulfan, which are all insecticides.

Novel Delivery Systems

A fundamental observation about delivery of bioaugmentation cultures has to be explained. Given all the trials that an introduced culture has to face, the delivery of the culture to a soil bioremediation project must be done correctly if it is to have a chance of success. In ex situ systems, this is best done by thorough mixing with the soil during the construction. It has been often observed that bacteria do not move appreciably through soil. The inoculum therefore must be distributed to nearly all sites immediately adjacent to the pollutants. This lack of movement is due to physical filtration by the solids in the soil and adsorption to soil particles. Therefore, the practice of addition of cultures to, say, a biopile is likely to be futile.

In groundwaters, similarly, there is evidence that bacteria do not travel great distances before becoming attached to a solid surface. Since appreciable biodegradation by an inoculum to groundwater would require the inoculum to travel substantial distances, it is not yet clear whether such an approach is valuable.

Delivery of bioaugmentation cultures in an immobilized form may offer more complete and/or more rapid degradation. In bioremediation applications, the immobilized matrix may also act as a bulking agent in contaminated soil, facilitating the transfer of oxygen, which is crucial for rapid hydrocarbon mineralization. Immobilization is known to reduce competition with indigenous microorganisms and to offer protection from predation and extremes of pH and toxic compounds in the contaminated soil. There is also evidence of increased biological stability, including plasmid stability, in immobilized cells.

BIOSTIMULATION

An alternative to increasing numbers of biodegrading microorganisms on the site is the practice of biostimulation. Biostimulation aims at enhancing the activities of indigenous microorganisms that are capable of degrading the offending contaminant. It is applicable to the bioremediation of oil-contaminated sites, often being viewed as helping nature, i.e., an extension of the natural remediation of soil and groundwater. In many cases, the addition of inorganic nutrients acts as a fertilizer to stimulate biodegradation by autochthonous microorganisms. Biostimulation in some cases involves the addition of inorganic nutrients to stimulate biodegradation by autochthonous microorganisms; in other cases, it is the intentional stimulation of resident xenobiotic-degrading bacteria by use of electron acceptors, water, nutrient addition, or electron donors.

Although not an approach to bioaugmentation as such, biostimulation leads to an increase in catabolic potential without increasing genetic diversity. In practice, it is a much simpler and less costly approach to bioremediation. The bioremediation community has diverging opinions on the subject. Many practitioners of bioremediation state that biostimulation is all that is required for the mineralization of naturally occurring hydrocarbons such as petroleum mixtures: 3.5 billion years of evolution has provided the genetic diversity required, and all that is needed to stimulate mineralization is the correct balance of carbon with nitrogen and phosphorus. However, it must be noted that bioaugmentation may be required for the biodegradation of more recalcitrant xenobiotic pollutants, where the evolutionary timescale may be limited to the years since the industrial revolution.

OXYGEN

Due to the limited solubility of oxygen in water, various attempts have been made to increase it. The most obvious route is the injection of pure oxygen rather than air, as this can increase the dissolved oxygen concentration several fold. More common has been the use of hydrogen peroxide.

OXYGEN AVAILABILITY AND TRANSPORT

As most of the bioremediation technologies are aerobic processes, due to the greater efficiency of aerobic biodegradation of organic compounds, the delivery of oxygen to soil and groundwater in ex situ and in situ bioremediation technologies is crucial to success. Indeed, oxygen should in most cases be regarded as the key component of a bioremediation that will ultimately determine the success or failure of the application. It has been shown on many occasions that hydrocarbon biodegradation will proceed without special soil amendments as long as oxygen is present.

The concentration of oxygen dissolved in water is low. Since moisture is critical to microbial activity, the essential interrelationship between oxygen and water is obvious. Oxygen transport from the open air to the pores of the soil bioremediation matrix to the pore water and thence to the microorganisms is a two-component process. Convection and diffusion are the two governing mechanisms, and oxygen transport is inextricably linked to the water content of the matrix since both convection and diffusion are dramatically reduced in water. Water saturation of soil pores slows oxygen transport to very low levels, and quickly, the rate of oxygen consumption by microorganisms exceeds the ability of convection and diffusion to replace it, resulting in deoxygenation and, if left unchecked, anaerobiosis.

The low rate of oxygen delivery is a major drawback in bioremediation. Soil contaminated with $10 \, m^3$ of hydrocarbons would require about 2 million m^3 (tons) of water saturated at 10 mg/L for its biodegradation.

CONVECTIVE AIR MOVEMENT

In ex situ bioremediation technologies, convection can be passive, making use of the buoyancy of hot air, or forced by using blowers or vacuum pumps. For example, in composting or biopile systems treating high levels of organic contamination, hot air can sometimes be seen rising from the top of the pile.

OXYGEN DELIVERY TO GROUNDWATER

There are several ways to deliver oxygen to groundwater during in situ bioremediation. The most appropriate method for a particular project will be dictated by the contaminants and their concentrations, which set the oxygen demand, and the hydrogeology of the site. Hydraulic conductivity has an enormous effect on the amount of oxygen that has to be supplied. For a given hydraulic gradient, oxygen supply rate varies by several orders of magnitude from low to high hydraulic conductivity. The simplest, least expensive method is to sparge air into the well bore by using porous bubbler devices such as those that are in common use in the wastewater treatment industry. Saturating water with air in this way will achieve a dissolved oxygen concentration of around 8–10 mg/L depending on salinity and temperature. Saturated water diffuses from the well bore into the subsurface at a rate determined by the hydraulic conductivity.

The very low diffusivity of oxygen in water compared to that in air threatens failure for many aquifer bioremediation projects. Alternatively, more drastic methods of aeration have been investigated. Of these, the injection of hydrogen peroxide (H_2O_2) has been the most successful. It is highly soluble in water and decomposes to water and oxygen. Therein lies the biggest problem with using hydrogen peroxide in situ: the rate of decomposition has to be controlled. Its decomposition is catalyzed by common soil components such as ferric iron and naturally occurring organic materials. For this reason, hydrogen peroxide is not a practical oxygen delivery method for the vadose zone. Too large amounts of oxygen generated are lost to the gas phase. In the saturated zone, more is dissolved in water. However, even then, if the decomposition cannot be controlled, the zone of influence around the injection well may be too limited.

OLEOPHILIC FERTILIZER

When the contaminant is oil or another hydrophobic pollutant, there is the danger that applied water-soluble sources of nitrogen and phosphorus will have poor contact with the material. The objective would be to encourage intimate contact between microorganisms, oil, and nutrients. The obvious way to facilitate this is the incorporation of surfactants into the formulation. To maximize the contact between oil and water, the surfactant of choice should minimize the interfacial tension between the two. When the oil droplets in water, or water droplets in oil, are sufficiently small that the system remains transparent, then the phase containing most of the surfactant and the dispersed droplets is called a microemulsion. Microemulsions are characterized by having ultralow interfacial tensions (less than 0.01 mN/m can readily be achieved in the laboratory with some fairly ordinary and inexpensive surfactants). One such oleophilic fertilizer was used in the beach cleanup operations for Exxon Valdez. It was designed to adhere to oil and is a microemulsion of a saturated solution of urea in oleic acid, containing tri(laureth-4)-phosphate and butoxy-ethanol.

APPLICATION OF NUTRIENTS

Nutrient additions can be calculated by various means. Knowledge of the composition of the average bacterium is the start point. Elemental assay of the dry mass of

Escherichia coli gives an approximate composition of the protoplasm of 50% carbon; 20% oxygen; 14% nitrogen; 8% hydrogen; 3% phosphorus; 2% potassium; 1% sulfur; 0.05% each calcium, magnesium, and chlorine; 0.2% iron; and a total of 0.3% trace elements, including manganese, cobalt, copper, zinc, and molybdenum. The nutrient requirements of the microbial cell are approximately the same as the cell composition except for carbon, which is supplied by the organic contaminant. A disadvantage of a commercial preparation containing mixed nutrients is that the ratio of various constituents cannot be modified according to requirements based on the carbon load created by the carbon-containing contaminant. In particular, the ratio of carbon to nitrogen to phosphorus seems to be important, and the ability to alter these components according to site contaminant concentrations would be important.

CO-METABOLISM AND ALTERNATIVE ELECTRON DONORS

Co-metabolism is defined as the degradation of a compound only in the presence of other organic material that serves as the primary energy source. Essentially, in co-metabolism, an enzyme for a natural substrate also transforms a pollutant compound. As an example, TCE can be transformed by methane monooxygenase and toluene monooxygenase. Either methane or toluene supplied as a co-substrate can be used for the bioremediation of TCE-contaminated groundwater.

The most favorable outcome of a co-metabolic event is that the transformation product can be mineralized, where the initial substrate was not. In other words, the product of the co-metabolism becomes an energy source for another microorganism. This, fortunately, often is the case. If not, then the next favored outcome is a reduction in toxicity of the transformation product.

Co-metabolism is more important in bioremediation than is generally appreciated. It is known that many microorganisms participate in co-metabolic processes, and co-metabolism is important for many transformations, including those of some PAHs, halogenated aliphatic and aromatic hydrocarbons, and pesticides. Aerobic or anaerobic co-metabolism is possible depending on the contaminants and the field situation. Other electron donors being used at Superfund groundwater bioremediation sites include acetate, lactate, benzoate, and methanol in all cases for the bioremediation of chlorinated compounds.

FURTHER READING

Aggarwal PK, Means JL, Hinchee RE. Formulation of nutrient solutions for in situ bioremediation. In: Hinchee RE, Olfenbuttel RF (eds.) *In Situ Bioreclamation. Applications and Investigations for Hydrocarbon and Contaminated Site Remediation*. Stoneham, MA: Butterworth-Heinemann; 1991:51–66.

Alexander M. *Biodegradation and Bioremediation*. 2nd ed. San Diego, CA: Academic Press; 1999.

Atagana HI, Haynes RJ, Wallis FM. 2003. Co-composting of soil heavily contaminated with creosote with cattle manure and vegetable waste for the bioremediation of creosote contaminated soil. *Soil Sed. Contam.* 12:885–899.

Balba MT, Al-Daher R, Al-Awadhi, N. Bioremediation of oil-contaminated desert soil: the Kuwaiti experience. *Environ. Int.* 1998;24:163–173.

Birke V, Burmeier H, Rosenau D. Permeable reactive barrier technologies for groundwater remediation in Germany: recent progress and new developments. *Fresenius Envi- Yon. Bull.* 2003;12:623–628.

Bollati A, Luzi, C. Appendix 3. In: Lecomte P, Mariotti C (ed.) *Handbook of Diagnostic Procedures for Petroleum – Contaminated Sites.* Chichester: John Wiley and Sons Ltd.; 1997:167–185.

Boving TB, Wang X, Brusseau ML. Solubilization and removal of residual trichloroethene from porous media: comparison of several solubilization agents. *J. Contam. Hydrol.* 2000;42:51–67.

Civilini M. Fate of creosote compounds during composting. *Microbiol. Ecru.* 1994;2:16–24.

Cookson J-T Jr. *Bioremediation Engineering: Design and Application.* New York: McGraw-Hill; 1995.

DePaoli DW. Design equations for soil aeration via bioventing. *Separations Technol.* 1996;6:165–174.

Dua M, Singh A, Sethunathan N, John AK. Biotechnology and bioremediation: successes and failures. *Appl. Microbiol. Biotechnol.* 2002;59:143–152.

Dupont RR. Fundamentals of bioventing applied to fuel contaminated sites. *Environ. Prog.* 1993;12:45–53.

Guerin TF. Co-composting of pharmaceutical wastes in soil. *Lett. Appl. Microbiol.* 2001;33:256–263.

Grasso D. *Hazardous Waste Site Remediation. Source Control.* Boca Raton, FL: CRC Press; 1993.

Hejazi RF, Husain T. Landfarm performance under arid conditions. 1. Conceptual framework. *Environ. Sci. Technol.* 2004;38:2449–2456.

Hejazi RF, Husain, T. Landfarm performance under arid conditions. 2. Evaluation of parameters. *Environ. Sci. Technol.* 2004;38:2457–2469.

Illmer P. Backyard composting: general considerations and a case study. In: Insam H, Riddech N, Klammer, S (ed.) *Microbiology of Composting.* Heidelberg: Springer-Verlag; 2002:3–4.

Jorgensen KS, Puustinen J, Suortti, AM. Bioremediation of petroleum hydrocarbon-contaminated soil by composting in biopiles. *Environ. Pollut.* 2000;107:245–254.

Kodres CA. Coupled water and air flows through a bioremediation soil pile. *Environ. Model. Software* 1999;14:37–47.

Kuyukina MS, Ivshina IB, Ritchkova MI, et al. Bioremediation of crude oil contaminated soil using slurry phase biological treatment and landfarming techniques. *Soil Sed. Contam.* 2003;12:85–99.

Laine MM, Jorgensen KS. Effective and safe composting of chlorophenol contaminated soil in pilot scale. *Environ. Sci. Technol.* 1997;31:371–378.

Layton AC, Muccini, M, Ghosh MM, Sayler GS. Construction of a bioluminescent reporter strain to detect polychlorinated biphenyls. *Appl. Environ. Microbiol.* 1998;64:5023–5026.

Martin E. Environmental protection. In: Lucas AG (ed.) *Modern Petroleum Technology*, 6th ed., vol. 2. Chichester: Downstream John Wiley and Sons; 2001:197–210.

Michel FC, Quensen J, Reddy CA. 2001. Bioremediation of a PCB-contaminated soil via composting. *Compost Sci. Utiliz.* 9:274–284.

Michel FC Jr, Reddy CA, Forney LJ. Microbial degradation and humification of the lawn care pesticide 2,4-dichlorophenoxyacetic acid during composting of yard trimmings. *Appl. Environ. Microbiol.* 1995;61:2566–2571.

Morgan P, Watkinson RJ. Hydrocarbon degradation in soils and methods for soil biotreatment. *Crit. Rev. Biotechnol.* 1989;8:305–333.

Namkoong W, Hwang EY, Park JS, Choi JY. Bioremediation of diesel contaminated soil with composting. *Environ. Pollut.* 2002;119:23–31.

Place MC, Coonfare CT, Chen ASC, Hoeppel RE, Rosansky SH. *Principles and Practice of Bioslurping*. Columbus, OH: Battelle Press; 2001.

Rittmann BE, Whiteman R. Bioaugmentation: a coming of age. *Water Qual. Int.* 1994;1:12–16.

Ronald MI, Atlas RMI, Philp J. *Bioremediation Applied Microbial Solutions for Real-World Environmental Cleanup*. Washington, DC: ASM Press; 2005.

Sasek V, Bhatt M, Cajthaml T, Malachova K, Lednicka, D. Compost-mediated removal of polycyclic aromatic hydrocarbons from contaminated soil. *Arch. Environ. Contam. Toxicol.* 2003;44:336–342.

Sayler GS, Ripp S. Field applications of genetically engineered microorganisms for bioremediation processes. *Curr. Opin. Biotechnol.* 2000;11:286–289.

Semple KT, Reid BJ, Fermor TR. Impact of composting strategies on the treatment of soils contaminated with organic pollutants. *Environ. Pollut.* 2001;112:269–283.

US Environmental Protection Agency. *Bioventing Principles and Practice, vol. 11. Bioventing Design. Manual. EPA/540/R-95/534A*. Washington, DC: U.S. Environmental Protection Agency; 1995.

Van Gestel K, Mergaert J, Swings J, Coosemans J, Ryckeboer, J. Bioremediation of diesel-contaminated soil by composting with biowaste. *Environ. Pollut.* 2003;125:361–368.

Zhu X, Venosa AD, Suidan MT, Lee K. *Guidelines for the Bioremediation of Oil-Contaminated Salt Marshes. [Online]*. Cincinnati, OH: National Risk Management Research Laboratory Office of Research and Development, U.S. Environmental Protection Agency; 2004. http://www.epa.gov.

6 Decomposition of Plant Cell Wall Structures

The primary cell wall of plants consists of cellulose, hemicellulose, pectin, and a small amount of protein. Cellulose provides the rigidity to support the plant structures, while hemicellulose and pectin form a matrix around the cellulose fibrils. Hemicellulose binds the cellulose microfibrils into a cohesive unit, while pectin forms a gel phase that contributes to the porosity of the cell wall and prevents collapse of the cellulose–hemicellulose framework. Lignin is added as secondary metabolism to the plant cell walls and contributes to mechanical stability of the plant tissues. The sequence of events in the degradation of fresh plant cell wall material is as follows: (1) protopectinase releases pectin from plant cell wall material, (2) pectin is degraded by pectinases, and (3) hemicellulose and lignin are degraded and the exposed cellulose is hydrolyzed by enzymes.

PROTOPECTINASE AND PECTINASE ACTIVITIES

The release of pectin from the cell wall is accomplished by protopectinases, and this action results in separation of cells. In terms of hydrolytic action, there are two types of protopectinases: an *A-type enzyme*, which releases pectin in soluble low-molecular weight segments, and a *B-type enzyme*, which releases high-molecular weight segments that have extensive polymerization. Protopectinases are produced by *Kluyveromyces fragilis* (yeast), *Aspergillus niger*, and *Bacillus subtilis* (Sakai et al., 1993).

Pectin, found in cell walls of plants, is a linear polymer of galacturonic acid bonded by α-1,4 sugar linkages, and some of the carboxyl groups of the sugar are esterified to methanol. In some cases, pectin has short sidechains of rhamnose, arabinose, or other neutral sugars. Because pectin may vary with cell type and plant species, the molecular mass of pectin ranges from 10,000 to 400,000. In the cell wall, pectin forms an interlocking gel that is attached to proteins and other structural components of the cell wall. The name commonly used to refer to enzymes that degrade pectin is *pectinase*, and these enzymes are primarily of the polygalacturonidase and esterase types. Pectinases are produced by various soil fungi and various bacteria, especially those that are plant pathogens. Commercially, there is an application using pectinases to remove turbidity in apple and other juices. It is of historical interest that *Clostridium* and *Bacillus* were used in a "retting" process to decompose pectin in flax straw with the release of textile fibers used in production of linen.

DEGRADATION OF HEMICELLULOSE

Hemicelluloses are polymeric sugar structures that account for about 25% of the primary cell wall of plants. These flexible polymers bind to cellulose microfibrils

DOI: 10.1201/9781003272618-8

with a composition that varies with plant species and cell type. Sugars present in hemicelluloses include xylose, glucose, mannose, arabinose, galactose, and methyl-glucuronic acid (Mishra and Singh, 1993). A major component in hemicelluloses is xylans (polymers of the pentose xylose), and the decomposition of xylans occurs by action of *endo-β*-1,4-xylanases or *exo-β*-1,4-xylanases. Release of other sugars from hemicellulose requires enzymes specific for bonds linking various sugars. Bacteria and fungi that hydrolyze various bonds in hemicellulose are broadly found in the environment. Pentoses or hexoses released from enzymatic hydrolysis of hemicellulose are consumed by microorganisms in the environment.

LIGNIN STRUCTURE

Lignin is formed in vascular plant cell walls by the oxidative coupling of several related phenylpropanoid precursors: coniferyl alcohol, sinapyl alcohol, and p-hydroxycinnamyl alcohol. Peroxidases or laccases in the plant cell wall oxidize these monomers by one electron, yielding transient resonance-stabilized phenoxy radicals that then polymerize in a variety of configurations. The possible ways that the precursors can couple can be portrayed on paper simply by drawing the conventional resonance forms of the phenoxy radicals and then by linking the most important of these in various pairwise combinations.

Lignin is covalently associated with hemicelluloses in the cell wall via numerous types of linkage. Among the most important ones are ether bonds between the benzylic carbon of lignin and the carbohydrate moiety, ester bonds between the benzylic carbon of lignin and uronic acid residues, and lignin-glycosidic bonds. In graminaceous plants, hydroxycinnamic acid residues are frequent in the lignin and are attached to hemicelluloses via ester linkages. The matrix of lignin and hemicellulose encrusts and protects the cellulose of the plant cell wall.

LIGNIN-DEGRADING MICROORGANISMS

In nature, lignin degradation is a slow process and takes a number of years. Bacteria, actinomycetes, yeasts, and fungi are known to be involved in lignin degradation. Of all naturally produced organic chemicals, lignin is probably the most recalcitrant one. This is consistent with its biological functions, which are to give vascular plants the rigidity they need to stand upright and to protect their structural polysaccharides (cellulose and hemicelluloses) from attack by other organisms. Lignin is the most abundant aromatic compound on earth and is second only to cellulose in its contribution to living terrestrial biomass. When vascular plants die or drop litter, lignified organic carbon is incorporated into the top layer of the soil. This recalcitrant material has to be broken down and recycled by microorganisms to maintain the earth's carbon cycle. Were this not so, all carbon would eventually be irreversibly sequestered as lignocellulose. Lignin biodegradation has diverse effects on soil quality. The microbial degradation of litter results in the formation of humus, and ligninolysis probably facilitates this process by promoting the release of aromatic humus precursors from the litter. These precursors include incompletely degraded lignin, flavonoids, terpenes, lignans, condensed tannins, and uberins. Undegraded lignocellulose, e.g., in the form of straw, has a deleterious effect on soil fertility because decomposing (as opposed to already decomposed) lignocellulose supports high populations of microorganisms that may produce phytotoxic metabolites. High microbial populations in undecomposed litter also compete with crop plants for soil nitrogen and other nutrients. By breaking down the most refractory component of litter, ligninolysis thus contributes to the removal of conditions that inhibit crop productivity.

Conditions that disfavor the biological breakdown of lignocellulose lead to soils with pronounced accumulations of litter. For example, the soils of coniferous forests may contain 50 years of accumulated litterfall, because the low pH of the litter and the lack of summer rainfall inhibit microbial activity. In mature forests of this type, woody material such as dead trunks and branches can constitute 50%–60% of the litter. Warm temperature, high moisture content, high oxygen availability, and high palatability of the litter to microorganisms all favor decomposition. The more highly lignified litter is, the less digestible it is and the more its decomposition depends on the unique organisms that can degrade lignocellulose.

BACTERIAL DEGRADATION

Much is not known regarding the degradation of lignin by bacteria. Pure culture studies on bacterial delignification are virtually absent because bacteria cannot grow on cellulose and lignin together. However, certain species of bacteria (e.g., *Aeromonas*, *Arthrobacterium*, *Flavobacterium*, *Pseudomonas*, and *Xanthomonas*) have the ability to degrade lignin. Species of *Pseudomonas* are the most efficient degraders of lignin. Non-filamentous bacteria mineralize less than 10% of synthetic lignin and can metabolize the low-molecular weight portion of lignin and the degradation products of lignin. Lignin degradation by thermophilic and anaerobic bacteria has not been demonstrated. Alkali lignin was utilized as the sole source of carbon from sulfate wastewater by species of *Corallina*, *Torula*, *Nocardia*, and *Pseudomonas*.

About 98% industrial kraft lignin degrades as the sole carbon source after 5 days of cultivation by *Aeromonas* spp. Cyanobacteria also plays an important role in the removal of lignin from paper mill effluents. Mixed cultures of bacteria, actinomycetes, and fungi in soil and compost can also mineralize lignin.

FUNGAL DEGRADATION

Fungi are the only microorganisms studied extensively for the degradation of lignin. Lignin degradation by fungi has been discussed by a number of researchers. Based on the nature of decay, the wood-rotting fungi are classified into three categories: soft-rot, brown-rot, and white-rot fungi. Soft-rot decay, caused by a number of molds of Ascomycetes and Imperfect Fungi, is known to degrade the major components of wood, including lignin. Fungi causing soft-rot decay include species of *Allescheria*, *Graphium*, *Monodictys*, *Paecilomyces*, *Papulospora*, and *Thielavia*. Two types of soft rot are recognized: Type I consists of cavities formed within secondary walls and type II relates to an erosion form of degradation. Better degradation of lignin by these fungi occurs in hardwood than in softwood. Xylariaceous Ascomycetes belonging to the genera *Daldinia*, *Hypoxylon*, and *Xylaria* are now grouped in type II.

These fungi occur on hardwood, and 53% of the weight loss in birch wood occurred within 2 months by *Daldinia concentrica*.

Brown-rot fungi include several species of Basidiomycetes and are most common in softwood. These fungi remove cellulose and hemicellulose from the wood, leaving the lignin as a crumbly brown residue. This is due to lignin demethylation, partial oxidation, and depolymerization. The brown color shows the presence of modified lignin in wood. Lignin can be degraded in softwood and hardwood by these fungi. All brown-rot fungi employ a Fenton-type catalytic system, producing hydroxyl radicals that attack wood components, but there are certain differences in wood decay. Based on the differences in the mechanism, brown-rot fungi are classified into two groups: one belonging to *Gloeophyllum trabeum* and the second including *Coniophora puteana* and *Poria (Postia) placenta*. *G. trabeum* accumulates oxalic acid, which can be used for the hydrolysis of polysaccharides and as a chelator for a $Fe(II)–H_2O_2$ system generating hydroxyl radicals. *P. placenta* demethoxylated spruce lignin, but there was no ring opening.

The only and most effective lignin degraders and/or mineralizers are the white-rot fungi or closely related litter-decomposing fungi, which include several hundred species of Basidiomycetes and a few Ascomycetes. These fungi are able to decompose hardwood more extensively than softwood and completely mineralize lignin and carbohydrate components of wood to CO_2 and water. Lignin and carbohydrates can be removed at the same proportional rate by some species of white-rot fungi. Selective species of other white-rot fungi remove lignin faster than cellulose. Many white-rot fungi colonize cell lumina, thus causing cell wall erosion. As decay progresses, the eroded areas coalesce and void areas filled with mycelia are formed. This process is known as *nonselective* or *simultaneous rot*; *Trametes versicolor* belongs to this category. Some white-rot fungi decompose lignin without loss of cellulose and create white-pocket rot; *Phellinus nigrolimitatus* belongs to this category. *Ganoderma applanatum* and *Heterobasidion annosum* produce both types of attack in the same wood. White-rot fungi occur more commonly on wood species of Angiosperms than on Gymnosperms. Syringyl units are degraded, whereas guaiacyl units are more resistant to degradation. Transmission electron microscopy has revealed partial removal of the middle lamella by *Ceriporiopsis subvermispora* and *Pleurotus eryngii* and removal of lignin from secondary cell walls by *Phlebia radiata*. In recent years, more taxonomically diverse fungi have been studied for lignin degradation. In general, it has been found that the physiological process of lignin degradation is fungus specific and is different from *Phanerochaete chrysosporium*.

Ecology of Fungal Lignocellulose Degradation

The organisms principally responsible for lignocellulose degradation are aerobic filamentous fungi, and the most rapid degraders in this group are Basidiomycetes. The ability to degrade lignocellulose efficiently is thought to be associated with a mycelial growth habit which allows the fungus to transport scarce nutrients, e.g., nitrogen and iron, over a distance into the nutrient-poor lignocellulosic substrate that constitutes its carbon source. It is curious in this regard that Actinomycetes have not evolved the capacity to degrade lignocellulose efficiently. It is possible that they have the ability to modify lignin somewhat, but no evidence has accumulated to show that they can degrade it.

FUNGAL LEAF LITTER DECAY

The processes by which fungi degrade leaf litter, as opposed to woody litter, are poorly understood. In some cases, leaves are colonized shortly after they fall by Basidiomycetes. For example, *Marasmius androsaceus* is an early colonizer and degrader of pine needles, a relatively recalcitrant and long-lived form of leaf litter. Older analyses indicate that conifer needles contain significant levels of lignin, but it remains to be shown whether the Basidiomycetes that are early colonizers of leaf litter are ligninolytic. In most cases, leaf litter decomposition is more complex, involving a succession of biodegradative activities that precede attack by lignocellulose degraders. The process typically begins with colonization by bacteria, Ascomycetes, and imperfect fungi that consume the least recalcitrant components present, e.g., sugars, starch, and low-molecular weight extractives. The cellulose present in non-lignified leaf tissues is then attacked by some of these organisms, but there is no evidence that lignin is degraded during this early stage of decay. Subsequently, the remaining lignified litter is modified by fauna such as earthworms, millipedes, slugs, and termites, which macerate lignocelluloses mechanically in a process that releases some digestible cellulose. Bacteria and fungi in the guts of these invertebrates then assist in the breakdown of this cellulose, but they do not degrade the lignin component appreciably. Instead, this mechanically modified lignocellulose is released relatively unchanged and becomes part of the soil organic matter. Fragmentation by animals significantly accelerates the degradation rate of the tougher types of litter such as tree leaves, but probably plays a lesser role in the degradation of soft herbaceous litter.

Finally, the modified but still lignified litter is colonized by Basidiomycetes that degrade it further. It is generally assumed that basidiomycete degraders of non-woody litter are ligninolytic, i.e., that they are more like white rotters than brown rotters, but so far little research has been done to confirm this view. The commercial edible mushroom *Agaricus bisporus* is the one litter decomposer whose degradative mechanisms have received some research attention. It degrades both cellulose and lignin, the former more rapidly, and contains ligninolytic enzymes. If *A. bisporus* is typical of other litter-decomposing Basidiomycetes, it is probably correct to infer that fungal ligninolysis is a significant process in non-woody litter. However, it remains unclear to what extent ligninolysis in litter plays the essential role that it does in wood by exposing trapped cellulose to fungal attack. Leaf litter is already finely milled by the time most Basidiomycetes colonize it, and certainly contains bioavailable cellulose, as shown by the fact that non-ligninolytic fungi can deplete cellulose during the comporting of litter.

Fungi that degrade lignin are faced with several problems. Since the polymer is extremely large and highly branched, ligninolytic mechanisms must be extracellular. Since it is interconnected by stable ether and carbon–carbon bonds, these mechanisms must be oxidative rather than hydrolytic. Since lignin consists of a mixture of stereoirregular units, fungal ligninolytic agents have to be much less specific than typical biological catalysts. Finally, the fact that lignin is insoluble in water limits its bioavailability to ligninolytic systems and dictates that ligninolysis is a slow process.

FUNGAL LIGNINOLYTIC MECHANISMS

Ligninolytic fungi are not able to use lignin as their sole source of energy and carbon. Instead, they depend on the more digestible polysaccharides in lignocellulosic substrates, and the primary function of ligninolysis is to expose these polysaccharides so that they can be cleaved by fungal cellulases and hemicellulases. In most fungi that have been examined, ligninolysis occurs during secondary metabolism, i.e., under nutrient limitation. With this approach, the fungus avoids synthesizing and secreting metabolically expensive ligninolytic agents when substrates more accessible than lignocellulose are present. The limiting nutrient for fungal growth in most woods and soils is probably nitrogen, and most laboratory studies of ligninolytic fungi have been done in nitrogen-limited culture media. However, a few ligninolytic fungi, e.g., some species of *Bjerkandera*, are ligninolytic even when sufficient nitrogen is present. Given the chemical recalcitrance of lignin, it is evident that white-rot fungi must employ unusual mechanisms to degrade it. Research has characterized several of these mechanisms in some detail and has shown that they all display one fundamental similarity: they depend on the generation of lignin free radicals which, because of their chemical instability, subsequently undergo a variety of spontaneous cleavage reactions.

LIGNIN PEROXIDASES

Lignin peroxidases (LiPs) were the first ligninolytic enzymes to be discovered. They occur in some frequently studied white-rot fungi, e.g., *P. chrysosporium*, *T. versicolor*, and *Bjerkandera* sp. but are evidently absent in others, e.g., *Dichomitus squalens*, *C. subvermispora*, and *Pleurotus ostreatus*. LiPs resemble other peroxidases such as the classical, extensively studied enzyme from horseradish, in that they contain ferric heme and operate via a typical peroxidase catalytic cycle. That is, LiP is oxidized by H_2O_2 to a two-electron deficient intermediate, which returns to its resting state by performing two one-electron oxidations of donor substrates. However, LiPs are more powerful oxidants than typical peroxidases are, and consequently, oxidize not only the usual peroxidase substrates such as phenols and anilines but also a variety of non-phenolic lignin structures and other aromatic ethers that resemble the basic structural unit of lignin. The simplest aromatic substrates for LiP are methoxylated benzenes and benzyl alcohols, which have been used extensively by enzymologists to study LiP reaction mechanisms. The H_2O_2-dependent oxidation of veratryl alcohol (3,4-dimethoxybenzyl alcohol) to veratraldehyde is the basis for the standard assay used to detect LiP in fungal cultures.

The LiP-catalyzed oxidation of a lignin substructure begins with the abstraction of one electron from the donor substrate's aromatic ring, and the resulting species, an aryl cation radical, then undergoes a variety of postenzymatic reactions. For example, dimeric model compounds that represent the major arylglycerol-b-aryl ether lignin structure undergo C_α–C_β cleavage upon oxidation by LiP.

Synthetic polymeric lignins are also cleaved at this position by the enzyme in vitro in a reaction that gives net depolymerization. These results strongly support a ligninolytic role for LiP, because C_α–C_β cleavage is a major route for ligninolysis in many white-rot fungi. Other LiP-catalyzed reactions that accord with fungal ligninolysis in vivo include aromatic ether cleavage at C_β- and C_α-oxidation without cleavage. It has been pointed out that ionization of the aromatic ring to give a cation radical is also what occurs when lignin model substrates are analyzed in a mass spectrometer, and indeed, the fragmentation pattern obtained by this procedure is similar to that obtained when LiP acts on lignin structures. There remains an unresolved problem with the proposal that LiP catalyzes fungal ligninolysis: LiP, like other enzymes, is too large to enter the pores in sound wood. If it initiates ligninolysis directly, LiP must therefore act at the surface of the secondary cell wall.

Fungal attack of this type is indeed found, but electron microscopic observations also indicate that white-rot fungi can remove lignin from the interior of the cell wall before they have degraded it enough for enzymes to penetrate. It has been proposed that LiP might circumvent the permeability problem by acting indirectly to oxidize low-molecular weight substrates that could penetrate the lignocellulosic matrix and act themselves as oxidants at a distance from the enzyme, but no convincing candidate for a diffusible LiP-dependent oxidant of this type has emerged so far. Notwithstanding these difficulties, LiP remains the only fungal oxidant known that can efficiently mimic, in vitro, the C_α–C_β cleavage reaction that is characteristic of ligninolysis by white-rot fungi such as *P. chrysosporium*. LiP must therefore be considered an important ligninolytic agent, but it may act in concert with other smaller oxidants that can penetrate and open up the wood cell wall.

MANGANESE PEROXIDASES

Manganese peroxidases (MnPs) may be the catalysts that provide these low-molecular weight oxidants. MnPs occur in most white-rot fungi, and are similar to conventional peroxidases, except that Mn(II) is the obligatory electron donor for reduction of the one-electron deficient enzyme to its resting state, and Mn(III) is produced as a result. This reaction requires the presence of bidentate organic acid chelators such as glycolate or oxalate, which stabilize Mn(III) and promote its release from the enzyme. The resulting Mn(III) chelates are small, diffusible oxidants that can act at a distance from the MnP active site. They are not strongly oxidizing and are consequently unable to attack the recalcitrant non-phenolic structures that predominate in lignin. However, Mn(III) chelates do oxidize the more reactive phenolic structures that make up approximately 10% of lignin. These reactions result in a limited degree of ligninolysis via C_α-aryl cleavage and other degradative reactions. It is an interesting possibility that MnP-generated Mn(III) might cleave phenolic lignin structures in this fashion to facilitate later attack by the bulkier but more powerful oxidant LiP.

CO-OXIDATION OF LIGNIN VIA PRODUCTION OF OXYRADICALS

The LiP- and MnP-catalyzed reactions just described cannot provide the only means by which fungi cleave polymeric lignin. LiP, despite its unique properties, is not essential because it is not produced by all white-rot fungi during ligninolysis. MnP-generated Mn(III) cannot be wholly responsible because white-rot fungi that lack LiP are nevertheless able to degrade the non-phenolic lignin structures that resist attack by chelated Mn(III). Other ligninolytic mechanisms must therefore exist. Recent work indicates that the production of diffusible oxyradicals by MnP may supply one such mechanism. In the presence of Mn(II), MnP promotes the peroxidation of unsaturated lipids, generating transient lipoxyradical intermediates that are known to act as potent oxidants of other molecules. The MnP/lipid peroxidation system, unlike MnP alone, oxidizes and cleaves non-phenolic lignin model compounds. It also depolymerizes both non-phenolic and phenolic synthetic lignins, which strongly supports a ligninolytic role for this system in vivo. Although lipid peroxidation has previously been implicated in a variety of biological processes, e.g., aging and carcinogenesis, we believe this is the first evidence that microorganisms may use it as a biodegradative tool.

LACCASES

Laccases are blue copper oxidases that catalyze the one-electron oxidation of phenolics and other electron-rich substrates. Most ligninolytic fungi produce laccases, *P. chrysosporium* being a notable exception. Laccases contain multiple copper atoms which are reduced as the substrates are oxidized. After four electrons have been received by a laccase molecule, the laccase reduces molecular oxygen to water, returning to the native state. The action of laccase on lignin resembles that of Mn(III) chelates, in that phenolic units are oxidized to phenoxy radicals, which can lead to degradation of some structures. In the presence of certain artificial auxiliary substrates, the effect of laccase can be enhanced so that it oxidizes non-phenolic compounds that otherwise would not be attacked, but it is not yet known whether natural versions of such auxiliary substrates function in vivo in lignin biodegradation, and indeed, the actual role of laccase has yet to be fully clarified.

PEROXIDE-PRODUCING ENZYMES

To support the oxidative turnover of the LiPs and MnPs responsible for ligninolysis, white-rot fungi require sources of extracellular H_2O_2. This need is met by extracellular oxidases that reduce molecular oxygen to H_2O_2 with the concomitant oxidation of a co-substrate. One such enzyme, found in *P. chrysosporium* and many other white-rot fungi, is glyoxal oxidase (GLOX). GLOX accepts a variety of 1–3 carbon aldehydes as electron donors. Some GLOX substrates, e.g., glyoxal and methylglyoxal, are natural extracellular metabolizers of *P. chrysosporium*. Another substrate for the enzyme, glycolaldehyde, is released as a cleavage product when the major arylglycerol-b-aryl ether structure of lignin is oxidized by LiP.

Aryl alcohol oxidases (AAOs) provide another route for H_2O_2 production in some white-rot fungi. In certain LiP-producing species of Bjerkandera, chlorinated anisyl alcohols are secreted as extracellular metabolizers and then reduced by a specific AAO to produce H_2O_2. It is noteworthy that, although many alkoxybenzyl alcohols are LiP substrates, chloroanisyl alcohols are not. The use of a chlorinated benzyl alcohol as an AAO substrate thus provides a strategy by which the fungus separates its ligninolytic and H_2O_2-generating pathways. A different approach is employed by some LiP-negative species of *Pleurotus*, which produce and oxidize a mixture of benzyl alcohols, including anisyl alcohol, to maintain a supply of H_2O_2. In yet another fungi, intracellular sugar oxidases might be involved in H_2O_2 generation.

DETECTION OF LIGNINOLYTIC ENZYMES IN COMPLEX SUBSTRATES

Once a fungus has been shown to degrade lignin in experiments with radiolabeled synthetic lignins, the question arises as to which ligninolytic enzymes the organism is expressing. If the degradation experiments have been done in defined liquid media, standard assays for LiP, MnP, laccase, and various H_2O_2-producing oxidases can be done with little difficulty. However, defined growth media that elicit the full expression of ligninolysis have not been developed for many fungi. Therefore, in experiments with the previously uninvestigated fungi that grow on litter, it is more pertinent to ask what ligninolytic enzymes are expressed in the natural growth substrate. This remains a difficult question because many peroxidases are easily inactivated by phenols or other inhibitors that occur in lignocellulosic substrates, and it is consequently difficult to assay these important ligninolytic enzymes reliably in solid-state cultures. Investigators must therefore turn to indirect methods for the detection of ligninolytic enzymes. One of these is to infiltrate a high-molecular weight lignin model compound into the lignocellulosic substrate and then to determine by subsequent product analysis whether the fungus cleaves it in the same way that purified LiP does. Another approach currently under development, and useful when the gene for the enzyme of interest has been sequenced, is to isolate fungal RNA from the substrate and use reverse transcription/polymerase chain reaction techniques to determine whether the gene for the enzyme is being expressed. These new research tools should help to alleviate our severe lack of knowledge about degradative mechanisms in litter-decomposing fungi.

ENZYMATIC DEGRADATION OF CELLULOSE

Cellulose synthesis by plants accounts for about 4×10^7 tons annually, making it the most abundant carbohydrate polymer on Earth. Cellulose is a linear polymer consisting of 8,000–12,000 glucose units bonded together by a β-1,4-glucosidic linkage. Microorganisms produce several different cellulases, including *endo*-1,4-β-glucanase, *exo*-1,4- β-glucanase, and 1,4-β-glucanase (Lynd et al., 2002, Singh and Hayashi, 1995). The best characterized systems for cellulose hydrolysis are with *P. chrysosporium* (a white-rot fungus), *P. placenta* (a brown-rot fungus), *Trichoderma reesei* (a soft-rot fungus), *Cellulomonas fimi* (an aerobic bacterium), *Clostridium thermocellum* (an anaerobic bacterium), and *Thermoactinomyces curvata* (an actinomycete). Cellulose hydrolysis by *P. chrysosporium* and *T. reesei* is attributed to multiple forms of *endo*-1,4-β-glucanases, *exo*-1,4-β-glucanases, and 1,4-β-glucanases, while *P. placenta* produces only *endo*-1,4-β-glucanases and 1,4-β-glucanases. Generally, cellulases produced by fungi are extracellular enzymes and are not bound to the mycelium. Cellulases produced by archaea and bacteria may be either free in the extracellular fluid or clustered in structures on the cell surface in a structure referred to as a cellulosome. Bacterial cellulosomes attach to insoluble cellulose substrates and facilitate the decomposition of cellulose for the benefit of the bacterium producing the cellulosome. The cellulosome contains numerous enzymes for cellulose, hemicellulose, and hydrolysis of other carbohydrate polymers with at least 15 enzymes in the cellulosome of *Ruminococcus albus*, over 90 cellulosomal enzymes produced by *Acetivibrio cellulolyticus*, and 50–60 enzymes in the cellulosome of *C. thermocellum* and *C. acetobutylicum* (Doi, 2008). In contrast, the thermophilic bacterium *Anaerocellum thermophilum* does not have cellulosomes but digests cellulose in switchgrass by free extracellular enzymes (Yang et al., 2009).

STARCH HYDROLYSIS

A common carbohydrate storage compound in plants is starch, which is a polymer of glucose consisting of linear and branched segments. Linear attachments of glucose are by α-1,4-glucosidic bonds, and formation of the branch is attributed to glucose attached by an α-1,6-glucosidic linkage. Many soil bacteria and fungi secrete α-amylase, α-1,6-glucosidase, and glucoamylase, which are capable of hydrolyzing starch to low-molecular weight molecules. The α-amylase attacks starch at the nonreducing end and releases maltose, while α-1,6-glucosidase hydrolyzes only the α-1,6-glucosidic linkage. Following extensive digestion of starch by α-amylase, a limit dextrin is produced where branches are terminated by a glucose unit attached to the linear segment by the α-1,6-glucosidic bond. Filamentous fungi such as *A. niger* may produce α-amylase, α-1,6-glucosidase, and glucoamylase. Glucoamylase hydrolyzes the α-1,6-glucosidic linkage and α-1,4-glucosidic linkage with equal efficiency. These extracellular enzymes work synergistically in the environment to rapidly degrade starch. *Glucoamylases* are carbohydrases that can attack numerous substrates in addition to starch. Glucoamylases will digest glycogen, a reserve carbohydrate polymer of animals, and polyglucose storage molecules in bacteria to glucose. *Clostridium* spp. are some of the few bacteria that produce glucoamylase, and this is significant in the environment because fungi do not grow in the anaerobic zones where clostridia flourish.

INULIN HYDROLYSIS

Inulin is the storage material found in tubers of Jerusalem artichoke, chicory, and dandelion roots (Vandamme and Derycke 1983). Inulin is a natural plant product consisting of a polyfructose molecule with a terminal glucose unit. The size of the molecule may vary with the plant species and a molecular weight of 3,500–5,500 is commonly produced. Inulinase is the enzyme that hydrolyzes the β-2,6-fructose bonds in linear polymer and β-2,1-fructose bond at branch points of the inulin structure. Inulinase is secreted by many different microorganisms, the most common of which are *A. niger*, *K. fragilis*, and *Arthrobacter ureafaciens*, which are molds, yeast, and bacteria respectively. There is the potential for use of inulin as a starting material for high-fructose syrups and as a fermentation substrate for ethanol production.

DECOMPOSITION OF DIVERSE BIOPOLYMERS INCLUDING ANIMAL FIBROUS PROTEINS

Various fibrous proteins found in animals are decomposed slowly in soil or aquatic environments by fungi and bacteria. Chitin, keratin, and silk have a molecular structure markedly distinct from that of soluble proteins, and this fibrous structure requires unique enzymes to degrade the fiber. Additionally, the interaction of insoluble structure of fibrous proteins with soluble enzymes contributes to the difficulty for their enzymatic hydrolysis. When an animal dies in the environment, the flesh and viscera are quickly decomposed by bacteria, but the skin, hair, and bones persist for some time.

CHITIN DIGESTION

Chitin is a polymer of *N*-acetylglucosamine found in the cell walls of fungi, exoskeleton of invertebrates, and insects. The most common is α-chitin with polymeric chains arranged in antiparallel configuration that are held together by hydrogen bonds to produce a relatively rigid sheet. A less common form is β-chitin, where the polymeric chains are arranged in parallel without hydrogen bonds. There is considerable range in the amount of chitin produced in marine environments. In the Atlantic Ocean, about 4.5 mg of chitin is produced per square meter annually by krill, and 1.5 g chitin is produced annually per square meter by lobsters in waters off of South Africa. There are many examples of microorganisms capable of degrading chitin with the production of chitosan, which is a deacylated chitin. Chitinolytic bacteria include *Cytophaga, Vibrio*, and *Streptomyces*, while examples of fungi-digesting chitin include *Mortierella, Trichoderma*, and *Penicillium*.

DECOMPOSITION OF KERATIN

The structural proteins of hair, horns, hooves, and wool are α-keratin. The individual proteins are long molecules with an α-helix. Wool keratin contains high levels of cysteine for the formation of disulfide cross bridges. The proteins in feathers, skin, claws, beaks, and scales of birds and reptiles are β-keratin because these proteins have a β-sheet structure. The α-keratin and β-keratin proteins require special enzymes for hydrolysis, and because of their compact molecular arrangement, they decay slowly. Keratinolytic enzymes are similar to serine metalloproteases and are produced by many pathogenic dermatophytes and mesophilic bacteria as well as numerous environmental microorganisms. Thermophilic bacteria and *Bacillus* species with keratinolytic activity are being considered for industrial processing systems. The quantity of keratin proteins from poultry processing plants primarily in the form of feathers is estimated to exceed 10,000 tons annually. Feathers are difficult to degrade because the proteins are often covered by a fine powder or oils from the birds to make the feathers non-wettable, and the β-sheet structure is a challenge for many enzymes. New feather-degrading bacteria are being isolated to find organisms that would be optimum for decomposition of bird wastes.

Additionally, pigments attached to feathers may influence the rate of keratin decomposition. Generally, the yellow-red feathers are attributed to carotenoids;

brown-black feathers, to melanins; and green feathers, to porphyrins. The bright red feathers of macaw parrots are attributed to a special chemical referred to as polyenal lipochrome. There is a suggestion that bacteria may have a role in evolution of bird plumage coloration because melanin-containing feathers are degraded before carotenoid-containing or unpigmented feathers. This would be especially important if feathers were degraded while attached to the bird and not just after they are released.

FIBROIN DECOMPOSITION

Another fibrous protein is *fibroin*, which contains lengthy regions of antiparallel β sheet with fibers fitting close together to produce a strong fiber. Silk spun by silkworms is an example of proteinaceous material with long stretches of β-sheet protein. Silk contains a great amount of glycine (Gly), alanine (Ala), and serine (Ser). Characteristic of the silk fiber is the following protein chain that accounts for a tight interaction between peptide segments in the β-sheet structure. Interspersed between the β sheets are folded protein regions attributed to relatively bulky amino acids, and these folded regions may contribute to the stretchiness of the silk fiber. The number of non-β-sheet structure in a silk fiber appears to be characteristic of the specific strain of silkworm. Enzymes that hydrolyze silk include the serine protease referred to as *elastin* and proteases produced by *Bacillus* or *Aspergillus*. The folded segments of silk fibroin are initially degraded, while the β sheets are more resistant to enzyme hydrolysis.

While the distribution of silk production by worms is limited in the environment, silk production by spiders is found worldwide. Webs from spiders may vary from ornate by orb weavers to mat-like appearance produced by various spiders. While spiderwebs persist when suspended above the ground, these natural fibers are readily digested by microorganisms in the soil. There are at least seven different types of silk produced by spiders, and each has distinctive mechanical properties (Rising et al., 2005). Silk from mulberry silkworms has a high concentration of serine, glycine, and alanine, while in spider silk these amino acids are in low concentration.

COLLAGEN BREAKDOWN

The most abundant protein in vertebrates is collagen because it is associated with various connective tissues including tendons and skin. Because collagen has a sequence of (glycine–any amino acid–proline) with about 40% of proline as hydroxyproline, the protein has a unique structure. About 14% of the lysine residues are hydroxylysine and these hydroxylysines are involved in crosslinking collagen fibers together (Watanabe, 2004). The enzyme, known as *collagenase*, is produced by various aerobic and anaerobic bacteria in the environment as well as terrestrial fungi. Collagen degradation by anaerobic oral bacteria can contribute to tooth loss in individuals with poor hygiene.

FURTHER READING

Alexandre G, Zhulin IB. Laccases are widespread in bacteria. *Trends Biotechnol.* 2000;18:41–42.

Beguin P, Aubert JP. The biological degradation of cellulose. *FEMS Microbiol. Rev.* 1994;13:25–58.

Cohen R, Hadar Y. The roles of fungi in agricultural waste conversion. In: *Fungi in Bioremediation*. Gadd GM (ed.) Cambridge: Cambridge University Press; 2001:305–334.

Doi RH. Cellulosomes from mesophilic bacteria. In: Wall JD, Harwood CS, Demain A (eds.) *Bioenergy*. Washington, DC: ASM Press;2008:97–107.

Leschine SB. Cellulose degradation in anaerobic environments. *Ann. Rev. Microbiol.* 1995;49:399–426.

Mishra P, Singh A. Microbial pentose utilization. *Adv. Appl. Microbiol.* 1993;39:91–153.

Paszczynski A, Crawford RL. Potential for bioremediation of xenobiotic compounds by white rot fungus *Phanerochaete chrysoporium*. *Biotechnol. Prog.* 1995;11:368–379.

Rising A, Nimmervoll H, Grip S, et al. Spider silk proteins – mechanical property and gene sequence. *Zool. Sci.* 2005;22:273–281.

Sakai T, Sakamoto T, Hallaert J, Vandamme EJ. Pectin, pectinase and protopectinase: production, properties, and applications. *Adv. Appl. Microbiol.* 1993;39:213–295.

Singh A, Hayashi K. Microbial cellulases: proteins architecture, molecular properties and biosynthesis. *Adv. Appl. Microbiol.* 1995;40:1–44.

Vandamme EJ, Derycke DG. Microbial inulinases: fermentation processes, properties and applications. *Adv. Appl. Microbiol.* 1983;29:139–176.

Vicuna R. Ligninolysis. A very peculiar microbial process. *Mol. Biotechnol.* 2000;14:173–176.

Watanabe K. Collagenolytic proteases from bacteria. *Appl. Microbiol. Biotechnol.* 2004;63:520–526.

Yang SJ, Kataeva I, Hamilton-Brehm SD, et al. Efficient degradation of lignocellulosic plant biomass, without pretreatment, by the thermophilic anaerobe "*Anaerocellum thermophilum*" DSM 6725. *Appl. Environ. Microbiol.* 2009;75:4762–4769.

7 Microorganisms and Metal Pollutants

Metals pose a very different pollution problem than organics. Metals cannot be degraded through biological, chemical, or physical means to an innocuous byproduct. More specifically, while the chemical nature of a metal can be changed through oxidation or reduction, the elemental nature of a metal remains the same. Consequently, metals are persistent and more difficult to remove from the environment. An important concept to define with respect to metals is bioavailability. A bioavailable metal is one that can be taken up by a microorganism, plant, or animal. Bioavailable metal usually consists of the ionic species that can be readily transformed into free ionic species in solution. Given this definition, metals can clearly exist in both bioavailable and unavailable forms in the environment, and it is only the bioavailable portion that can exert toxicity on microbes, plants, or animals. As a result, the total metal in a sample does not necessarily reflect the degree of biological metal toxicity, making it difficult to accurately assess the extent of risk posed by metal contamination. Only recently have investigators begun to try to elucidate the ecological significance of bioavailable metal concentrations.

Because of the toxicity and the ubiquity of metals in the environment, microorganisms have developed multiple ways of dealing with both essential and unwanted toxic metals. Levels of essential metals have to be carefully regulated to ensure sufficient supply while avoiding toxicity.

This process is often referred to as metal homeostasis, as opposed to resistance. All organisms need to maintain homeostasis of different essential metals such as copper, iron, manganese, and zinc to maintain cell functioning. The most common way microorganisms deal with excess metal is to pump the metal ions out of their cells while simultaneously restricting metal uptake. In addition, some microorganisms have mechanisms to sequester and immobilize metals, whereas others actually enhance metal solubility in the environment.

CAUSE FOR CONCERN

Concern over metal pollution was once primarily related to mining activity and industrial waste. Now reports of metal-related contamination can be found almost daily in the news, including reports on mercury in fish and arsenic in drinking water. The environmental levels of metals in many locations around the world continue to increase, in some cases to toxic levels, due to contributions from a wide variety of industrial and domestic sources. For example, anthropogenic emissions of lead, cadmium, vanadium, and zinc exceed those from natural sources by up to 100-fold.

Metal-contaminated environments pose serious health and ecological risks. Metals, such as aluminum, antimony, arsenic, cadmium, lead, mercury, and silver, cause adverse effects including heart disease, liver damage, cancer, neurological and

DOI: 10.1201/9781003272618-9

cardiovascular disease, central nervous system damage, encephalopathy, hypophosphatemia, and sensory disturbances. The problem of mercury pollution came into focus in Minamata Bay, Japan, after the discovery of high levels of methylmercury in fish and shellfish that resulted in thousands of poisonings and hundreds of deaths. The mercury contamination originated from a chemical factory that generated small amounts of highly toxic and bioavailable methylmercury during its manufacturing process, which was disposed of into Minamata Bay, and ultimately accumulated in fish. It is also likely that microbial activity in the sediment converted elemental mercury that was disposed of into the bay into methylmercury.

Lead is a second metal of concern because lead poisoning of children is common and leads to behavioral problems resulting from impaired mental function and even semi-permanent brain damage. The Centers for Disease Control (CDC) and the Agency for Toxic Substances and Disease Registry (ATSDR) estimate that 10% of children in the United States have blood lead levels greater than 10 µg/dL, a potentially toxic level. Historians have speculated that the decline of the Roman Empire may have been due in part a decrease in the mental skills of the ruling class as a result of lead poisoning from wine stored in pottery lined with lead and from lead water pipes. Although contamination of drinking water supplies and concentration of metals in edible fish are of particular concern, soils and sediments are the major sinks for metal-containing wastes as the production of domestic and industrial wastes increases.

METAL POLLUTION

Anthropogenic Sources

Metal pollution results when human activity disrupts normal biogeochemical activities or results in disposal of concentrated metal wastes. Sometimes a single metal is involved, but more often mixtures of metals are present. Mining; ore refinement; nuclear processing; and the industrial manufacture of batteries, metal alloys, electrical components, paints, preservatives, and insecticides are examples of processes that produce metal byproducts. Examples of specific metal contaminants include arsenic, copper, and zinc salts that have been used extensively as pesticides in agricultural settings; silver salts that are used to treat skin burns; and lead, which is utilized in the production of batteries, cable sheathing, pigments, and alloys. Other examples include mercury compounds that are used in electrical equipment, paints, thermometers, and fungicides and as preservatives in pharmaceuticals and cosmetics. Triorganotin compounds, such as tributyltin chloride and triphenyltin chloride, can be used as antifouling agents in marine paints because of their toxicity to plankton and bacteria. The extent of metal pollution becomes even more obvious when one considers the amount of waste generated in metal processing.

Thus, while metals are ubiquitous in nature, human activities have caused metals to accumulate in soil. Such contaminated soils provide a metal sink from which surface waters, groundwaters, and the vadose zone can become contaminated. Contaminated soil contributes to high metal concentrations in the air through metal volatilization and creation of windborne dust particles. In addition, industrial emissions and smelting activities cause release of substantial amounts of metals into the atmosphere.

NATURAL SOURCES

Naturally occurring high metal concentrations can also be found as a result of the weathering of parent materials that contain high levels of metals. One metalloid currently receiving attention in countries around the world is arsenic. The concern is contamination of groundwater which serves as the source of drinking water and, in some cases, irrigation water. This contamination is most often from naturally occurring arsenic in the parent minerals that make up the soils and subsurface in these areas. Regardless of the source, metals are of concern because they cannot be degraded, and therefore, accumulate in the environment, which results in the potential for increased exposure and toxicity over time.

MICROBIAL METAL TRANSFORMATIONS

OXIDATION-REDUCTION

Many microbial transformations of metals occur due to their use as terminal electron acceptors in anaerobic respiration (metal reduction) or their use as an energy substrate in which the metal is oxidized. A number of metals and metalloids are subject to redox cycling in the environment including iron, manganese, selenium, and arsenic. For example, the oxidation of arsenite (As(III)) to arsenate (As(V)) can be described by the following chemical reaction:

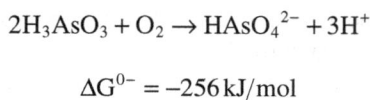

$$2H_3AsO_3 + O_2 \rightarrow HAsO_4^{2-} + 3H^+$$

$$\Delta G^{0-} = -256\,kJ/mol$$

Arsenite oxidation can occur as an abiotic process, but microorganisms play an important role in arsenite oxidation in natural systems. As(III) oxidation can also serve as a detoxification reaction since As(III) is up to 50 times more toxic to bacterial cells than As(V) in most biological systems (Silver et al., 2002). A variety of arsenite-oxidizing bacteria have been identified including both chemoautotrophs and heterotrophs (Ehrlich, 2002).

Reduction of arsenate to arsenite occurs by one of two mechanisms—dissimilatory reduction or detoxification. In detoxification, reduction of arsenate is not coupled to respiration and does not provide energy for the bacterium—it is simply reduced to arsenite and then exported out of the cell. In dissimilatory reduction, As(V) is used as the terminal electron acceptor during anaerobic respiration. An example of such a reaction is the growth of *Bacillus arsenicoselenatis* using lactate as the electron donor and As(V) as the TEA. The reaction can be described by the following equation (Oremland et al., 2002):

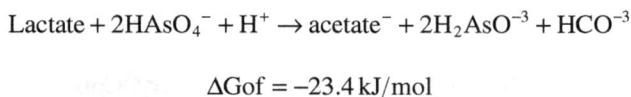

$$\text{Lactate} + 2HAsO_4^- + H^+ \rightarrow \text{acetate}^- + 2H_2AsO^{-3} + HCO^{-3}$$

$$\Delta Gof = -23.4\,kJ/mol$$

Studies of the rates of arsenate reduction by this mechanism show half-lives averaging around 30 hours (Inskeep et al., 2002). For dissimilatory reduction of As(V) to occur, however, there must be high enough arsenic concentrations to support growth, and strict anaerobic conditions may be required. Environments such as sediments, hot springs, and freshwater and marine systems can support such conditions.

In addition to some metals being directly reduced during anaerobic respiration, metals can also be reduced indirectly through reaction with other reduced products such as sulfides. In fact, for some metals, such as uranium, this may actually be a major mechanism for their reduction in the environment. The oxidation and reduction of metals can have profound practical implications. For example, in subsurface geological formations, metals are often found in a reduced state as, for example, pyrite (FeS_2). Pyrite is often associated with metal ore deposits. Pyrite is stable until it is exposed to oxygen by mining activities, e.g., strip mining. Upon the introduction of oxygen, a combination of autoxidation and chemoautotrophic microbial oxidation of iron and sulfur results in the production of large amounts of acid. Acid, in turn, facilitates metal solubilization, resulting in a metal-rich acidic leachate called acid mine drainage. Acid mine drainage is a problem associated with many types of mining activity including subsurface mining, where metal deposits become exposed to atmospheric oxygen; strip mining, where large expanses of land are exposed to oxygen; and mine tailing wastes, which are large deposits of processed or spent ore. Microbially induced corrosion of metal pipes and fuel and storage tanks is a second significant problem of concern. Corrosion occurs due to cooperation between two groups of bacteria, the anaerobic chemoheterotrophic sulfate-reducing bacteria (SRB) and the aerobic chemoautotrophic iron-oxidizing bacteria. These two groups of bacteria work together to create an environmental niche on pipe surfaces that is favorable for their simultaneous activity, even though one group requires oxygen and the other does not (Figure 7.1a and b).

Methylation

The microbial methylation of metals not only results in increased metal mobility because some organometals are volatile but also because in some cases it can change the toxicity of the metal. Methylation involves the transfer of methyl groups (CH_3) to metals and metalloids, e.g., lead, mercury, arsenic, and selenium. The resulting organometal is more lipophilic than the metal species. This results in the potential of bioaccumulation and biomagnification in food webs.

Methylation of mercury occurs in the sediments of lakes, rivers, and estuaries, where organic matter concentrations are high and redox conditions are favorable for the activity of SRB, the primary generators of methylmercury (Drott et al., 2007). The most important intracellular agent of mercury methylation is believed to be methylcobalamine (CH_3CoB_{12}), a derivative of vitamin B12. Methylation reactions can be summarized as follows:

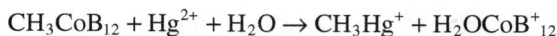

$$CH_3CoB_{12} + Hg^{2+} + H_2O \rightarrow CH_3Hg^+ + H_2OCoB^+_{12}$$

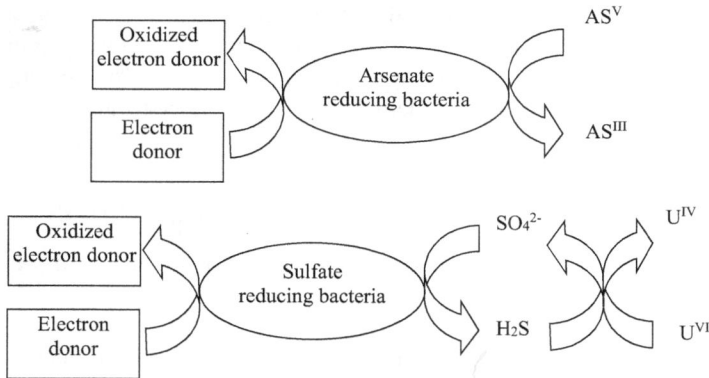

FIGURE 7.1 (a) Direct reduction through use as a terminal electron acceptor in anaerobic respiration. (b) Indirect reduction following abiotic reactions with reduced chemical species such as sulfides.

Methylcobalamine methylmercury:

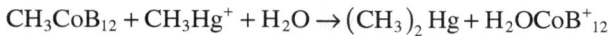

$$CH_3CoB_{12} + CH_3Hg^+ + H_2O \rightarrow (CH_3)_2 Hg + H_2OCoB^+{}_{12}$$

Methylcobalamine dimethylmercury:

The dominant product formed is the salt of the methylmercuric ion, CH_3Hg^+ (methylmercury), because the volatile dimethylmercury $((CH_3)_2Hg)$ forms at a much slower rate. Since methylation actually increases the toxicity of mercury, methylation of mercury may facilitate diffusion of both methylmercury and dimethylmercury from the cell more easily than Hg^{2+}. In contrast, methylation of selenium directly decreases its toxicity of selenium.

Mercury is used extensively in the electrical industry, instrument manufacturing, electrolytic processes, and chemical catalysis. Mercury salts and phenylmercury compounds are also used as fungicides and disinfectants. Approximately 10,000 metric tons of mercury are produced worldwide annually. Fossil fuel burning releases an additional 3,000 metric tons. Methylmercury compounds are highly lipophilic and neurotoxic. Several outbreaks of mercury poisoning have occurred throughout history. In Minamata Bay, Japan, release of mercury-containing effluents by a chemical processing plant resulted in serious illness in people who consumed fish with elevated levels of mercury. Another example is the Great Lakes in the United States, which until the 1970s had relatively uncontrolled releases of polychlorinated biphenyls (PCBs), dioxins, and mercury. For a period of time, parts of the lakes were closed to fishing, but the problem has improved due to restricted use and release of the organic and metal pollutants. At this time, health advisories are in effect that make recommendations about the type and amounts of fish that can be safely consumed.

Arsenic is another example of a metal that is methylated as a resistance mechanism. It is methylated by some bacteria, such as *Rhodopseudomonas palustris*, and many fungi, such as *Scopulariopsis brevicaulis* to mono-, di-, and trimethylarsine, volatile forms of arsenic. Arsenic poisonings have occurred in the past when fungi growing on damp wallpaper converted and volatilized arsenate (used as a coloring agent) in the wallpaper. Illness occurred upon inhalation of the resulting methylated arsenic species. There is also growing evidence that microbial methylation of soil arsenic has led to elevated levels of methylated arsenic in rice from some parts of the world (Somenahally et al., 2011).

PHYSICOCHEMICAL METHODS OF METAL REMEDIATION

The remedial methods used to treat contaminated soil or sediments may be broadly divided into two main categories:

- Methods aimed at preventing movement of metals to the immediate surroundings, also called immobilization.
- Methods aimed at metal removal.

The goal in metal immobilization is to reduce metal solubility. Two immobilization strategies include pH alteration and addition of organic matter. Since metal solubility decreases with increasing pH, metal solubility should be reduced when site pH is raised. Liming is sometimes used to increase soil pH causing precipitation of contaminating metals as calcic and phosphoric metal-containing minerals. Amendment with organic matter can also aid in metal immobilization as a result of the electrostatic attraction between metals and organic particles. The addition of organic matter may involve the addition of highly organic waste material, such as biosolids. Often sites containing high levels of toxic metals have little or no vegetation. Revegetation of such sites, while sometimes difficult to achieve, is a good way to increase organic matter content.

Metal removal from soils or sediments can be achieved by excavation (which simply moves the problem to another location) or by using soil washing techniques. Soil washing methods rely on chemicals to facilitate metal removal. Washing with acidic solutions or chelating agents, e.g., ethylenediaminetetraacetic acid (EDTA) or nitrilotriacetic acid (NTA), solubilizes metals, enhancing removal from the system. One problem with the use of these chemical agents is the residual toxicity left by the washing agent after treatment. Newer biodegradable chelating agents show promise for metal removal. For example, 84% of nickel in a spent catalyst was removed with the biodegradable chelating agent [S,S]-ethylenediaminedisuccinic acid (EDDS) (Chauhan et al., 2012), while studies show the degree of metal removal by chelating agents can be metal specific (Tandy et al., 2004). Researchers are also looking at biological alternatives to these chemicals. For example, biosurfactants have been explored as alternative "green" soil washing agents for metal-contaminated soils (Maier and Soberon-Chavez, 2000).

While metal removal by excavation may be appropriate when the area of contamination is small or there is immediate risk to human health, increasing cost and shrinking landfill space emphasize the need for cheaper, environmentally friendly alternatives. Following excavation, contaminated soils must be stored in a hazardous waste containment facility or incinerated. In some situations, however, excavation can exacerbate the problem. For example, excavation of sediments, called dredging, can actually result in increased metal toxicity. Metal sediments are often anaerobic and the metals within the sediment exist in an immobile, reduced state. Exposure to oxidizing conditions results in metal oxidation and increased metal solubility, increasing both bioavailability and transport. There is current discussion about whether physically removing metal-containing sediments is more detrimental than leaving them in place.

Incineration of soils can be used to remove metals from soils. However, incineration is not only expensive and impractical for large volumes of soil but also releases metals to the atmosphere only to be deposited elsewhere. In addition, such thermal treatment of soil also destroys important soil properties, destroying soil structure and soil biota.

The nonbiological remediation of aquatic systems, including surface water, groundwater, and wastewater, is fairly straightforward, albeit costly. Metals are removed and concentrated from contaminated waters through flocculation, complexation, and/or precipitation. Lime addition precipitates metals as metal hydroxides. Chelating agents complex metals and can be recovered with a change in pH.

A developing approach for the removal of organic and metal contaminants from water is the use of permeable reactive barriers (PRBs) containing materials such as zero-valent iron. In the case of metal contaminants, the goal is to convert the metal into a less toxic form and/or immobilize it within the PRB as the contaminated water passes through. For example, a zero-valent or oxidized iron barrier would promote the sorption and precipitation of arsenic within the barrier, resulting in the water exiting the PRB having decreased levels of arsenic. This approach has tremendous potential for containment and/or remediation; however, in practice, it has faced major challenges including microbial and chemical fouling of the PRBs that prematurely shorten their effective lifetime. Continued advances in PRB materials and coatings may improve their longevity and applications in the future.

MICROBIAL APPROACHES IN THE REMEDIATION OF METAL-CONTAMINATED SOILS AND SEDIMENTS

The goals of microbial remediation of metal-contaminated soils and sediments are to

- immobilize the metal in situ to reduce metal bioavailability and mobility;
- remove the metal from the soil.

There are several proposed methods for microbial remediation of metal-contaminated soils including microbial leaching, microbial surfactants, microbially induced metal volatilization, and microbial immobilization and complexation.

Certain microorganisms, such as *Acidithiobacillus ferrooxidans*, can facilitate the removal of metals from soil through metal solubilization or leaching via the same acidification process as seen with the formation of acid mine drainage. Generally used in the recovery of economically valuable metals from ores, bioleaching has also been used to recover copper, lead, zinc, and uranium from tailings. The process uses acidophilic iron and sulfur oxidizers (e.g., *Acidithiobacillus*, *Leptospirillum*) and is considered to be environmentally friendly (Rawlings, 2002). These microorganisms can participate in both direct bioleaching and indirect bioleaching of metals from a variety of ores. Copper is the major metal recovered using bioleaching.

There are two commercial-scale approaches for bioleaching. The first is used primarily for copper and involves recycling leach liquor through a copper sulfide ore body. This can be done in situ or on ore heaps placed on pads on the ground. In situ bioleaching can occur either in a spent mine or can be applied to a new unmined ore body. However, in situ bioleaching requires suitable hydrologic conditions to allow efficient collection of the leachates and also to ensure that leachates do not go off-site. Heap bioleaching or dump bioleaching usually involves mining the ore, crushing it and then placing it in piles on an irrigation pad. The leach liquor is applied to the top of the heap and percolates through the ore, collecting metals. The metal-laden leach liquor is collected from the bottom of the pile, processed to remove the metals, and then recycled onto the top of the pile. The second commercial-scale approach for bioleaching involves the use of a series of continuous flow bioreactors, a much more costly process. This process is usually used for high value metals such as gold.

However, the principle is the same—the bioreactors are filled with ore and leach liquor is cycled through the bioreactors to remove the metals from the ore. Metals recovered by leaching can be concentrated by complexation with chelating agents or precipitation with lime. Bioleaching also has potential in the removal of metals from contaminated soils and metal-containing sludges. Unfortunately, this aspect of microbial leaching has received little attention.

Microorganisms can also increase metal solubility for recovery through the production of surfactants. Because of their small size, biosurfactants are a potentially powerful tool in metal remediation. Bacterial surfactants are water-soluble, low-molecular weight molecules (<1,500) that can move relatively freely through soil pores. In addition to their small size, biosurfactants have a high affinity for metals so that, once complexed, contaminating metals can be removed from the soil by soil flushing. Some surfactants, such as the rhamnolipid produced by *Pseudomonas aeruginosa*, show specificity for certain metals, such as cadmium and lead (Ochoa-Lozae et al., 2011). Biosurfactant specificity allows the optimization of removal of a particular metal. Related to biosurfactants, the higher molecular weight (~10^6) bioemulsifiers such as emulsan, produced by *Acinetobacter calcoaceticus*, can also aid in metal removal and are increasingly being looked at as a potential application for metal recovery (Gutnick and Bach, 2002).

Like leaching, methylation of metals can increase metal bioavailability and toxicity. Some methylated metals are more lipophilic than their nonmethylated counterparts. In spite of the possible increased toxicity, many microorganisms still volatilize metals to facilitate their removal from the immediate environment. Because methylation enhances metal removal, methylation of certain metals has been used as a remediation strategy. Mercury is another metal commonly methylated by microorganisms. However, mercury is susceptible to bioaccumulation in the food chain, posing serious health risks to the human population, and therefore, removal of mercury by volatilization would not be an acceptable approach.

Immobilization strategies include metal sequestration which takes advantage of the ability of some microorganisms to produce metal-complexing polymers (both extracellular and intracellular) or to convert metals to a less soluble form. Recall that exopolymers have high affinities for various metals. The overall approach in microbial metal sequestration is to introduce the polymer-producing microorganism into the contaminated soil and allow the organism to grow and replicate, thereby increasing the amount of polymer present in the soil and increasing the number of organisms producing the polymer. Microbial metal sequestration has been shown to be effective in laboratory studies but has yet to be proven effective in the field. A second immobilization strategy is to create reducing or anaerobic conditions which results in the reduction and precipitation of metals. For example, the reduction of sulfate to sulfide under anaerobic conditions can lead to the formation of metal sulfide precipitates that are immobile. Likewise, the reduction of uranium (uranium(VI)) to a less-soluble form (uranium (IV)) has been demonstrated to dramatically lower concentrations of dissolved uranium in groundwater in field-scale studies.

It should be noted that this approach is very metal specific, since reduced forms of some metals (e.g., arsenic) are actually more soluble than their oxidized counterparts. Also, this approach would require that reducing conditions be maintained at the site in order to prevent reoxidation of the sequestered metals. Although immobilization strategies are generally more economical than removal strategies and appear to have tremendous potential for many sites, there is not yet sufficient evidence to confirm the long-term effectiveness of immobilization.

MICROBIAL APPROACHES IN THE REMEDIATION OF METAL-CONTAMINATED AQUATIC SYSTEMS

Microbially facilitated removal of metals from water is based on the ability of microorganisms to complex and precipitate metals, resulting in both detoxification and removal from the water column. Specific interactions for metal removal include metal binding to microbial cell surfaces and exopolymer layers, intracellular uptake, metal volatilization and metal precipitation via microbially facilitated metal redox reactions. Although these microbial mechanisms can effectively remove metals from contaminated aquatic systems, it is important to note that the metals are not destroyed and still have to be disposed of properly.

Wetland treatment is a cost-effective and efficient method for removal of metals from contaminated waters, such as acid mine drainage. Metal reductions are often greater than 90% (Scholz and Xu, 2002). Wetland remediation is based on microbial adsorption of metals, metal bioaccumulation, bacterial metal oxidation, and sulfate reduction. The high organic matter content of wetlands provided by high plant and algal growth encourages both the growth of sulfate-reducing microorganisms and metal sorption to the organic material. Although these various processes contribute to the removal of toxic metals from the water column, the metals are not destroyed. Consequently, wetlands are constantly monitored for any environmental change that may adversely affect metal removal. For example, a decrease in pH may solubilize precipitated metals or a disturbance of the wetland sediment may change the redox conditions and oxidize reduced metals. Wetlands are resilient systems, and as long as new vegetative growth and organic inputs occur, wetlands can effectively remove metals for an indefinite period of time.

The most common treatment for metal-contaminated waters is with microbial biofilms. Many microorganisms, including *Pseudomonas, Arthrobacter, Bacillus, Citrobacter, Streptomyces*, and the yeasts *Saccharomyces* and *Candida*, produce exopolymers as part of their growth regime. Metals have high affinities for these anionic exopolymers. Microbial biofilms may be viable or nonviable when used in remediation. In general, the biofilm is immobilized on a support as contaminated water is passed through the support. Often, a mixture of biofilm-producing organisms grows on these supports, providing a constant supply of fresh biofilm. For example, live *Citrobacter* spp. biofilms are used to remove uranium from contaminated water. Both *Arthrobacter* spp. biofilms and biomass (non-living) are used in recovery of cadmium, chromium, copper, lead, and zinc from wastewaters. Non-living *Bacillus* spp. biomass preparations effectively bind cadmium, chromium, copper, mercury, and nickel, among other metals. Microbial biofilms are likewise used in the removal of metals from domestic wastewater. In domestic waste treatment, the important biofilm-producing organisms include *Zoogloea, Klebsiella*, and *Pseudomonas* spp. Complexed metals are removed from the wastewater via sedimentation before release from the sewage treatment plant.

FURTHER READING

Atlas RM, Philp JC. *Bioremediation: Applied Microbial Solutions for Real-World Environmental Cleanup.* Washington, DC: ASM Press; 2005.

Barton LL, Plunkett RM, Thomson BM. Reduction of metals and nonessential elements by anaerobes. In: Ljungdahl LG, Adams MW, Barton LL, Ferry JG, Johnson MJ (eds.) *Biochemistry and Physiology of Anaerobic Bacteria.* New York: Springer; 2003:220–235.

Chauhan G, Pant KK, Nigam KDP. Extraction of nickel from spent catalyst using biodegradable chelating agent EDDS. *Ind. Eng. Chem. Res.* 2012;51:10354–10363.

Drott A, Lambertsson L, Bjorn E, Skyllberg U. Importance of dissolved neutral mercury sulfides for methyl mercury production in contaminated sediments. *Environ. Sci. Technol.* 2007;41:2270–2276.

Ehrlich HL. Bacterial oxidation of As(III) compounds. In: Frankenberger WT Jr (ed.) *Environmental Chemistry of Arsenic.* New York: Marcel Dekker, Inc.; 2002:313–327.

Gadd GM. Fungi and yeast for metal accumulation. In: Ehrlich HL, Brierley CL (eds.) *Microbial Mineral Recovery.* New York: McGraw-Hill; 1990.

Gutnick DL, Bach H. Engineering bacterial biopolymers for the biosorption of heavy metals; new products and novel formulations. *Appl. Microbiol. Biotechnol.* 2000;54:451–460.

Inskeep WP, McDermott TR, Fendorf, S. Arsenic (V)/(III) cycling in soils and natural waters: chemical and microbiological processes. In: Frankenberger WT Jr (ed.) *Environmental Chemistry of Arsenic.* New York: Marcel Dekker; 2002:183–215.

Lloyd JR. Mechanisms and environmental impact of microbial metal reduction. In: Gadd GM, Semple KT, Lappin-Scott HM (eds.) *Micro-organisms and Earth Systems – Advances in Geomicrobiology.* Cambridge: University Press; 2005:273–302.

Maier RM, Soberon-Chavez G. Pseudomonas aeruginosa rhamnolipids: biosynthesis and potential applications. *Appl. Microbiol. Biotechnol.* 2000;54:625–633.

Ochoa-Loza FJ, Artiola JF, Maier RM. Stability constants for the complexation of various metals with a rhamnolipid biosurfactant. *J. Environ. Qual.* 2001;30:479–485.

Oremland RS, Newman DK, Kail BW, Stolz JF. Bacterial respiration of arsenate and its significance in the environment. In: Frankenberger WT Jr (ed.) *Environmental Chemistry of Arsenic.* New York: Marcel Dekker, Inc.; 2002:273–295.

Rawlings DE. Heavy metal mining using microbes. *Ann. Rev. Microbiol.* 2002;56:65–91.

Robinson NJ, Whitehall SK, Cavet JS. Microbial metallothioneins. *Adv. Microbial Physiol.* 2001;44:184–216.

Scholz M, Xu J. Performance comparison of experimental constructed wetlands with different filter media and macrophytes treating industrial wastewater contaminated with lead and copper. *Bioresour. Technol.* 2002;83:71–79.

Silver S, Phung LT, Rosen BP. Arsenic metabolism: resistance, reduction, and oxidation. In: Frankenberger WT Jr (ed.) *Environmental Chemistry of Arsenic.* New York: Marcel Dekker; 2002: 247–272.

Somenahally AS, Hollister EB, Yan W, Gentry TJ, Loeppert RH. Water management impacts on arsenic speciation and iron-reducing bacteria in contrasting rice-rhizosphere compartments. *Environ. Sci. Technol.* 2011;45:8328–8335.

Tandy S, Bossart K, Mueller R, et al. Extraction of heavy metals from soils using biodegradable chelating agents. *Environ. Sci. Technol.* 2004;38:937–944.

Volesky B. *Biosorption of Heavy Metals.* Boca Raton, FL: CRC Press; 1990.

8 Bioleaching

Biomining is the generic term that describes the processing of metal containing ores and concentrates using microbiolocgal technology. Biomining has a particular application as an alternative to traditional physical-chemical methods of mineral processing in a variety of niche areas. These include deposits where the metal values are low, where the presence of certain elements (e.g., arsenic) would lead to smelter damage, or where environmental considerations favor biological treatment options.

Bioleaching is a process described as being "the dissolution of metals from their mineral source by certain naturally occurring microorganisms" or "the use of microorganisms to transform elements so that the elements can be extracted from a material when water is filtered through it." In the case of copper, copper sulfide is microbially oxidized to copper sulfate and metal values are present in the aqueous phase. Remaining solids are discarded. "Biooxidation" describes the microbiological oxidation of host minerals which contain metal compounds of interest. As a result, metal values remain in the solid residues in a more concentrated form. In gold mining operations, biooxidation is used as a pretreatment process to (partly) remove pyrite or arsenopyrite. This process is also called "biobeneficiation" where solid materials are refined and unwanted impurities are removed. The terms "biomining," "bioextraction," and "biorecovery" are also applied to describe the mobilization of elements from solid materials mediated by bacteria and fungi. "Biomining" concerns mostly applications of microbial metal mobilization processes in large-scale operations of mining industries for an economical metal recovery.

Although the terms bioleaching and biooxidation are often used interchangeably, there are distinct technical differences between these process technologies. Bioleaching refers to the use of microorganisms, principally *Thiobacillus ferrooxidans*, *Leptospirillum ferrooxidans* and thermophilic species of *Sulfobacillus*, *Acidianus*, and *Sulfolobus*, to leach a metal of value such as copper, zinc, uranium, nickel, and cobalt from a sulfide mineral. Bioleaching place the metal values of interest in the solution phase during oxidation. These solutions are handled for maximum metal recovery and the solid residue is discarded. Minerals biooxidation refers to a pretreatment process that uses the same bacteria as bioleaching to catalyze the degradation of mineral sulfides, usually pyrite or arsenopyrite, which host or occlude gold, silver, or both. Biooxidation leaves the metal values in the solid phase and the solution is discarded.

PRINCIPLES OF MICROBIAL METAL LEACHING

LEACHING MECHANISMS

Mineralytic effects of bacteria and fungi on minerals are based mainly on three principles, namely acidolysis, complexolysis, and redoxolysis. Microorganisms are able to mobilize metals by (1) the formation of organic or inorganic acids (protons), (2) oxidation and reduction reactions, and (3) the excretion of complexing agents.

DOI: 10.1201/9781003272618-10

Sulfuric acid is the main inorganic acid found in leaching environments. It is formed by sulfur-oxidizing microorganisms such as thiobacilli. A series of organic acids are formed by bacterial (as well as fungal) metabolism resulting in organic acidolysis, complex and chelate formation. A kinetic model of the coordination chemistry of mineral solubilization has been developed which describes the dissolution of oxides by the protonation of the mineral surface as well as the surface concentration of suitable complex forming ligands such as oxalate, malonate, citrate, and succinate. Proton-induced and ligand-induced mineral solubilization occurs simultaneously in the presence of ligands under acidic conditions.

Models of Leaching Mechanisms

Originally, a model with two types of mechanisms which are involved in the microbial mobilization of metals has been proposed: (1) Microorganisms can oxidize metal sulfides by a "direct" mechanism obtaining electrons directly from the reduced minerals. Cells have to be attached to the mineral surface and a close contact is needed. The adsorption of cells to suspended mineral particles takes place within minutes or hours. This has been demonstrated using either radioactively labeled *T. ferrooxidans* cells grown on $NaH_{14}CO_3$ or the oxidative capacity of bacteria attached to the mineral surface. Cells adhere selectively to mineral surfaces occupying preferentially irregularities of the surface structure. In addition, a chemotactic behavior to copper, iron, or nickel ions has been demonstrated for *L. ferrooxidans*. Genes involved in the chemotaxis were also detected in *T. ferrooxidans* and *Thiobacillus thiooxidans*. (2) The oxidation of reduced metals through the "indirect" mechanism is mediated by ferric iron (Fe^{3+}) originating from the microbial oxidation of ferrous iron (Fe^{2+}) compounds present in the minerals. Ferric iron is an oxidizing agent and can oxidize, e.g., metal sulfides, and it is (chemically) reduced to ferrous iron which, in turn, can be microbially oxidized again. In this case, iron has a role as electron carrier. It was proposed that no direct physical contact is needed for the oxidation of iron.

In many cases, it was concluded that the "direct" mechanism dominates over the "indirect" mostly due to the fact that "direct" was equated with "direct physical contact." This domination has been observed for the oxidation of covellite or pyrite in studies employing mesophilic *T. ferrooxidans* and thermophilic *Acidianus brierleyi* in bioreactors which consisted of chambers separated with dialysis membranes to avoid physical contact. However, the attachment of microorganisms on surfaces is not an indication per se for the existence of a direct mechanism. The term "contact leaching" has been introduced to indicate the importance of bacterial attachment to mineral surfaces. The following equations describe the "direct" and the "indirect" mechanism for the oxidation of pyrite:

Direct:

$$c\ 2FeS_2 + 7O_2 + 2H_2O \underline{\text{ Thiobacilli }} 2FeSO_4 + 2H_2SO_4 \qquad (8.1)$$

Indirect:

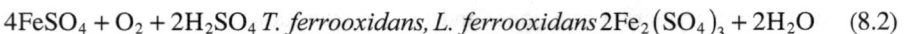

$$4FeSO_4 + O_2 + 2H_2SO_4 \underline{\textit{ T. ferrooxidans, L. ferrooxidans }} 2Fe_2(SO_4)_3 + 2H_2O \qquad (8.2)$$

$$FeS_2 + Fe_2(SO_4)_3 \underline{\text{Chemical oxidation}} FeSO_4 + 2S \qquad (8.3)$$

$$2S + 3O_2 + H_2O \underline{\text{T. thiooxidans}} 2H_2SO_4 \qquad (8.4)$$

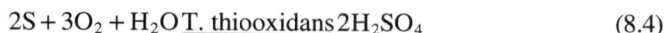

All facts have been combined and a mechanism has been developed which is characterized by the following features: (1) cells have to be attached to the minerals and in physical contact with the surface; (2) cells form and excrete exopolymers; (3) these exopolymeric cell envelopes contain ferric iron compounds which are complexed to glucuronic acid residues and these are part of the primary attack mechanism; (4) thiosulfate is formed as intermediate during the oxidation of sulfur compounds; and (5) sulfur or polythionate granules are formed in the periplasmatic space or in the cell envelope.

Thiosulfate and traces of sulfite have been found as intermediates during the oxidation of sulfur. Sulfur granules (colloidal sulfur) have been identified as energy reserves in the exopolymeric capsule around cells of *T. ferrooxidans* during growth on synthetic pyrite films. "Footprints" of organic films containing colloidal sulfur granules are left on the mineral surface upon detachment of the bacteria.

FACTORS INFLUENCING BIOLEACHING

The leaching effectiveness depends largely on the efficiency of the microorganisms and on the chemical and mineralogical composition of the ore to be leached. The maximum yields of metal extraction can be achieved only when the leaching conditions correspond to the optimum growth conditions of the organism.

NUTRIENTS

Microorganisms used for metal extraction from sulfide materials are chemolithoautotrophic bacteria, and therefore, only inorganic compounds are required for growth. In general, the mineral nutrients are obtained from the environment and from the material to be leached. For optimum growth, iron and sulfur compounds may be supplemented together with ammonium, phosphate, and magnesium salts.

Metal oxidation mediated by acidophilic microorganisms can be inhibited by a variety of factors such as organic compounds, surface active agents, solvents, or specific metals: The presence of organic compounds (yeast extract) inhibited pyrite oxidation of *T. ferrooxidans*. Certain metals present in bioleaching environments can inhibit microbial growth, thereby reducing leaching efficiencies. For instance, arsenic added to cultures inhibited *Sulfolobus acidocaldarius* grown on pyrite and *T. ferrooxidans* grown on arsenopyrite. Addition of copper, nickel, uranium, or thorium adversely influenced iron(II) oxidation by *T. ferrooxidans* with uranium and thorium showing higher toxicities than copper and nickel. Silver, mercury, ruthenium, and molybdenum reduced the growth of *Sulfolobus* grown on a copper concentrate. Industrial biocides such as tetra-*n*-butyltin, isothiazolinones, N-dimethyl-N′-phenyl-N′-(fluorodichloro-methylthio)-sulfamide, or 2,2′-dihydroxy-5,5′-dichlorophenylmethane (dichlorophen) reduced the leaching of manganese oxides by heterotrophic microorganisms. Biocides were externally added as selective inhibitors to suppress unwanted organisms and to improve manganese leaching efficiencies. At low concentrations of ~5 mg/L, however, manganese mobilization was increased by 20%.

O_2 and CO_2

An adequate supply of oxygen is a prerequisite for good growth and high activity of the leaching bacteria. In the laboratory, this can be achieved by aeration, stirring, or shaking. On a technical scale, particularly in the case of dump or heap leaching, sufficient supply with oxygen may cause some difficulties. Carbon dioxide is the only carbon source required, but there is no need for addition of CO_2.

Gaseous compounds can show inhibitory effects on metal leaching: Aqueous-phase carbon dioxide at concentration >10 mg/L was inhibiting growth of *T. ferrooxidans* on pyrite–arsenopyrite–pyrrhotite ore. Optimal concentrations of carbon dioxide were found to be in the range of 3–7 mg/L. There are reports on the stimulation of bacterial leaching and the increase of leaching rates by supplementing leaching fluids with carbon dioxide. Concentrations of 4% (v/v) carbon dioxide in the inlet gas of a fermenter showed maximum growth rates of *T. ferrooxidans* and maximum iron(II), copper, and arsenic oxidation.

pH

The adjustment of the correct pH value is a necessary condition for the growth of the leaching bacteria and is decisive for the solubilization of metals. pH values in the range of 2.0–2.5 are optimum for the bacterial oxidation of ferrous iron and sulfide. At pH values below 2.0, a considerable inhibition of *T. ferrooxidans* will occur but *T. ferrooxidans* may be adapted to even lower pH values by increasing addition of acid.

Standard test methods have been developed to determine leaching rates of iron from pyrite mediated by *T. ferrooxidans*. Metal bioleaching in acidic environments is influenced by a series of different factors. Physicochemical as well as microbiological factors of the leaching environment are affecting rates and efficiencies. In addition, properties of the solids to be leached are of major importance. As examples, pulp density, pH, and particle size were identified as major factors for pyrite bioleaching by *S. acidocaldarius*. Optimal conditions were 60 g/L, 1.5, and ~20 μm respectively. The influence of different parameters such as activities of the bacteria itself, source energy, mineralogical composition, pulp density, temperature, and particle size were studied for the oxidation of sphalerite by *T. ferrooxidans*. Best zinc dissolution was obtained at low pulp densities (50 g/L), small particle sizes, and temperatures of approximately 35°C.

Pulp Density

Pulp densities of 20 g/L delayed the onset of bioleaching of pyrite derived from coal. Increasing pulp densities from 30 to 100 g/L decreased rates of pyrite oxidation in *Sulfolobus* cultures. For fungi such as *Aspergillus niger*, optimal pulp densities for maximum metal leaching efficiencies were found to be in the range of 30–40 g/L. Quartz particles at pulp densities of 80 g/L almost completely inhibited the oxidation of covellite by *T. ferrooxidans* especially in the absence of iron(II).

Temperature

The optimum temperature for ferrous iron and sulfide oxidation by *T. ferrooxidans* is between 28°C and 30°C. At lower temperatures, a decrease in metal extraction will occur, but even at 4°C bacterial solubilization of copper, cobalt, nickel, and zinc was observed. At higher temperatures (50°C–80°C), thermophilic bacteria can be used for leaching purposes.

Mineral Substrate

The mineralogical composition of the leaching substrate is of primary importance. At high carbonate content of the ore or gangue material, the pH in the leaching liquid will increase and inhibition or complete suppression of bacterial activity occurs. Low pH values, necessary for the growth of the leaching bacteria, can be achieved by external addition of acid, but this may not only cause the formation and precipitation of gypsum but will also affect the cost of the process. The rate of leaching also depends on the total surface of the substrate. A decrease in the particle size means an increase in the total particle surface area so that higher yields of metal can be obtained without a change in the total mass of the particles. A particle size of about 42 µm is regarded as the optimum.

An enlargement of the total mineral surface area can also be obtained by an increase in pulp density. An increase in the pulp density may result in an increase in metal extraction but the dissolution of certain compounds which have an inhibitory or even toxic effect on the growth of leaching bacteria will increase as well.

It has been demonstrated that the addition of small amounts of amino acids (cysteine in this case) resulted in an increased pyrite corrosion by *T. ferrooxidans* as compared to controls without addition. It is suggested that the microorganisms may profit from weakening and breakup of chemical bonds mediated by the formation of the cysteine–pyrite complex. This might also be the case under natural conditions by the excretion of cysteine-containing metabolites. An inexpensive alternative to increase metal recovery from ore heaps is by the addition of sulfur-containing amino acids such as cysteine.

During bioleaching processes, co-precipitation of metals with mineral phases such as jarosites can reduce leaching efficiencies. In addition, the precipitation of compounds present in the leachates on the minerals to be leached can make the solid material inaccessible for bacterial leaching. Organic solvents such as flotation or solvent extraction agents, which are added for the downstream processing of leachates from bioleaching, might also lead to inhibition problems. Isopropylxanthate and LIX 984 (used as flotation agent and solvent extraction agent respectively) prevented the oxidation of pyrite and chalcopyrite by *T. ferrooxidans*. This fact is of special importance when spent leaching liquors are recycled for a reuse.

Heavy Metals

The leaching of metal sulfides is accompanied by an increase in metal concentration in the leachate. In general, the leaching organisms, especially the *Thiobacilli*, have a high tolerance to heavy metals and various strains may even tolerate 50 g/L Ni, 55 g/L Cu, or 112 g/L Zn. Different strains of some species may show completely different sensitivities to heavy metals. Very often, it is possible to adapt individual strains to higher concentrations of metals or to specific substrates by gradually increasing the concentration of metals or substrates.

Surfactants and Organic Extractants

Surfactants and organic compounds used in solvent extraction generally have an inhibitory effect on the leaching bacteria, mainly because of a decrease in the surface tension and reduction of the mass transfer of oxygen. Solvent extraction is currently preferred for the concentration and recovery of metals from pregnant solution. When bacterial leaching and solvent extraction are coupled, the solvents become enriched in the aqueous phase and have to be removed before the barren solution is recirculated to the leaching operation.

Other metabolites excreted by *Thiobacillus* might also enhance metal leaching efficiencies: Wetting agents such as mixtures of phospholipids and neutral lipids are formed by *T. thiooxidans*. As a consequence, growth of *T. thiooxidans* on sulfur particles is supported by the excretion of metabolites acting as biosurfactants which facilitate the oxidation of elemental sulfur. It was also hypothesized that *Thiobacillus caldus* is stimulating the growth of heterotrophic organisms in leaching environments by the excretion of organic compounds and is supporting the solubilization of solid sulfur by the formation of surface active agents. Metal solubilization might also be facilitated by microbial metabolites excreted by organisms other than *Thiobacillus* which are part of microbial consortia found in bioleaching operations. Microbial surfactants, which show large differences in their chemical nature, are formed by a wide variety of microorganisms. In the presence of biosurfactants which lead to changes in the surface tension, metal desorption from solids might be enhanced resulting in an increased metal mobility in porous media. It has been suggested that this metabolic potential can be practically used in the bioremediation of metal-contaminated soils. However, there is some evidence that surface active compounds as well as organic solvents are inhibitory to bioleaching reactions and prevent bacterial attachment. The external addition of Tween reduced the oxidation of chalcopyrite by *T. ferrooxidans*. It was concluded that the need of the microorganisms for surfactants is met by their own formation. In contrast, it was reported that the addition of Tween 80 increased the attachment of *T. ferrooxidans* on molybdenite and the oxidation of molybdenum in the absence of iron(II).

LEACHING TECHNIQUES

In Vitro Methods

The bioleaching of minerals is a simple and effective technology for the processing of sulfide ores and is used on a technical scale mainly for the recovery of copper and uranium. The effectiveness and economics of microbial leaching processes depend highly on the activity of the bacteria and on the chemical and mineralogical composition of the ore. Therefore, processes tested on individual types of ores cannot be transferred to other ones. Before a technical application is possible, the optimum leaching conditions have to be elaborated for each type of ore.

Percolator Leaching

The first experiments on bacterial leaching were carried out in air-lift percolators. In the simplest case, the percolator consists of a glass tube provided in its bottom part with a sieve plate and filled with ore particles. The ore packing is irrigated or flooded with a nutrient inoculated with bacteria. The leach liquor trickling through the column is pumped up by compressed sterile air to the top of the column for recirculation. Simultaneously, the stream of air takes care of the aeration of the system. To monitor the course of the leaching process, liquid samples are taken at intervals and the state of the leaching process is determined on the basis of pH measurements, microbiological investigations, and chemical analysis of the metals that have passed into solution.

Submerged Leaching

Because the oxygen supply is often inadequate and the surface ratio is unfavorable, percolator leaching is not very efficient, fairly slow, and series of experiments lasting 100–300 days are not unusual. Therefore, percolator leaching has been substantially displaced by submerged leaching using fine-grained material (particle size $<100\,\mu m$) which is suspended in the leaching liquid and kept in motion by shaking or stirring. Higher rates of aeration and a more accurate monitoring and control of various parameters favor the growth and the activity of the bacteria so that the reaction times are considerably shortened and the metal extraction substantially increases. Suspension leaching can be carried out in Erlenmeyer flasks or, in a more sophisticated manner, in a bioreactor. Besides mechanically stirred systems, an air-lift reactor has been proved suitable for the treatment of ore concentrates, industrial waste products, and for the bio-desulfurization of coal.

Column Leaching

Column leaching operates on the principle of percolator leaching and is used as a model for heap or dump leaching processes. Depending on their size, the columns may be made of glass, plastic, lined concrete, or steel. Their capacities range from several kilograms to a few tons. At various distances, most column systems have devices for taking samples or for installing special instruments for measuring temperature, pH, humidity, oxygen, or carbon dioxide. This gives information about what has to be expected in heap or dump leaching and how the leaching conditions can be optimized.

Industrial Leaching Processes

Currently, bioleaching is used on an industrial scale for the treatment of low-grade ores which generally contain metal concentrations below 0.5% (w/w). The simplest way of conducting microbial leaching is to pile the material in heaps, allow water to trickle through the heap, and collect the seepage water (leachate). Since the bacterial oxidation of sulfides is much slower than other biotechnical processes, the leachate is recirculated. There are three main procedures in use: dump leaching, heap leaching, and underground leaching.

Dump Leaching

Dump leaching is the oldest process. The size of the dumps varies considerably and the amount of ore may be in the range of several hundred thousand tons of ore. The top of the dump is sprinkled continuously or flooded temporarily. Depending on the ore, the lixiviant may be water, acidified water, or acid ferric sulfate solution from other leaching operations on the same mining property. Before recirculation, the leachate may pass through an oxidation basin in which the bacteria and ferric iron are regenerated.

Heap Leaching

This procedure is mainly used for fine-grained ores that cannot be concentrated by flotation. The leaching is practiced in large basins containing up to 12,000 tons of ore. The procedure is similar to that of dump leaching. In some heap leaching operations, pipes are placed in strategic positions within the heaps during its construction to provide the deeper portions of the heap with sufficient amounts of oxygen.

Underground Leaching

Underground leaching is usually done in abandoned mines. Galleries are flooded or unmined ore or mine waste in side tunnels is sprinkled or washed under pressure. The water is collected in deeper galleries and shafts and is then pumped to a processing plant at the surface. The best known application of this procedure is at the Stanrock uranium mine at Elliot Lake in Ontario, Canada.

Ore deposits that cannot be mined by conventional methods because they are too low grade or because they are too small can be leached in situ. Solutions containing the appropriate bacteria are injected into boreholes in the fractured orebody. After a sufficient time for reaction, the leachate is pumped from neighboring wells or collected in drifts. The procedure requires sufficient permeability of the orebody and impermeability of the gangue rock so that any seepage of the pregnant leaching solution is prevented.

Tank Leaching

Considering the high yields in metal extraction by submerged leaching, the change from shake flasks to bioreactors was tested very early. Tank leaching was found to be most effective for the treatment of ore concentrates and more than 80% of the total zinc was extracted from a zinc sulfide concentrate. Tank leaching is more expensive to construct and to operate than dump, heap, or in situ leaching processes. But the rate of metal extraction is much higher, and currently, this technique is successfully used for bioleaching of refractory gold ores.

In Situ Bioleaching

In situ bioleaching has been commercially used as a scavenger technology for nearly 30 years to extract uranium and copper from depleted underground mines. When conventional mining is completed, the underground workings are blasted to fragment the ore and overburden material establishing permeability. Shafts are left intact to allow for aeration of the fragmented ore and to recover the metal-bearing solutions from sumps. Acidified leach solutions, applied to the top surface of the entire rubblized ore zone, percolate through the fragmented ore. The leaching bacteria become established and facilitate metal extraction. Metal-rich solutions, recovered in sumps, are pumped to the surface for metal recovery. Ultimate metal recovery from in situ operations is dependent on the degree of ore fragmentation and uniform irrigation of the fragmented ore zone. Poor metal recoveries can often be traced to leach solutions following preferential flow paths and not contacting the ore uniformly.

BACTERIAL ATTACHMENT ON MINERAL SURFACES

It is known that the formation of extracellular polymeric substances plays an important role in the attachment of thiobacilli to mineral surfaces such as sulfur, pyrite, or covellite. Extraction or loss of these exopolymers prevents cell attachment resulting in decreased metal leaching efficiencies. It was concluded that a direct contact between bacterial cells and solid surfaces is needed and represents an important prerequisite for an effective metal mobilization. Interactions between microorganisms and the mineral surface occur at two levels. The first level is a physical sorption because of electrostatic forces. Due to low pH usually occurring in leaching environments, microbial cell envelopes are positively charged leading to electrostatic interactions with the mineral phase. The second level is characterized by chemical sorption where chemical bonds between cells and minerals might be established (e.g., disulfide bridges). In addition, extracellular metabolites are formed and excreted during this phase in the near vicinity of the attachment site. Low-molecular weight metabolites excreted by sulfur oxidizers include acids originating from the TCA cycle, amino acids, or ethanolamine, whereas compounds with relatively high molecular weights include lipids and phospholipids. In the presence of elemental sulfur, sulfur-oxidizing microorganisms from sewage sludge form a filamentous matrix similar to a bacterial glycocalyx suggesting the relative importance of these extracellular substances in the colonization of solid particles.

COPPER BIOLEACHING

Sulfide copper ores are predominantly recovered by means of the flotation process and pyrometallurgical smelting of the flotation concentrate. A strong increase in the number of copper leaching operations (chemical and biological leaching) in the 1990s and at the beginning of the twenty-first century increased the share of leaching operations from 10% to 20% of total copper production. The conventional leaching of oxide ores with dilute sulfuric acid is increasingly being substituted by a growing share of bioleaching of sulfide ores. An important reason for this change is the decreasing resource of oxide copper ore in comparison to low-grade sulfide ore. Three different processes for bioleaching of sulfide ores can be distinguished:

- Heap or dump/stockpile bioleaching with secondary, mostly low-grade sulfide ores, which contain minerals such as chalcocite (Cu_2S) and covellite (CuS).
- Heap bioleaching of low-grade primary copper sulfides such as chalcopyrite ($CuFeS_2$; pilot or demonstration scale).
- Stirred-tank bioleaching with copper concentrates (pilot or demonstration scale).

Most important for copper bioleaching to date is heap or dump/stockpile bioleaching with secondary copper ores. About 80% of the bioleached copper originates from projects with secondary copper ores. In terms of the remaining 20%, low-grade primary ore is increasingly bioleached via dump/stockpile leaching. Heap leaching of any copper ore has a recognized potential of expansion.

The concept behind bioleaching is the microbial production of primary products responsible for the oxidation/solubilization of mineral sulfides, releasing the metals of interest in the leaching solution before a further recuperation by electrochemical/biosulfidation methods. The most studied microorganism involved in sulfide ores bioleaching is the aerobic, mesophilic, and extremophile acidic bacteria *Acidithiobacillus ferrooxidans*. This chemolithotrophic bacterium oxidizes reduced forms of iron (Fe^{2+}) and sulfur as electrons and energy source, using carbon dioxide (CO_2) as a carbon source. During the chalcopyrite oxidation process (Eq. 8.5), *A. ferrooxidans* oxidizes both the ferrous ions (Fe^{2+}) (Eq. 8.6) as well as the elemental sulfur (S^0) (Eq. 8.7) generated as a byproduct, resulting in the formation of ferric ions (Fe^{3+}) and H^+, that attack the chalcopyrite ($CuFeS_2$) in a loop effect.

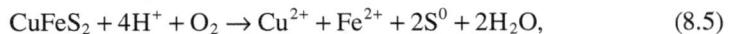

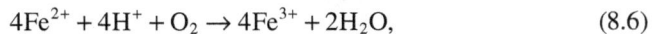

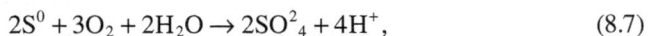

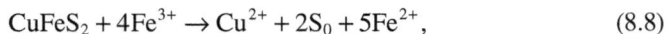

$$CuFeS_2 + 4H^+ + O_2 \rightarrow Cu^{2+} + Fe^{2+} + 2S^0 + 2H_2O, \qquad (8.5)$$

$$4Fe^{2+} + 4H^+ + O_2 \rightarrow 4Fe^{3+} + 2H_2O, \qquad (8.6)$$

$$2S^0 + 3O_2 + 2H_2O \rightarrow 2SO^2_4 + 4H^+, \qquad (8.7)$$

$$CuFeS_2 + 4Fe^{3+} \rightarrow Cu^{2+} + 2S_0 + 5Fe^{2+}, \qquad (8.8)$$

A high copper extraction rate could be achieved with bioleaching using moderate thermophilic bacteria at approximately 50°C or thermophilic archaea (genera *Acidianus, Metallosphaera, Sulfolobus*) up to 80°C.

Uranium Bioleaching

Uranium is a widespread metal in nature, occurring in granites and various other mineral deposits. Uranium is used mainly as fuel in nuclear power stations, although some uranium compounds are also used as catalysts and staining pigments. In recent years, there has been an upsurge in uranium exploration across the globe, driven by a renewed interest in nuclear energy and higher uranium prices. For chemical in situ leaching of uranium, an acid or alkaline digestion is applied depending on the acid-consuming characteristics of the rocks in the deposit. For this purpose, an oxidizing solution with complexing agents is introduced in the ore deposit via bore holes and the uranium-enriched solution is pumped to the surface for further processing. In most cases, uranium occurs as uraninite, which requires an oxidative leach for its extraction. The uranium ore is extracted through mechanical means, such as blasting, drilling, pneumatic drilling, picks, and shovels, and then transported to the surface. Due to very low concentration of uranium in the rock, immense amounts of rock have to be moved and processed in order to get a few kilograms of natural uranium. This results in enormous heaps. For instance, with a concentration of 0.1% of uranium, 1,000 tons of radioactive waste have to be dumped onto heaps to get just 1 ton of natural uranium.

In practice, the iron is usually obtained from pyrite—which is either associated with or added to the ore, and a high redox potential is achieved by adding oxidizing agents such as pyrolusite (MnO_2) and sodium chlorate ($NaClO_3$). However, these are expensive reagents, and they create environmental problems by releasing harmful impurities into the leaching circuit. Recent studies have evaluated process flowsheets that encompass conventional atmospheric leaching using chemical oxidants for these types of orebodies. Two process options that were investigated involved the use of MnO_2 and SO_2/air as the oxidant. These studies showed that the costs associated with the purchase and transport of chemical reagents are high, and the use of potentially cheaper SO_2/air system creates technical challenges in the subsequent ion exchange process used to recover the uranium.

In the leaching of uranium, the bacteria do not directly attack the uranium mineral. But they generate Fe(III) from pyrite and soluble Fe(II). Fe(III) readily attacks minerals incorporating U(IV) to converting it to U(VI) which is soluble in dilute sulfuric acid. The biooxidation is about 10^5–10^6 times faster than the chemical oxidation. The uranium solubilization by indirect mechanism can be described as follows:

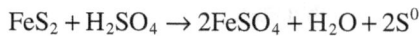

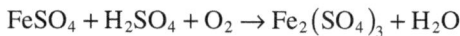

$$FeS_2 + H_2SO_4 \rightarrow 2FeSO_4 + H_2O + 2S^0$$

$$FeSO_4 + H_2SO_4 + O_2 \rightarrow Fe_2(SO_4)_3 + H_2O$$

Fe(II) can be re-oxidized by microbes to Fe(III) which takes part in the oxidation process again. Sulfur formed is simultaneously oxidized depending on the species to H_2SO_4 which aids (oxidizing agent) the dissolution of uranium as follows:

$$2S^0 + 3O_2 + 2H_2O \rightarrow 2H_2SO_4$$

Uraninite reacts with ferric sulfate to produce soluble uranyl sulfate and ferrous sulfate:

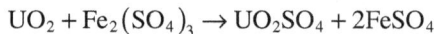

$$UO_2 + Fe_2(SO_4)_3 \rightarrow UO_2SO_4 + 2FeSO_4$$

This reaction requires a source of iron and a high redox potential.

In situ bioleaching of uranium ores is a procedure in which insoluble UO_2 is oxidized to water-soluble uranyl ions $(UO_2)^{2+}$ by means of microorganisms such as A. ferrooxidans. In the process, U(IV) is oxidized to U(VI), and in a coupled redox reaction, Fe(III) is reduced to Fe(II). The oxidizing agent Fe(III) for UO_2 is provided by the microbial Fe(II) oxidation as in copper biooxidation.

GOLD BIOLEACHING

Gold is not leached biologically because oxidation of the gold, which exists in the metallic state, does not take place. However, sulfidic iron and, perhaps, the arsenic matrix, in which the gold is either bound in the crystal lattice or enclosed as a particle, is biologically oxidized (biooxidation). The liberation of the original refractory gold is facilitated by extracting solubilized oxidized mineral components. Afterward, the gold can be attacked by cyanide leaching.

Because the leachable gold deposits that occurred close to surfaces were preferentially exploited in the past and are now going to be depleted, it can be assumed that the future production of gold coming from refractory or low-grade sulfide ores will increase significantly. Refractory gold ores are those which contain gold that cannot be satisfactorily recovered by basic gravity concentration or cyanidation and yield gold recoveries of less than 80%.

Many of the newly developed, more deeply situated gold deposits have to be assumed as refractory in terms of mineralogy, because the gold is encased in sulfide minerals. In order to prepare these refractory ores for cyanide leaching, a pretreatment that includes oxidation of the sulfides is required. For many years, the roasting of the sulfide ore was the only economical process for a pretreatment of sulfide ores prior to cyanide leaching. Another process that can be used economically for the pretreatment of rich sulfide ores and concentrates is the pressure oxidation in autoclaves. In the 1990s, biooxidation as an alternative pretreatment process, particularly for low-grade sulfide gold ores, was introduced. Biooxidation takes place in large-scale tank reactors set up in series in which several process parameters such as temperature, pH, O_2, and CO_2 are controlled. Optimum conditions for the metal sulfide-oxidizing bacteria are regulated. Because of the relatively slow process kinetics, the residence time in the bioreactor is up to a few days. In contrast to pressure oxidation, the general advantages of biooxidation projects are the relatively low capital expenditures and the ease of operability.

There are three different technical processes distinguished:

- Biooxidation in heaps for low-grade, refractory gold ores (e.g., applied by Newmont Mining Corporation).
- Biooxidation in stirred tanks for refractory gold ores with higher content of gold (e.g., applied by Gold Fields, BIOX® process).
- Coating of inert tailings material with sulfidic gold concentrates and its biooxidation in heaps (e.g., applied by GeoBiotics).

Biooxidation in Heaps

The reaction rate is relatively slow (months up to years for a heap), but the operating expenses and capital costs are very low. Diminished gold extraction in comparison to the other processes—between 50% and 75% gold recovery—is observed. Likewise, the long operation time, because of the long-term capital commitment, is considered a drawback in the case of the subsequently applied heap cyanide leach process. The process can be profitable especially for low-grade sulfide gold ores.

Biooxidation with Stirred Tanks

This process is characterized by a rapid reaction rate. The capital costs and operating expenses are substantially higher in comparison to the other processes, so that the application can only be used for high-grade gold ores and gold concentrates. This is considered at the moment to be the only biooxidation technology that is widely applied at industrial scale.

Biooxidation with Thin Layer Technology (Coating)

This process lies with respect to performance indicators and the technical-economic parameters between the stirred-tank and heap leaching process.

MICROBES INVOLVED IN GOLD BIOLEACHING

Although gold is relatively inert, it is solubilized by the formation of complexes with certain biogenic products. Environmental concern over the use of large amounts of cyanide for the leaching of gold from ores has prompted studies of potential biogenic agents for gold dissolution. Cyanide is produced by a wide range of fungi, strains of *Chromobacterium violaceum* and by several species of *Pseudomonas*. The localized biogenesis of cyanide in gold deposits by cyanogenic microorganisms might be used to extract gold with minimal environmental risks. Cyanogenic bacteria have been shown to solubilize gold in laboratory experiments. Several strains of *C. violaceum* solubilized gold at levels up to $215\,mg^{-1}$ of culture medium. The solubilized form of gold was identified as $[Au(CN)_2]^-$. Gold solubilization was correlated with cyanide production, being most pronounced during the stationary phase of growth. The amino acids serine, asparagine, histidine, aspartic acid, glycine, and alanine form soluble complexes with gold, apparently by forming Au-N bonds in alkaline media. Bacteria and fungi that excrete these amino acids solubilize up to a few mg of Au/1 in laboratory tests. Oxidative reagents in combination with amino acids increase gold yields over those with amino acids alone. A mutant strain of *Bacillus* is able to grow at pH 8.5–9.0 and excrete peroxides, and amino acids extracted about 20% of the gold.

Screening tests with microorganisms might result in discovery of novel gold lixiviants. Further investigations into controlling the microbiological generation of cyanide in situ in heap leaching environments might lead to processes for localized low-level production of cyanide for extraction of gold from ores.

LEACHING OF METALS WITH FUNGI

Almost all knowledge of biohydrometallurgy accumulated up to now deals with chemolithoautotrophic microorganisms of the genus *Thiobacillus*. Several facts account for this disregard of fungi as a possible means for leaching: (1) heterotrophic microorganisms need a lot of organic carbon source for growth and for the production of great amounts of leaching agents; (2) many biohydrometallurgists are not very familiar with the handling of fungi; (3) the turnover of the leaching material is considerably slower with fungi in comparison to the "classic" substrate of *T. ferrooxidans*, pyrite; and (4) there seems to be less experience concerning genetic approaches.

However, on closer inspection, the following counter arguments must be emphasized: (1) a lot of mentioned metal-containing materials increase the pH of the medium and are therefore not a suitable environment for the commercially important acidophilic *Thiobacillus* species; (2) the ability of fungal leaching agents (organic acids) to form complexes with the metal ions is of advantage in order to increase the solubility of metal ions in neutral environments and in order to reduce the toxicity of heavy metal ions; (3) a great part of the mentioned leaching materials contains no energy source for the growth of chemolithoautotrophic bacteria which depend on the oxidation of sulfur or reduced iron and sulfur compounds; (4) tightened environmental standards and subsequently increasing costs for environmental precautions will force the industry to apply more and more non-polluting metal winning processes and to revalue conventional recycling methods for secondary raw materials for their ecologic sustainability; (5) the recycling of secondary raw materials will become more and more important because of the exhaustion of natural resources; (6) the genetics of yeasts are well known and there are also a growing number of publications concerning the transformation of filamentous fungi; (7) if we had as much knowledge about the causal connections of the leaching processes with fungi as we have with autotrophic bacteria, optimizing strategies could be applied much more efficiently. For these reasons, and if cheap organic wastes like whey permeate or molasses are used as growth substrates, leaching processes with heterotrophic microorganisms should be applicable in special niches where autotrophic bacteria are not feasible and conventional pyro- or hydrometallurgical recycling techniques have environmental, economic, or operational disadvantages.

Additionally, ecological implications of fungal leaching are of importance. Fungi contribute to weathering and mineralization of metal containing materials not only in specialized niches such as mining areas but also in a wide diversity of habitats. Knowledge about these processes is important in order to avoid the contamination of the environment with heavy metals solubilized from deposited wastes. The present article will give a brief survey of several diverse subjects which are of importance for leaching processes with fungi. Some of the treated topics may at the first impression appear to be far from biotechnology. But only a profound knowledge of the fundamentals gives us the opportunity to optimize and control a biotechnological process.

COMPARISON TO *THIOBACILLUS* SPP.

While chemolithoautotrophic bacteria of the genus *Thiobacillus* only need carbon dioxide and a reduced iron or sulfur compound for growth, chemoorganoheterotrophic fungi are dependent on an organic carbon and energy source. Nevertheless, the fixation of carbon dioxide also exists in fungi: the high efficiency observed when citric acid is excreted by *A. niger* can only be obtained because high amounts of carbon dioxide are fixed by anaplerotic pathways. In addition, the ability to gain energy from the oxidation of thiosulfate can be found with fungi. This oxidation of elemental sulfur or reduced sulfur compounds leads to the solubilization of metals. However, the rates of this process are much less efficient than with thiobacilli. Chemolithoautotrophic bacteria and fungi show strikingly different ways of solubilizing metals from solid materials. Chemolithoautotrophic bacteria solubilize solid metal compounds by two mechanisms: (1) in order to gain energy, the leaching material is solubilized by enzymically catalyzed reactions which imply a physical contact between the bacterium and the leaching material and lead to the destruction of the mineral; (2) the end products of these reactions, namely sulfuric acid, supplying protons, and ferric iron, supplying oxidizing capacity and protons, contribute also to the solubilization process. Redox processes play a major role. Contrary to that, fungi extract metals only by the excretion of metabolites like protons, organic acids, amino acids, peptides, and proteins; and redox processes play a minor role. Enzymically catalyzed solubilization of a metal compound has not been confirmed with fungi up to now. However, a cell-particle contact may also enhance indirect leaching processes. This is observed when citric acid excretion is stimulated by the adsorption of filter dust or when iron is reduced by oxalic acid produced by a fungus.

$T.$ $ferrooxidans$ must transfer 2 mol of electrons from ferrous iron to oxygen in order to produce 1 mol of ATP. This results in a cell yield of 0.5 g of dry weight. Fungi can synthesize 1 mol of ATP and about 2 g of dry weight from the transfer of 0.66 mol of electrons from glucose to oxygen. This corresponds to the oxidation of 0.027 mol of glucose. In other words, the turnover of the growth substrate ferrous iron by $T.$ $ferrooxidans$ is 74 times faster than that of glucose by a fungus. The leaching materials dealt with in this article contain in most cases no energy source for chemolithoautotrophic bacteria. Therefore, ferrous iron, for instance, has to be added. In this case, the efficiency of the metabolism of $T.$ $ferrooxidans$ and of a fungus to produce leaching equivalents are placed in the same order of magnitude: 1 mol of ferric iron produced by $T.$ $ferrooxidans$ is able to produce a maximum of 3 mol of protons or to oxidize 1 mol of a metal ion. From this, it can be deduced that between 1 and 1.5 mol of a metal ion, which is twofold positively charged after the reaction, can be solubilized. A fungus can produce up to 1.3 mol of citric acid from 1 mol of sucrose; from this amount of citric acid, 3.9 mol of protons are available which can solubilize up to 1.95 mol of a twofold positively charged metal ion. However, the time in which the citric acid is produced is 10 days for a fungal leaching process compared to less than 1 hour for the oxidation of 1 mol of ferrous iron by $T.$ $ferrooxidans$.

The temperature for optimal growth and activity of the commercially important $Thiobacillus$ species and of most leaching active fungi is around 30°C. The pH range in which growth and leaching activities occur lies between 1.0 and 5.5 with an optimum around 2.4 for the commercially important $T.$ $ferrooxidans$. But attempts have also been made to leach with autotrophic bacteria in a near-neutral environment. The pH optimum for the most frequently used genus $Penicillium$ is between 2.0 and 8.0.

Important for a future application is the property of fungi to grow on a wide variety of substrates including organic carbon containing wastes from agriculture, food industry, or biotechnological processes. However, whether a fungus produces great amounts of metabolites from these substrates or not has to be tested with each specific substrate.

Two qualities are necessary for a fungus which is to be used in leaching processes: it must excrete considerable amounts of metabolites (in most cases, organic acids) and it must be resistant to heavy metals. The question arises if a screening program should take both criteria into consideration simultaneously. If this is done, strains would be excluded which are not resistant to heavy metals but are able to produce large amounts of metabolites. Such strains could, for example, be employed in two-step processes. Resistance to heavy metals and excretion of organic acids are only in few cases causally connected. An example is the production of oxalic acid by *A. niger*, *Penicillium spinulosum*, and other fungi. Contrary to the induction of the excretion of a metabolite by a heavy metal ion, *A. niger*, for instance, produces great amounts of citric acid only if the medium is deficient of a certain metal, namely manganese.

There is—with the exception of weathering—no causal connection between growth of fungi and the solubilization of metal compounds with fungi as it is the case with thiobacilli. Therefore, two different methods have to be applied in order to develop highly efficient fungal leaching processes: as a first step, screening programs must be carried out applying direct as well as indirect criteria; second, the mechanisms of the excretion of metabolites must be elucidated and subsequently optimized.

Indirect criteria of screening programs are based on the assumption that fungi which are able to solubilize non-metal compounds of low solubility, for example, $CaHPO_4$, are also suitable for metal leaching processes. However, great care must be taken, because acid production by fungi is very strongly dependent on medium composition and conditions of cultivation.

FUNGAL METABOLISM

Generally, fungi acidify the nutrient medium during growth. This acidification can largely be put down to four processes: (1) the excretion of protons via the proton-translocating plasma membrane ATPase (the excretion of protons can even be used as a parameter for the determination of growth; (2) the absorption of nutrients in exchange for protons; (3) the excretion of organic acids; and (4) the acidification brought about through the carbon dioxide produced by the respiratory activity of the fungus.

Although these processes occur also within the frame of a "normal" metabolism, they usually end at a certain point. The production of great amounts of metabolites is only possible if there is an "imbalance" of the metabolism. Therefore, it is of utmost importance to know the reasons of such a disorganization. The best known case of metabolite excretion, caused by an imbalance of metabolism, is the citric acid production by *A. niger*. The main effect is a manganese deficiency which induces the synthesis of a protease. Subsequently, this leads to an increased protein turnover and to an elevation of the intracellular ammonium concentration. The ammonium then counteracts the citrate inhibition of the phosphofructokinase.

A short selection of other fungi capable of producing organic acids in amounts relevant for biotechnologists are *Yarrowia lipolytica* (citric acid), *Mucor* spp. (fumaric acid, gluconic acid), *Rhizopus* spp. (lactic acid, fumaric acid, gluconic acid), *A. niger* (citric acid, oxalic acid, gluconic acid), *Aspergillus* spp. (citric acid, malic acid, tartaric acid, ketoglutaric acid, itaconic acid, aconitic acid), *Penicillium* spp. (citric acid, tartaric acid, ketoglutaric acid, malic acid, gluconic acid), *Schizophyllum commune* (malic acid), and *Paecilomyces variotii* (malic acid). The nature and amount of organic acids excreted by fungi are mainly influenced by pH of the medium (production of oxalic acid by *A. niger*, *Penicillium simplicissimum*), the buffering capacity of the medium, the carbon source, and the presence or absence of certain heavy metals and trace elements (Mn, Fe, Cu, Mn).

Additional factors which are important for the excretion of, for instance, citric acid by *A. niger* are the balances of nitrogen and phosphate. Operational parameters of growth such as stirrer rate and aeration rate are also of importance because they influence the oxygen supply and the form of growth (pellets or filamentous).

The fungal cell is necessarily exposed to many environmental influences. The first one is the alteration of the activity of enzymes and of the regulation of metabolic pathways by heavy metals, the second one deals with environmental influences on transport processes through the plasma membrane, and the third one with the influence of reactive (an)organic surfaces. It is well known that citric acid production by *A. niger* is strongly influenced by metal ions. For example, an excess of copper ions neutralizes the inhibition of the aconitase by iron. Another example is the stimulating influence of manganese ions on enzymes of the glycolysis, the TCA cycle, and the pentose phosphate cycle which occurs during the leaching of manganese ores. It is obviously impossible to control the concentrations of every metal ion which is liberated during a leaching process. However, the addition of specific metal ions in order to stimulate the production of metabolites should be possible. If metabolites are to be excreted, they must pass the plasma membrane. Because the intracellular pH of fungal cells ranges from slightly acidic to neutral, most of the common metabolites occur in an ionized form. For this reason, the membrane permeability for these molecules is rather low, and some sort of driving force for the excretion must be assumed. The plasma membrane of fungi contains a proton-translocating ATPase which is a key enzyme for the fungal cell. It manages, or at least participates, in the control of growth, uptake of nutrients by secondary transport processes, control of intracellular and extracellular pH, germination of spores, and control of polarity of growth. Because the main product of its activity is a proton motive force consisting of a proton gradient (outside acidic) and a transmembrane potential gradient (inside negative), it becomes clear that interactions have to be expected between the environment present in leaching cultures and this enzyme or its products. Possible interactions include the dissipation of the proton gradient by leaching materials with high buffering capacity, a hyperpolarization of the transmembrane potential gradient by the neutralization of the extruded protons, a depolarization of the potential gradient by uptake of metal ions, direct influences on the activity of the H^+-ATPase, and changes in membrane permeability. These influences may also affect the excretion of organic acids. It was shown that citric acid excretion in a leaching culture with *P. simplicissimum* is decreased if the H^+-ATPase is inhibited by specific inhibitors, for example, diethylstilbestrol, sodium orthovanadate, and miconazole.

Therefore, the investigation of interactions between the environment in leaching cultures and the H+-ATPase seems to be important. The presence of inorganic or organic surfaces may also greatly influence the metabolic activity of the cells. For example, the presence of clay minerals stimulated the excretion of organic acids by chemoorganoheterotrophic bacteria.

ABIOTIC REACTIONS

Four mechanisms of the solubilization of solid metal compounds can be observed in leaching processes with fungi: acidolysis, complexolysis, redoxolysis, and the mycelium functioning as a "sink." The first three processes occur through metabolites excreted by the fungus. The fourth process can be observed if the fungus accumulates the metal ion from the solution and—by disturbing the equilibrium between solid and dissolved metal—causes the continuous solubilization of the metal. By far, the most important mechanism is the acidolysis. The oxygen atoms covering the surface of a metal compound are protonated very fast. The protons and the oxygen combine with water and the metal is therefore detached from the surface. The rate-limiting steps are the removal of the metal ion from the surface and the production of the metabolite by the fungus. On the other hand, protons are also able to catalyze solubilization reactions without their neutralization. The term complexolysis means that a metal ion is solubilized due to the complexing capacity of a molecule. Reactions of this kind are slower than solubilization with protons. Complexes with high stability constants are formed, for example, by oxalic acid and iron, magnesium, and aluminum, by citric acid and calcium and magnesium, as well as by tartaric acid and calcium, magnesium, iron, silicium, and aluminum. Also, certain amino acids form stable complexes with metal ions. Complexation plays an important role in enhancing the solubility of a metal ion which has been solubilized via acidolysis. Because it is the final aim of leaching processes to obtain metal ions in solution, this is an important characteristic property of fungal leaching processes. Furthermore, the complexation of a heavy metal often reduces its toxicity for the fungus.

Fungal Leaching Techniques

For the excretion of great amounts of organic acids, one of the most critical factors is the availability of oxygen. This means that oxygen may be needed as a substrate for synthesizing an organic acid but also that oxygen plays a role in the regulation of the metabolic pathways involved in acid formation. Because the supply with sufficient oxygen and carbon source as well as the maintenance of a defined metabolic imbalance is essential, an efficient leaching process with a fungus is only thinkable in a bioreactor at the moment.

The characteristics of various fungal growth forms and acid production processes necessitate the use of specially constructed bioreactors. For leaching processes, a construction as simple as possible is desirable. One possibly promising technique would be the use of biphasic media, for instance, a mixture of dextran and polyethylene glycol.

Before a process is scaled up, it must be carefully considered if a one-step or a two-step process should be employed. Some useful criteria are the tolerance of the fungus to the material and the metal ions, the possibly necessary presence of the material in order to stimulate the formation and excretion of acid, the advantages or disadvantages of binding of the leaching material and the solubilized metal ions to the fungal biomass, and the possible use of higher temperatures in a two-step process.

The winning of copper from copper sulfide by chemolithoautotrophic bacteria is "the most commonly practiced ore leaching process and is responsible for more than 15% of the world's primary copper production." Contrary to that, leaching of copper with fungi was only done on a shake-flask scale. However, there are many copper-containing leaching materials which cannot be treated with *Thiobacillus* spp. Several investigations were carried out to extract copper with fungi from ores which contain carbonates, silicates, and oxides. The fungi belonged to the genera *Aspergillus* or *Penicillium*. The leaching of copper sulfides with *A. niger* and *Trichoderma harzianum* has been described as possible but not feasible. Metabolites produced by *A. niger* and *P. variotii* (citric and oxalic acid) were able to leach copper from an oxidized copper ore. Increasing the inoculum had no positive effect on the solubilization of copper. A considerable amount of copper was solubilized with *Penicillium notatum* at a pH near neutral from elemental copper and pyrite; within 30 days, copper could be solubilized from manganese nodules if the manganese were reduced by oxalic acid produced, for example, by *A. niger*.

Uranium in the form of oxides, phosphate, or thorites was leached from different uranium-bearing rocks (0.018%–4.40% uranium). The fungi used were *Aspergillus ochraceus* and *Penicillium funiculosum*. The mechanism of uranium solubilization was attributed to the formation of stable organo-uranyl complexes with citric acid and glutamic acid. The complexes had molecular weights between 2,000 and 2,900 which indicated the existence of polyelectrolytes. The fungi also accumulated the uranium. The dissipation of the proton-motive force promoted the uranium uptake.

A lot of results concerning the solubilization of silicates with fungi have been collected from investigations about weathering. The treatment of silicate minerals will become more and more important, because the increasing depletion of high-grade ores will make it necessary to win metals from the abundant low-grade silicate minerals. Possible applications include the destruction of the crystal lattice of silicate minerals in order to solubilize the metals contained in the mineral (e.g., aluminum, potassium, lithium, iron, uranium, copper, nickel, zinc) and the removal of silicon from low-grade bauxite ores in order to enhance the recovery rate of aluminum won via the Bayer process or from magnesite ores.

Several fungal species are able to solubilize zinc from solid materials. Some of these results came from research on the solubilization of micronutrients from metal compounds (oxides, carbonates, silicates, hydroxides, chromates, and others) in soils.

Nickel was leached from lateritic ores (0.9%–3.1% nickel) with several species of the genera *Penicillium* (*P. simplicissimum, P. funiculosum*, and others) and *Aspergillus*. *P. simplicissimum* was adapted to tolerate up to 30,000 mg/L nickel. Abiotic leaching experiments with 15 organic acids revealed that citric acid was the most suitable acid for this leaching process (up to 90% of the nickel extracted with 0.5 M citric acid after 15 days). With *Penicillium* sp., up to 70% of the nickel were extracted. However, the extraction rate was strongly dependent on the type of the ore (silicatic or limonitic). *A. niger* was also tested but was much more sensitive to nickel and additionally accumulated nickel. The manganese of ferromanganese nodules can be reduced by oxalic acid which is produced by *Penicillium* spp. or *Aspergillus* spp. The solubilization of manganese concomitantly releases nickel.

The leaching of aluminum is closely connected to the solubilization of silicates. Citric acid produced by *P. simplicissimum* was used for extracting aluminum from basalt rock. A two-stage process for the microbial winning of aluminum was described. The lixiviant was a spent medium from *A. niger* containing citric and oxalic acid acidified with inorganic acids to pH 0.5. The leaching process itself was carried out at 90°C–100°C. Heating the minerals to 600°C increased the yield and inhibited the leaching of iron which is troublesome at the recovery of aluminum from the pregnant solution. After 3–5 hours, more than 90% of the aluminum was extracted from thermally activated genthite and vermiculite. Other minerals yielded between 60% and 90%. The aluminum oxide Al_2O_3 is also susceptible to solubilization with organic acids produced by *A. niger.*

The solubilization of iron with fungi is possible in two ways: by reducing ferric iron to ferrous iron (catalyzed by an excreted metabolite) and by solubilizing ferric iron with protons or complexing agents. Both reactions can be carried out with several fungi *(A. niger, Fusarium* spp., *Actinomucor* spp., *Alternaria* spp.) and are supposed to be non-enzymatic by means of excreted metabolites such as oxalic acid, malate, oxaloacetate, and pyruvate. The reduction process is stimulated in acidic and anaerobic environments.

The leaching of manganese ores (mainly oxides) may serve to win the manganese or to extract other valuable metals like copper, nickel, and cobalt from these ores. Because the manganese in these ores is trivalent or tetravalent, a reduction must take place to get the manganese soluble. Fungi from the genera *Aspergillus, Penicillium,* and *Pichia* can reduce manganese only non-enzymatically via excreted metabolites.

FURTHER READING

Brierley CL. Bacterial succession in bioheap leaching. *Hydrometallurgy.* 2001;59:249–255.
Brierley CL. *Biomining: Theory, Microbes and Industrial Processes.* Rawlings DE (ed.) Berlin Heidelberg New York: Springer; 1997.
Brierley JA. Heap leaching of gold-bearing deposits: Theory and operational description. In: Rawlings DE (ed.) *Biomining: Theory, Microbes, and Industrial Processes.* Berlin Heidelberg New York: Springer; 1997:103–115.
Crundwell FK. How do bacteria interact with minerals? *Hydrometallurgy.* 2003;71:75–81.
Dew DW, Lawson EN, Broadhurst JL. The BIOX™ process for biooxidation of gold-bearing ores or concentrates. In: Rawlings DE (ed.) *Biomining.* Berlin Heidelberg New York: Springer; 1997:45–80.
Rawlings DE, Barrie Johnson D. *Biomining.* Berlin Heidelberg New York: Springer; 2006.
Ritchie AIM. Optimization of biooxidation heaps. In: Rawlings DE (ed.) *Biomining.* Berlin Heidelberg New York: Springer;1997:201–226.
Schnell HA. Bioleaching of copper. In: Rawlings DE (ed.) *Biomining.* Berlin Heidelberg New York: Springer; 1997:21–43.
van Aswegen PC, van Niekerk J, Olivier W. In: Rawlings DE, Barrie Johnson D (eds.) *Biomining.* Berlin Heidelberg New York: Springer; 2007.

9 Microorganisms and Organic Pollutants

INTRODUCTION

Global release of industrial and agricultural chemicals has resulted in widespread environmental pollution. The energy production industry alone generates large amounts of waste during the processing of coal and oil, and also nuclear energy production. As just one example, the United States Environmental Protection Agency (USEPA) estimates that more than 1 million underground storage tanks, predominantly employed for gasoline storage, have been in service in the United States. There have been over 510,000 confirmed accidental releases from these underground storage tanks. This type of contamination is known as a point source. In contrast, the application of pesticides and fertilizers over vast land areas can result in what is called nonpoint source contamination. It has been reported that >90% of the monitored streams and >55% of shallow groundwater sites on agricultural and urban lands are contaminated with pesticides (Gilliom et al., 2006).

Groundwater contamination is a critical issue from two perspectives. First, groundwater constitutes approximately 97% of all available fresh water on Earth. Second, there is a hydrologic interchange between surface and subsurface water systems, such that there is a conduit for moving groundwater contamination into surface waters.

THE OVERALL PROCESS OF BIODEGRADATION

Biodegradation is the breakdown of organic contaminants that occurs due to microbial activity. As such, these organic contaminants can be considered as a microbial food source or substrate. Biodegradation of any organic compound can be thought of as a series of biological degradation steps or a pathway that ultimately results in the oxidation of the parent compound. Complete biodegradation or mineralization involves oxidation of the parent compound to form carbon dioxide and water, a process that provides both carbon and energy for growth and reproduction of cells. The series of degradation steps constituting mineralization is similar whether the carbon source is a simple sugar such as glucose or plant polymer such as cellulose or a pollutant molecule. Each degradation step in the pathway is catalyzed by a specific enzyme made by the degrading cell. Enzymes are most often found within a cell, but are also made and released from the cell to help initiate degradation reactions. Enzymes found external to the cell are known as extracellular enzymes. Extracellular enzymes are important in the degradation of macromolecules such as the plant polymer cellulose. Macromolecules must be broken down into smaller subunits outside the cell

DOI: 10.1201/9781003272618-11

to allow transport of the smaller subunits into the cell. Biotransformation may stop at any step in the biodegradation pathway if the appropriate enzyme, either internal or extracellular, is not present. In fact, lack of appropriate biodegrading enzymes is one common reason for persistence of organic contaminants, particularly those with unusual chemical structures that the existing enzymes do not recognize. Thus, contaminant compounds that have structures similar to those of natural substrates are normally easily degraded. Those that are quite dissimilar to natural substrates are often degraded slowly or not at all. Some organic contaminants are only partially degraded by environmental microorganisms. This can result from absence of the appropriate degrading enzyme as mentioned earlier. A second type of incomplete degradation is co-metabolism in which a partial oxidation of the substrate occurs but the energy derived from the oxidation is not used to support microbial growth. The process occurs when organisms possess one or more enzymes that coincidentally can degrade a particular contaminant in addition to its target substrate. Thus, such enzymes are nonspecific.

Co-metabolism can occur during periods of active growth or can result from the interaction of resting (non-growing) cells with an organic compound. Co-metabolism is difficult to measure in the environment, but has been demonstrated for some environmental contaminants. For example, the industrial solvent trichloroethene (TCE; also known as trichloroethylene) can be oxidized co-metabolically by methanotrophic bacteria that grow on methane as a sole carbon source (Suttinun et al., 2013). TCE is of great interest for several reasons. It is one of the most frequently reported contaminants at hazardous waste sites, it is a suspected carcinogen, and it is generally resistant to biodegradation. The first step in the oxidation of methane by methanotrophic bacteria is catalyzed by the enzyme methane monooxygenase. This enzyme is so nonspecific that it can also co-metabolically catalyze the first step in the oxidation of TCE when both methane and TCE are present. The bacteria receive no energy benefit from this co-metabolic degradation step. The subsequent degradation steps may be catalyzed spontaneously by other bacteria, or in some cases, by the methanotroph. This is an example of a co-metabolic reaction that may have great significance in remediation. Research is currently investigating the application of these methanotrophs, as well as co-metabolizing microorganisms that grow on toluene, ethylene, propylene, propane, butane, and even ammonia, to TCE-contaminated sites.

Partial or incomplete degradation can also result in polymerization or synthesis of compounds that are more complex and stable than the parent compound. This occurs when initial degradation steps, often catalyzed by extracellular enzymes, create reactive intermediate compounds. These highly reactive intermediate compounds can then combine with each other or with other organic matter present in the environment. This is illustrated in Figure 9.1, which shows some possible polymerization reactions that occur with the herbicide propanil during biodegradation. These include formation of dimers or larger polymers, which are quite stable in the environment. Stability is due to low bioavailability (high sorption and low solubility), lack of degrading enzymes, and the fact that some of these residues become chemically bound to the soil organic matter fraction.

$$CH_4 \qquad\qquad Cl_2C=CHCl$$

Methane   Methane   TCE
monooxygenase

$$CH_3OH \qquad\qquad Cl_2C\text{-}CHCl$$

$$H_2CO$$

$$HCOOH$$

$$CO_2 \qquad\qquad\qquad\qquad\qquad CO_2 + Cl^-$$

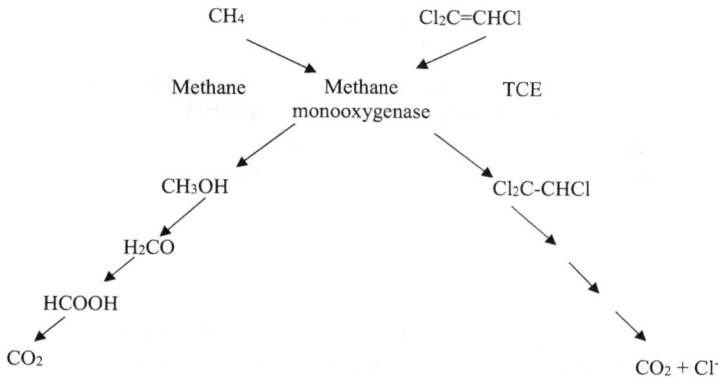

FIGURE 9.1 The oxidation of methane by methanotrophic bacteria is catalyzed by the enzyme methane monooxygenase. The same enzyme can act non-specifically on trichloroethene (TCE). Subsequent TCE degradation steps may be catalyzed spontaneously, by other bacteria, or in some cases by the same methanotroph (Pepper et al., 2006)

CONTAMINANT STRUCTURE, TOXICITY, AND BIODEGRADABILITY

The vast majority of the organic carbon available to microorganisms in the environment is material that has been photosynthetically fixed (plant material). Of concern are environments that receive large additional inputs of carbon from agriculture or industry (petroleum products, organic solvents, pesticides). Although many of these chemicals can be readily degraded because of their structural similarity to naturally occurring organic carbon, the amounts added may exceed the existing carrying capacity of the environment. Carrying capacity is defined here as the maximum level of microbial activity that can be expected under the existing environmental conditions. Microbial activity may be limited by both biological and physical-chemical factors. These factors include low numbers of microbes, insufficient oxygen or nutrient availability, and suboptimal temperature or water availability (Miller and Herman, 1997).

GENETIC POTENTIAL

The onset of contaminant biodegradation generally follows a period of adaptation or acclimation of indigenous microbes, the length of which depends on the contaminant structure. The efficient cycling of plant-based organic matter by soil microorganisms can promote the rapid degradation of organic contaminants that have a chemical structure similar to those of natural soil organic compounds. Previous exposure to a contaminant through repeated pesticide applications or through frequent spills will create an environment in which a biodegradation pathway is maintained within an adapted community. Adaptation of microbial populations most commonly occurs by induction of enzymes necessary for biodegradation followed by an increase in the population of biodegrading organisms (Leahy and Colwell, 1990).

Naturally occurring analogues of certain contaminants may not exist, and previous exposure may not have occurred. Degradation of these contaminants requires a second type of adaptation that involves a genetic change such as a mutation or a gene transfer. This results in the development of new metabolic capabilities. The time needed for an adaptation requiring a genetic change, or for the selection and development of an adapted community, is not yet predictable, but it may require weeks to years or may not occur at all (van der Meer, 2006).

TOXICITY

Chemical spills and engineered remediation projects, such as landfarming of petroleum refinery sludges, can involve extremely high contaminant concentrations. In these cases, toxicity of the contaminant to microbial populations can slow the remediation process. One common type of toxicity is that associated with non-ionic organic contaminants such as petroleum hydrocarbons or organic solvents. This toxicity is mainly due to a nonspecific narcotic-type mode of action, which is based on the partitioning of a dissolved contaminant into the lipophilic layer of the cell membrane, which causes a disruption of membrane integrity (Sikkema et al., 1995). This effect is important because, due to hydrophobic interactions, the cell membrane is a major site of organic contaminant accumulation in microorganisms. In addition, functional groups such as halogens and even the molecular weight of a compound influence its toxicity to microbial cells (Kenawy et al., 2007).

Models have been developed that relate bioconcentration (the accumulation of a hydrophobic contaminant by a cell or organism) and toxicity to the physicochemical attributes or descriptors of the organic contaminant. These models are referred to as quantitative structure-activity relationship (QSAR) models. A number of models have been developed based on different attributes such as structure, functional groups, and metabolic pathways of degradation (Pavan and Worth, 2006). Specific descriptors can include hydrophobicity and molecular connectivity (which represents the surface topography of a compound). As might be expected, no one QSAR works for all compounds, in fact some studies show that such models correctly predict biodegradability 68%–91% of the time depending on the model used and the set of contaminants used to test the models (Tunkel et al., 2005; Pavan and Worth, 2006). Refinement of these models is currently an international effort because the use of QSAR is expected to increase in the future due to (1) the high costs associated with experimental determination of contaminant persistence, bioconcentration, and toxicity and (2) international pressure to reduce the use of animal testing.

BIOAVAILABILITY

For a long time, biodegradation was thought to occur if the appropriate microbial enzymes were present. As a result, most research focused on the actual biodegradative process, specifically the isolation and characterization of biodegradative enzymes and genes. There are, however, two steps in the biodegradative process. The first is the uptake of the substrate by the cell, and the second is the metabolism or degradation of the substrate. Assuming the presence of an appropriate metabolic pathway, degradation of a contaminant can proceed rapidly if the contaminant is available in a water-soluble form. However, degradation of contaminants with limited water solubility or those that are strongly sorbed to soil or sediments can be limited due to their low bioavailability (Maier, 2000).

Growth on an organic compound with limited water solubility poses a unique problem for microorganisms. Most microorganisms obtain substrate from the surrounding aqueous phase, but the opportunity for contact between the degrading organism and an organic compound with low water solubility is limited. Such a compound may be present in a liquid or solid state, both of which can form a two-phase system with water. Liquid hydrocarbons can be less or more dense than water, forming a separate phase above or below the water surface. For example, polychlorinated biphenyls (PCBs) and chlorinated solvents such as TCE are denser than water, and form a separate phase below the water surface. Solvents less dense than water, such as benzene and other petroleum constituents, form a separate phase above the water surface. There are three possible modes of microbial uptake of a liquid organic:

- Utilization of the solubilized organic compound.
- Direct contact of cells with the organic compound. This can be mediated by cell modifications, such as fimbriae, or cell surface hydrophobicity, which increase attachment of the cell to the organic compound.
- Direct contact with fine or sub-micrometer size substrate droplets dispersed in the aqueous phase.

The mode that predominates depends largely on the water solubility of the organic compound. In general, direct contact with the organic compound plays a more important role (modes 2 and 3) as water solubility decreases.

Some microbes can enhance the rate of uptake and biodegradation as a result of production of biosurfactants or emulsifiers. There are two effects of biosurfactants. First, they can effectively increase the aqueous solubility of the hydrocarbon through formation of micelles or vesicles that associate with hydrocarbons. Second, they can facilitate attachment of cells to the hydrocarbon by making the cell surface more hydrophobic and thus better able to stick to a separate oil phase. This makes it possible to achieve greatly enhanced biodegradation rates in the presence of biosurfactants (Herman et al., 1997). For organic compounds in the solid phase, e.g., waxes, plastics, or polyaromatic hydrocarbons (PAHs), there are only two modes by which a cell can take up the substrate:

1. Direct contact with the substrate.
2. Utilization of solubilized substrate.

Available evidence suggests that for solid-phase organic compounds, utilization of solubilized substrate is most important. Thus, low water solubility has a greater impact on degradation of solid-phase organic compounds than on liquid-phase organics.

Another factor that affects bioavailability of an organic compound is sorption of the compound by soil or sediment (Novak et al., 1995). Depending on the sorption mechanism, organic compounds can be weakly (hydrogen bonding, van der Waals forces, hydrophobic interactions) or strongly (covalent bonding) bound to soil. Sorption of weakly bound or labile residues is reversible, and when a sorbed residue is released back into solution, it becomes available for microbial utilization.

Bioavailability can also be reduced by the diffusion of contaminants into soil matrix microsites that are inaccessible to bacteria because of pore-size exclusion (Alexander, 1995). There is evidence that the proportion of labile residues made available by desorption decreases with the length of time the residues are in the soil. Thus, as contaminants age and become sequestered more deeply within inaccessible microsites, bioavailability, and therefore biodegradation, can be expected to decrease. Finally, some contaminants may be incorporated into soil organic matter by the catalytic activity of a wide variety of oxidative enzymes that are present in the soil matrix. The incorporation of contaminants into soil organic matter is called humification, a process that is usually irreversible and that may be considered as one factor in the aging process. These bound or humified residues are released and degraded only very slowly as part of the normal turnover of humic material in soil.

CONTAMINANT STRUCTURE

Steric Effects

Some types of contaminant structures can lead to low degradation rates even if the contaminant structure is similar to naturally occurring molecules. The presence of branching or functional groups can slow degradation by changing the chemistry of the degradation reaction site. The reaction site is the contact area between a degradative enzyme and the contaminant substrate where a transformation step occurs. When the reaction site is blocked by branching or a functional group, contact between the contaminant and enzyme at the reaction site is hindered. This is known as a steric effect. Branching or functional groups can also affect transport of the substrate across the cell membrane, especially if the transport is enzyme assisted. Steric effects usually increase as the size of the functional group increases (Pitter and Chudoba, 1990).

Electronic Effects

Functional groups may also contribute electronic effects that hinder biodegradation by affecting the interaction between the contaminant and the enzyme. Functional groups can be electron donating (e.g., CH_3) or electron withdrawing (e.g., Cl), and therefore, can change the electron density of the reaction site. In general, functional groups which add to the electron density of the reaction site increase biodegradation rates, and functional groups that decrease the electron density of the reaction site decrease biodegradation rates.

ENVIRONMENTAL FACTORS AFFECTING BIODEGRADATION

A number of parameters influence the survival and activity of microorganisms in any given environment. One factor that has great influence on microbial activity is organic matter, the primary source of carbon for heterotrophic microorganisms in most environments. Surface soils have a relatively high and variable organic matter content, and therefore are characterized by high microbial numbers and diverse metabolic activity. In contrast, the subsurface unsaturated zone and saturated zone usually have a much lower content and diversity of organic matter, resulting in lower microbial numbers and activity. Exceptions to this rule are some areas of the saturated zone that have high flow or recharge rates, which can lead to numbers and activities of microorganisms similar to those found in surface soils. Occurrence and abundance of microorganisms in an environment are determined not only by available carbon but also by various physical and chemical factors. These include oxygen availability, nutrient availability, temperature, pH, salinity, and water activity. Inhibition of biodegradation can be caused by a limitation imposed by any one of these factors, but the cause of the persistence of a contaminant is sometimes difficult to determine. Perhaps, the most important factors controlling contaminant biodegradation in the environment are oxygen availability, organic matter content, nitrogen availability, and contaminant bioavailability. Interestingly, the first three of these factors can change considerably depending on the location of the contaminant.

REDOX CONDITIONS

Redox conditions are very important in determining the extent and rate of contaminant biodegradation. For most contaminants, aerobic biodegradation rates are much higher than anaerobic biodegradation rates. For example, petroleum-based hydrocarbons entering the aerobic zones of freshwater lakes and rivers are generally susceptible to microbial degradation, but oil accumulated in anaerobic sediments can be highly persistent. Oxygen is especially important for degradation of highly reduced hydrocarbons such as the alkanes. For example, low-molecular weight alkanes such as methane do not degrade anaerobically. Higher molecular weight alkanes, such as hexadecane ($C_{16}H_{34}$), can occur, but degradation is very limited, and usually is only found in historically petroleum-contaminated sites. In contrast, there are some highly chlorinated compounds (e.g., perchloroethene (PCE)) that are recalcitrant under aerobic conditions, but amenable to biotransformation under anaerobic conditions.

ORGANIC MATTER CONTENT

Surface soils have large numbers of microorganisms. Bacterial numbers commonly range from 10^7 to 10^{10} per gram of soil with somewhat lower fungal numbers, 10^5–10^6 per gram of soil. In contrast, microbial populations in deeper regions such as the deep vadose zone and groundwater region are often lower by two orders of magnitude or more. This large decrease in microbial numbers with depth is primarily due to differences in organic matter content. Both the vadose zone and the groundwater region have low amounts of organic matter. One result of low total numbers of microorganisms is that a low population of contaminant degraders may be present initially. Thus, biodegradation of a particular contaminant may be slow until a sufficient biodegrading population has been built up. A second reason for slow biodegradation in the vadose zone and groundwater region is that because a low amount of organic matter is present, the organisms in this region are often dormant. This can cause their response to an added carbon source to be slow, especially if the carbon source is a contaminant molecule that has low bioavailability, or to which the organisms have not had prior exposure.

NITROGEN

Microbial utilization of organic contaminants, particularly hydrocarbons composed primarily of carbon and hydrogen, creates a demand for essential nutrients such as nitrogen and phosphorus. Thus, biodegradation can often be improved simply by the addition of nitrogen fertilizers. This is particularly true in the case of biodegradation of petroleum oil spills in which nitrogen shortages can become acute. In general, microbes have an average C:N ratio within their biomass of about 5:1–10:1 depending on the type of microorganism. Therefore, a ratio of approximately 100:10:1 (C:N:P) is often used in such sites (Wang and Bartha, 1990).

OTHER ENVIRONMENTAL FACTORS

Temperature

Hydrocarbon degradation has been reported to occur at a range of temperatures from close to freezing to more than 30°C. Bacteria can adapt to temperature extremes in order to maintain metabolic activity; however, seasonal temperature fluctuations in the natural environment have been shown to affect the rate at which degradation occurs (Palmisano et al., 1991). For example, the degradation rates of hexadecane and naphthalene in a river sediment were reduced approximately 4.5-fold and 40-fold respectively in winter (0°C–4°C) compared with summer (8°C–21°C) samples.

pH

In soils, the rate of hydrocarbon degradation is influenced by pH with the highest rates generally observed at neutral pH. However, microorganisms have been isolated from historically contaminated sites that have adapted to growth on hydrocarbons even at very acidic pH levels (pH 2–3). It has been observed that the diversity of these microorganisms is lower than that of their counterparts that grow at neutral pH (Uyttebroek et al., 2007).

Salinity

In typical terrestrial or freshwater ecosystems, co-contamination with moderate-to-high levels of salinity tends to slow hydrocarbon degradation (Ulrich et al., 2009). In marine ecosystems, hydrocarbons are frequently introduced naturally from oil seeps and natural gas deposits as well as anthropogenically from oil tanker spills and discharges; therefore, marine environments tend to contain microbial populations adapted for degradation of hydrocarbons under the salinity levels typically found in these ecosystems.

Water Activity

Optimal conditions for activity of aerobic soil microorganisms occur when between 38% and 81% of the soil pore space is filled with water (also referred to as percent saturation). In this range of water content, water and oxygen availability are maximized. At higher water contents, the slow rate of oxygen diffusion through water limits oxygen replenishment, thereby limiting aerobic activity. At lower water contents, water availability becomes limiting. It is because optimal activity really depends upon a combination of factors including water content and available pore space. Available pore space is measured as bulk density, which is defined as the mass of soil per unit volume (g/cm^3). This means that in any given soil, increasing bulk density indicates increasing compaction of the soil. In a soil that is loosely compacted (lower bulk density), a water saturation of 70% represents more water (more filled small pores and pore throats) than in a highly compacted soil. Therefore, in a soil with low bulk density, oxygen diffusion constraints become important at lower water saturation than for highly compacted soils.

BIOREMEDIATION

The objective of bioremediation is to exploit naturally occurring biodegradative processes to clean up contaminated sites. There are several types of bioremediations. In situ bioremediation is the in-place treatment of a contaminated site. Ex situ bioremediation may be implemented to treat contaminated soil or water that is removed from a contaminated site. Biostimulation, which is the modification of environmental conditions (e.g., addition of oxygen and nitrogen) to enhance the biodegradation activity of indigenous microorganisms, is often used to increase the speed and effectiveness of bioremediation. In contrast, intrinsic bioremediation or natural attenuation is the indigenous level of contaminant biodegradation that occurs without any stimulation or treatment. All of these types of bioremediations continue to receive increasing attention as viable remediation alternatives for several reasons. These include generally good public acceptance and support, good success rates for some applications, and a comparatively low cost of bioremediation when it is successful. As with any technology, there are also drawbacks. Success can be unpredictable because a biological system is being used. A second consideration is that bioremediation rarely restores an environment completely. Often the residual contamination left after treatment is strongly sorbed and not available to microorganisms for degradation. Over a long period of time (years), these residuals can be slowly released. There is little research concerning the fate and potential toxicity of such released residuals, and therefore, there is both public and regulatory concern about the importance of residual contamination. Although it is often not thought of as bioremediation, domestic sewage waste has been treated biologically for many years with good success. Interestingly, even sewage treatment is undergoing re-examination in the light of detection of trace levels of endocrine disrupting compounds (EDCs) in treated wastewater. These compounds mimic hormone activities in mammalian endocrine systems and arise from pharmaceuticals and personal care products (PPCPs) that are in sewage but are not completely removed by conventional drinking and wastewater treatment plants (Synder et al., 2003). Removal of EDCs and PPCPs is incomplete for two reasons: diverse chemical structures that require acclimation and low concentrations which may fail to induce biodegradation pathways. In application of bioremediation to problems other than sewage treatment, it must be kept in mind that biodegradation is dependent on the pollutant structure and bioavailability. Therefore, bioremediation success will depend on the type of pollutant or pollutant mixture present and the type of microorganisms present. The first successful application of bioremediation outside sewage treatment was the cleanup of oil spills, and success in this area is now well documented. In the past few years, many new bioremediation technologies have emerged that are being used to address other types of pollutants including (USEPA, 2001):

* volatile organic compounds (including chlorinated VOCs)
* PAHs
* pesticides, herbicides
* explosives

Several key factors are critical to successful application of bioremediation: environmental conditions, contaminant and nutrient availability, and the presence of degrading microorganisms. If biodegradation does not occur, the first thing that must be done is to isolate the factor limiting bioremediation, and sometimes this can be a very difficult task. Initial laboratory tests using soil or water from a polluted site can usually determine whether degrading microorganisms are present and whether there is an obvious environmental factor that limits biodegradation, such as extremely low or high pH or lack of nitrogen and/or phosphorus. However, sometimes the limiting factor is not easy to identify. Often pollutants are present as mixtures, and one component of the pollutant mixture can have toxic effects on the growth and activity of degrading microorganisms. Low bioavailability due to sorption and aging is another factor that can limit bioremediation and can be difficult to evaluate in the environment.

ADDITION OF OXYGEN OR OTHER GASES

One of the most common limiting factors in bioremediation is availability of oxygen. Oxygen is an element required for aerobic biodegradation. In addition, oxygen has low solubility in water and a low rate of diffusion (movement) through both air and water. The combination of these three factors makes it easy to understand that inadequate oxygen supplies will slow bioremediation. Several technologies have been developed to overcome a lack of oxygen. A typical bioremediation system used to treat a contaminated aquifer as well as the contaminated zone above the water table contains a series of injection wells or galleries, and a series of recovery wells that comprise a two-pronged approach to bioremediation. First, the recovery wells remove contaminated groundwater, which is treated above ground, in this case using a bioreactor containing microorganisms that are acclimated to the contaminant. This would be considered ex situ treatment. Following bioreactor treatment, the clean water is supplied with oxygen and nutrients, and then, it is reinjected into the site. The reinjected water provides oxygen and nutrients to stimulate in situ biodegradation. In addition, the reinjected water flushes the vadose zone to aid in the removal of the contaminant for aboveground bioreactor treatment. This remediation scheme is a very good example of the use of a combination of physical, chemical, and biological treatments to maximize the effectiveness of the remediation treatment.

Bioventing is a technique used to add oxygen directly to a site of contamination in the vadose zone (unsaturated zone). In bioventing alone, air is injected at very low flow rates into the contaminated zone to promote biodegradation. Alternatively, in some cases, flow rates can be increased to combine soil vapor extraction technology and bioremediation. In this case, extracted vapor-phase contaminants are treated aboveground either biologically or chemically, and in addition, in situ bioremediation is stimulated.

The bioventing zone includes the vadose zone and contaminated regions just below the water table. A series of wells have been constructed around the zone of contamination. To initiate bioventing, a vacuum is drawn on these wells to force accelerated air movement through the contamination zone. This effectively increases the supply of oxygen throughout the site and thus the rate of contaminant biodegradation. If the rate of air movement is increased further, contaminants are volatilized and removed as air is forced through this system. This contaminated air can be treated biologically by passing the air through aboveground soil beds in a process called biofiltration (Jutras et al., 1997).

In contrast, air sparging is used to add oxygen to the saturated zone. In this process, an air sparger well is used to inject air under pressure below the water table. The injected air displaces water in the soil matrix, creating a temporary air-filled porosity. This causes oxygen levels to increase, resulting in enhanced biodegradation rates. In addition, volatile organics will volatilize into the airstream, and can be removed by a vapor extraction well. Methane is another gas that can be added with oxygen in extracted groundwater and reinjected into the saturated zone. Methane is used specifically to stimulate methanotrophic activity and co-metabolic degradation of chlorinated solvents. Methanotrophic organisms produce the enzyme methane monooxygenase to degrade methane, and this enzyme also co-metabolically degrades several chlorinated solvents. Co-metabolic degradation of chlorinated solvents is presently being tested in field trials to determine the usefulness of this technology.

NUTRIENT ADDITION

A common bioremediation treatment is the addition of nutrients, in particular nitrogen and phosphorus. Many contaminated sites contain organic wastes that are rich in carbon but contain minimal amounts of nitrogen and phosphorus. Injection of nutrient solutions takes place from an aboveground batch feed system. The goal of nutrient injection is to optimize the ratio of carbon, nitrogen, and phosphorus (C:N:P) in the site to approximately 100:10:1. However, sorption of added nutrients can make it difficult to achieve the optimal ratio accurately.

SEQUENTIAL ANAEROBIC-AEROBIC DEGRADATION

The rapid biodegradation of many priority pollutants requires both anaerobic and aerobic stages. As already discussed, aerobic conditions favor the biodegradation of compounds with fewer halogen substituents, and anaerobic conditions favor the biodegradation of compounds with a high number of halogen substituents. However, complete biodegradation of highly halogenated aliphatics under anaerobic conditions often does not take place. Therefore, some researchers have proposed the use of a sequential anaerobic and aerobic treatment. Initial incubation under anaerobic conditions is used to decrease the halogen content, and subsequent addition of oxygen creates aerobic conditions to allow complete degradation to proceed aerobically. This approach was used successfully to treat a groundwater plume containing TCE, DCE, vinyl chloride, and petroleum hydrocarbons (Morkin et al., 2000).

ADDITION OF SURFACTANTS

Surfactant addition has been proposed as a technique for increasing the bioavailability and hence biodegradation of contaminant. Surfactants can be synthesized chemically and are also produced by many microorganisms in which case they are called biosurfactants. Surfactants work similar to industrial and household detergents that effectively remove oily residues from machinery, clothing, or dishes. Individual contaminant molecules can be "solubilized" inside surfactant micelles. These micelles range from 5 to 10 nm in diameter. Alternatively, surfactant molecules can coat oil droplets and emulsify them into solution—a property that makes them useful in dispersants for "breaking up" oil spills. In addition, surfactants can enhance the ability of microbes to stick to oil droplets. There have been extensive laboratory and field tests performed with both synthetic and biosurfactants. While these materials definitely increase the bioavailability of organic contaminants, they do not always stimulate biodegradation. However, enough successful tests have been performed to indicate that, if chosen carefully, surfactants can be used to enhance the remediation process (Maier and Soberon-Chavez, 2000; Martienssen and Schirmer, 2007).

ADDITION OF MICROORGANISMS OR DNA

If appropriate biodegrading microorganisms are not present in soil or if microbial populations have been reduced because of contaminant toxicity, specific microorganisms can be added as "introduced organisms" to enhance the existing populations. This process is known as bioaugmentation. Scientists are also capable of creating "superbugs," organisms that can degrade pollutants at extremely rapid rates. Such organisms can be developed through successive adaptations under laboratory conditions or can be genetically engineered.

Although bioaugmentation, with naturally occurring or engineered organisms, has been demonstrated to increase contaminant degradation in numerous lab-based studies, it generally has not been successful for remediation in actual field sites (Gentry et al., 2004; Stroo et al., 2013). The problem is that introduction of a microorganism to a contaminated site may fail for at least two reasons. First, the introduced microbe often cannot establish a niche in the environment. In fact, these introduced organisms often do not survive in a new environment beyond a few weeks. Second, there are difficulties in delivering the introduced organisms to the site of contamination, because microorganisms, like contaminants, can be strongly sorbed by solid surfaces.

Currently, very little is known about microbial transport and establishment of environmental niches. These are areas of active research, and in the next few years, scientists may gain a further understanding of microbial behavior in soil ecosystems. However, until we discover how to successfully deliver and establish introduced microorganisms, bioaugmentation will not be a viable bioremediation option for the majority of contaminated sites. If bioaugmentation does not work, one way to take advantage of the superbugs that have been developed is to use them in bioreactor systems under controlled conditions. Extremely efficient biodegradation rates can be achieved in bioreactors that are used in aboveground treatment systems. Another bioaugmentation strategy is to add specific genes that can confer a specific degradation capability to indigenous microbial populations. The addition of degradative genes relies on the delivery and uptake of the genetic material by indigenous microbes. There are two approaches that can be taken in delivery of genes. The first is to use microbial cells to deliver the DNA via conjugation. The second is to add "naked" DNA to the soil to allow uptake via transformation. This second approach may reduce the difficulty of delivery since DNA alone is much smaller than a whole cell.

BIODEGRADATION OF ORGANIC POLLUTANTS

ALIPHATICS

There are several common sources of aliphatic hydrocarbons that enter the environment as contaminants. These include straight-chain and branched-chain structures found in petroleum hydrocarbons; the linear alkyl benzenesulfonate (LAS) detergents; and the one- and two-carbon halogenated compounds such as chloroform and TCE that are commonly used as industrial solvents.

ALKANES

Aerobic Conditions

Because of their structural similarity to fatty acids and plant paraffins, which are ubiquitous in nature, many microorganisms in the environment can utilize n-alkanes (straight-chain alkanes) as a sole source of carbon and energy. In fact, it is easy to isolate alkane-degrading microbes from any environmental sample. As a result, alkanes are usually considered to be the most readily biodegradable type of hydrocarbon. Biodegradation of alkanes occurs with a high biological oxygen demand (BOD) using one of the two pathways shown. The more common pathway is the direct incorporation of one atom of oxygen onto one of the end carbons of the alkane by a monooxygenase enzyme resulting in the formation of a primary alcohol. Alternatively, a dioxygenase enzyme can incorporate both oxygen atoms into the alkane to form a hydroperoxide. The end result of both pathways is the production of a primary fatty acid.

Fatty acids are common metabolites found in all cells. They are used in the synthesis of membrane phospholipids and lipid storage materials. The common pathway used to catabolize fatty acids is known as β-oxidation, a pathway that cleaves off consecutive two-carbon fragments. Each two-carbon fragment is removed by coenzyme A as acetyl-CoA, which then enters the tricarboxylic acid (TCA) cycle for complete mineralization to CO_2 and H_2O. It becomes apparent that if one starts with an alkane that has an even number of carbons, the two-carbon fragment acetyl-CoA will be the last residue. If one starts with an alkane with an odd number of carbons, the three-carbon fragment propionyl-CoA will be the last residue. Propionyl-CoA is then converted to succinyl-CoA, a four-carbon molecule that is an intermediate of the TCA cycle.

In general, midsize straight-chain aliphatics (n-alkanes C_{10} to C_{18} in length) are utilized by microbes more readily than n-alkanes with either shorter or longer chains. Long-chain n-alkanes are utilized more slowly because of low bioavailability resulting from extremely low water solubilities. For example, the water solubility of decane (C_{10}) is 0.052 mg/L, and the solubility of octadecane (C_{18}) is almost 10-fold less (0.006 mg/L). Solubility continues to decrease with increasing chain length. In contrast, short-chain n-alkanes have higher aqueous solubility, e.g., the water solubility of butane (C_4) is 61.4 mg/L, but they are toxic to cells. Short-chain alkanes are toxic to microorganisms because their increased water solubility results in increased uptake of the alkanes, which are then dissolved in the cell membrane. The presence of these short alkanes within the cell membrane can alter the fluidity and integrity of the cell membrane.

The toxicity of short-chain n-alkanes can be mediated in some cases by the presence of free-phase oil droplets. Protection occurs because the short-chain alkanes partition into the oil droplets. This results in reduced bioavailability because the aqueous phase concentration is decreased. Thus, n-alkane degradation rates will differ depending on whether the substrate is present as a pure compound or in a mixture of compounds. Biodegradability of aliphatics is also negatively influenced by branching in the hydrocarbon chain. The degree of resistance to biodegradation depends on both the number of branches and the positions of methyl groups in the molecule.

Alkenes are hydrocarbons that contain one or more double bonds. The majority of alkene biodegradability studies have used 1-alkenes as model compounds (Britton, 1984). The initial step in 1-alkene degradation can involve attack at the terminal or a subterminal methyl group as described for alkanes. Alternatively, the initial step can be attack at the double bond, which can yield a primary or secondary alcohol or an epoxide. Each of these initial degradation products is further oxidized to a primary fatty acid, which is degraded by β-oxidation.

Anaerobic Conditions

In comparison to aerobic conditions, aliphatic hydrocarbons, which are highly reduced molecules, are degraded slowly, if at all, anaerobically. This is supported by the fact that hydrocarbons in natural underground reservoirs of oil (which are under anaerobic conditions) are not degraded despite the presence of microorganisms. The current view is that low-molecular weight alkanes (e.g., methane) do not energetically support anaerobic degradation. This is because they must be activated or functionalized prior to degradation. In contrast, higher molecular weight aliphatics have been shown to undergo degradation using a unique pathway, where the first step in biodegradation involves activation through the addition of a four-carbon oxygen-containing moiety, fumarate, into the alkane. Once oxygen has been introduced into the molecule through the addition of fumarate, it is mineralized (Widdel and Rabus, 2001). Aliphatics that are already activated, including both alkenes and aliphatics containing oxygen (aliphatic alcohols and ketones), are readily biodegraded anaerobically.

Halogenated Aliphatics

Chlorinated solvents such as trichloroethene (TCE) and perchloroethene (PCE) have been extensively used as industrial solvents. As a result of improper use and disposal, these solvents are among the most frequently detected types of organic contaminants in groundwater. The need for efficient and cost-effective remediation of solvent-contaminated sites has stimulated interest in the biodegradation of these C_1 and C_2 halogenated aliphatics.

Aerobic Conditions

Under aerobic conditions, halogenated aliphatics are generally degraded more slowly than aliphatics without halogen substitution. For example, although 1-chloroalkanes ranging from C_1 to C_{12} are degraded as a sole source of carbon and energy in pure culture, they are degraded more slowly than their non-chlorinated counterparts. The presence of two or three chlorines bound to the same carbon atom inhibits aerobic degradation (Janssen et al., 1990). For example, while TCE is degraded under aerobic conditions, PCE is not. These results can be explained by the decreasing electronic effects of the chlorine atom on the enzyme-carbon reaction center as the alkane chain length increases.

Biodegradation of halogenated aliphatics occurs by one of two basic types of reactions. Substitution is a nucleophilic reaction (the reacting species brings an electron pair) in which the halogens on a mono- or dihalogenated compound are substituted by a hydroxy group. Oxidation reactions are catalyzed by a select group of monooxygenase and dioxygenase enzymes that have been reported to oxidize highly chlorinated C_1 and C_2 compounds (e.g., TCE). These monooxygenase and dioxygenase enzymes are produced by bacteria and oxidize a variety of non-chlorinated compounds including methane, ammonia, toluene, and propane (Bhatt et al., 2007).

Anaerobic Conditions

In some very limited instances, C_1 chlorinated aliphatics such as chloromethane and dichloromethane can serve as a source of carbon and energy to support growth; however, chlorinated aliphatics are generally metabolized under anaerobic conditions primarily through two processes: (1) co-metabolism and (2) use as a terminal electron acceptor to support growth, a process called halorespiration. Both of these processes usually result in partial transformation of the substrate rather than complete mineralization. In general, the process of removing chlorines under anaerobic conditions is referred to as reductive dehalogenation. Reductive dehalogenation can be mediated by reduced transition metal complexes found in coenzymes such as vitamin B_{12} or in enzymes that can participate in dehalogenation (Fennel et al., 2004).

It is now clearly recognized that reductive dehalogenation is a very important process in contaminated environments for highly chlorinated aliphatics and chlorinated compounds in general. This is because anaerobic conditions favor the degradation of highly chlorinated compounds, whereas aerobic conditions favor the degradation of mono- and di-substituted halogenated compounds. Recall that TCE and PCE are among the most common groundwater contaminants. Under aerobic conditions, PCE is inert, and while TCE can be co-metabolized aerobically, this process requires optimization of the electron donor to TCE ratio. However, under anaerobic conditions, halorespiration of both TCE and PCE can occur quite rapidly resulting in lesser chlorinated metabolites that become amenable for aerobic biodegradation. If appropriate populations of microorganisms (e.g., *Dehalococcoides* spp.) are present, TCE and PCE can even be completely dechlorinated to ethane through use of the chlorinated organic as a terminal electron acceptor during metabolism of a corresponding electron donor (e.g., H_2 or a C_1 or C_2 organic compound such as ethanol).

ALICYCLICS

Alicyclic hydrocarbons are major components of crude oil, 20%–70% by volume. They are commonly found elsewhere in nature as components of plant oils and paraffins, microbial lipids and pesticides (Tridgill, 1984). Various components can be simple, such as cyclopentane and cyclohexane, or complex, such as trimethylcyclopentane and various cycloparaffins. The use of alicyclic compounds in the chemical industry, and the release of alicyclics to the environment through industrial processes, other than oil processing and utilization, is more limited than for aliphatics and aromatics. Consequently, the issue of health risks associated with human exposure to alicyclics has not reached the same level of importance as for the other classes of compounds, especially the aromatics.

Aerobic Conditions

It is difficult to isolate pure cultures that degrade alicyclic hydrocarbons using enrichment techniques. Although microorganisms with complete degradation pathways have been isolated, alicyclic hydrocarbon degradation is thought to occur primarily by commensalistic and co-metabolic reaction. In a series of reactions, one organism converts cyclohexane to cyclohexanol co-metabolically during the oxidation of propane, but is unable to further transform the compound. A second organism that is unable to oxidize cyclohexane to cyclohexanol can perform the subsequent transformations including lactonization, ring opening, and mineralization of the remaining aliphatic compound (Perry, 1984).

Cyclopentane and cyclohexane derivatives that contain one or two OH, C=O, or COOH groups are readily metabolized, and such degraders are easily isolated from environmental samples. In contrast, degradation of alicyclic derivatives containing one or more CH_3 groups is inhibited. This is reflected in the decreasing rate of biodegradation for the following series of alkyl derivatives of cyclohexanol: cyclohexanol > methylcyclohexanol > dimethylcyclohexanol (Pitter and Chudoba, 1990).

Anaerobic Conditions

Anaerobic biodegradation of complex mixtures of alicyclic compounds in gas condensate has been demonstrated under methanogenic and sulfate-reducing conditions. Gas condensate is the mixture of hydrocarbons in raw natural gas. It is a mixture containing primarily aliphatics (C_2–C_{12}), cyclopentanes, cyclohexanes, and aromatics (benzene, toluene, ethylbenzene, xylene (BTEX)). In examining the fate of the alicyclic components of gas condensate, sulfate-reducing conditions were found to support the anaerobic biodegradation of unsubstituted cyclopentanes and cyclohexanes as well as those with one methyl or ethyl substitution.

AROMATICS

Aromatic compounds contain at least one unsaturated ring system with the general structure C_6R_6, where R is any functional group. Benzene (C_6H_6) is the parent hydrocarbon of this family of unsaturated cyclic compounds.

Compounds containing two or more fused benzene rings are called polyaromatic hydrocarbons (PAHs; also known as polycyclic aromatic hydrocarbons). Aromatic hydrocarbons are natural products; they are part of lignin and are formed as organic materials are burned, for example, in forest fires. However, the addition of aromatic compounds to the environment has increased dramatically through activities such as fossil fuel processing, and utilization and burning of wood and coal. The quantity and composition of the aromatic hydrocarbons are of major concern when evaluating a contaminated site because several components of the aromatic fraction have been shown to be carcinogenic to humans. Aromatic compounds also have demonstrated toxic effects toward microorganisms.

UNSUBSTITUTED AROMATICS

Aerobic Conditions

A wide variety of bacteria and fungi can carry out aromatic transformations, both partial and complete, under a variety of environmental conditions (Johnsen et al., 2005). Under aerobic conditions, the most common initial transformation is a hydroxylation that involves the incorporation of molecular oxygen. The enzymes involved in these initial transformations are either monooxygenases or dioxygenases. In general, prokaryotic microorganisms transform aromatics by an initial dioxygenase attack to cis-dihydrodiols. The cis-dihydrodiol is rearomatized to form a dihydroxylated intermediate, catechol. The catechol ring is cleaved by a second dioxygenase either between the two hydroxyl groups (ortho pathway) or next to one of the hydroxyl groups (meta pathway) and further degraded to completion.

Most eukaryotic microorganisms do not mineralize aromatics; rather, they are processed for detoxification and excretion. This is done by an initial oxidation with a cytochrome P-450 monooxygenase, which incorporates one atom of molecular oxygen into the aromatic compound and reduces the second to water, resulting in the formation of an arene oxide. This is followed by the enzymatic addition of water to yield a trans-dihydrodiol.

Alternatively, the arene oxide can be isomerized to form phenols, which can be conjugated with sulfate, glucuronic acid, and glutathione. These conjugates are similar to those formed in higher organisms, which are used in the elimination of aromatic compounds. A small group of eukaryotes, the ligninolytic fungi, can completely mineralize aromatic compounds in a process known as ligninolytic degradation. The lignin structure is based on two aromatic amino acids, tyrosine and phenylalanine. In order to degrade an amorphous aromatic-based structure such as lignin, the white-rot fungi release nonspecific extracellular enzymes such as laccase or H_2O_2- dependent lignin peroxidase. These enzymes generate oxygen-based free radicals that react with the lignin polymer to release residues that are taken up by the cell and degraded. Since the lignin structure is based on an aromatic structure and the initial enzymes used to degrade lignin are nonspecific, the white-rot fungi are able to use the same activity to degrade a variety of aromatic contaminants. The most famous of the white-rot fungi is *Phanerochaete chrysosporium*, which has been demonstrated to degrade a variety of aromatic compounds.

Often the capacity for aromatic degradation is plasmid mediated (Ghosal et al., 1985). Plasmids can carry both individual genes and operons encoding partial or complete biodegradation of an aromatic compound. An example of the latter is the NAH7 plasmid, which codes for the entire naphthalene degradation pathway. The NAH7 plasmid was obtained from *Pseudomonas putida*, and contains genes that encode the enzymes for the first 11 steps of naphthalene oxidation. This plasmid or closely related plasmids are frequently found in sites that are contaminated with PAHs (Ahn et al., 1999). This plasmid has also been used to construct a luminescent bioreporter gene system. Here, the lux genes that cause luminescence have been inserted into the nah operon in the NAH plasmid. When the nah operon is induced by the presence of naphthalene, both naphthalene-degrading genes and the lux gene are expressed. As a result, naphthalene is degraded and the reporter organism luminesces. Such reporter organisms are currently being used as sensors to study the temporal and spatial activity of the reporter in soil systems. In general, aromatics composed of one, two, or three condensed rings are transformed rapidly and often completely mineralized, whereas aromatics containing four or more condensed rings are transformed much more slowly, often as a result of a co-metabolic attack. This is due to the limited bioavailability of these high-molecular weight aromatics. Such PAHs have very limited aqueous solubility, and sorb strongly to particle surfaces in soil and sediments. However, it has been demonstrated that chronic exposure to aromatic compounds will result in increased transformation rates because of adaptation of an indigenous population to growth on aromatic compounds.

Anaerobic Conditions

Like aliphatic hydrocarbons, aromatic compounds can be completely degraded under anaerobic conditions. Anaerobic mineralization of aromatics produces benzoyl CoA as the common degradation intermediate. If the aromatic is oxygenated such as for benzoate, biodegradation occurs rapidly and even at rates comparable to aerobic conditions. However, under anaerobic conditions, a mixed microbial community works together even though each of the microbial components requires a different redox potential. For example, mineralization of benzoate can be achieved by growing an anaerobic benzoate degrader in co-culture with a methanogen and sulfate reducer. The initial transformations in such a system are often carried out fermentatively, and this results in the formation of aromatic acids, which in turn are transformed to methanogenic precursors such as acetate, carbon dioxide, and formate. These small molecules can then be utilized by methanogens. Such a mixed community is called a consortium. It is not known how this consortium solves the problem of requiring different redox potentials in the same vicinity in a soil system. Clearly, higher redox potentials are required for degradation of the more complex substrates, such as benzoate, leaving smaller organic acid or alcohol molecules that are degraded at lower redox potentials. To ultimately achieve degradation may require that the organic acids and alcohols formed at higher redox potential be transported by diffusion or by movement with water (advection) to a region of lower redox potential. On the other hand, it may be that biofilms form on the soil surface and that redox gradients are formed within the biofilm allowing complete degradation to take place.

SUBSTITUTED AROMATICS

One group of aromatics of special interest is the chlorinated aromatics. These compounds have been used extensively as solvents and fumigants (e.g., dichlorobenzene) and wood preservatives (e.g., pentachlorophenol (PCP)), and are parent compounds for pesticides such as 2,4-dichlorophenoxyacetic acid (2,4-D) and DDT. The difficulty for aerobic microbes in the degradation of chlorinated aromatics is that the common intermediate in aromatic degradation is catechol. Catechol formation requires two adjacent unsubstituted carbons so that hydroxyl groups can be added. Chlorine substituents can block these sites. Some bacteria solve this problem by removing a chlorine using a dehalogenase or monooxygenase enzyme. Chlorinated phenols are particularly toxic to microorganisms. In fact, phenol itself is very toxic and is used as a disinfectant. Chlorination adds to toxicity, which increases with the degree of chlorination.

Methylated aromatic derivatives, such as toluene, constitute another common group of substituted aromatics. These are major components of gasoline and are commonly used as solvents. These compounds can initially be attacked either on the methyl group or directly on the ring. This can be compared to anaerobic degradation of toluene. Alkyl derivatives of aromatics are attacked first at the alkyl chain, which is shortened by β-oxidation to the corresponding benzoic acid or phenylacetic acid, depending on the number of carbon atoms. This is followed by ring hydroxylation and cleavage.

DIOXINS AND PCBs

Dioxins and dibenzofurans are created during waste incineration and are part of the released smoke stack effluent. Once thought to be one of the most potent carcinogens known, 2,3,7,8-tetrachlorodibenzo-p-dioxin (TCDD) is associated with the manufacture of 2,4-D and 2,4,5-trichlorophenoxy acetic acid (2,4,5-T), hexachlorophene, and other pesticides that have 2,4,5-T as a precursor. Current thinking is that TCDD is less dangerous in terms of carcinogenicity and teratogenicity than once thought, but that non-cancer risks including diabetes, reduced IQ, and behavioral impacts may be more important. The structure of TCDD and its low water solubility, 0.002 mg/L, result in great stability of this molecule in the environment.

Biphenyl is the unchlorinated analogue or parent compound of the PCBs, which were first described in 1881. The PCBs consist of different chlorine-substituted biphenyls of which, in theory, there are 209 possible isomers. Only approximately 100 actually exist in commercial formulations. The aqueous solubility of biphenyl is 7.5 mg/L, and any chlorine substituent decreases the water solubility. In general, the water solubility of monochlorobiphenyls ranges from 1 to 6 mg/L, compared to 0.08–5 mg/L for dichlorobiphenyls. In contrast, for hexachlorobiphenyl, the aqueous solubility is just 0.00095 mg/L. By 1930, because of their unusual stability, PCBs were widely used as nonflammable heat-resistant oils in heat transfer systems, as hydraulic fluids and lubricants, as transformer fluids, in capacitors, as plasticizers in food packaging materials, and as petroleum additives. PCBs were used as mixtures of variously chlorinated isomers and marketed under various trade names, e.g., Aroclor (USA), Clophen (Germany), Phenoclor (Italy), Kanechlor (Japan), Pyralene (France), and Soval (Russia).

Past use of PCBs has resulted in their accumulation in the environment from waste dumps and spills and as a result of PCB manufacturing processes. Although some PCB degradation occurs, it is limited by low bioavailability, by the recalcitrance of highly chlorinated PCB congeners under aerobic conditions, and by incomplete degradation under anaerobic conditions. The extensive research that has been performed to understand PCB degradation has suggested several strategies for promoting biodegradation. Of these, the most promising is the use of a sequential anaerobic–aerobic process first to allow the removal of chlorines using halorespiration and then allow mineralization of the less chlorinated congeners.

HETEROCYCLIC COMPOUNDS

Heterocyclics are cyclic compounds containing one or more heteroatoms (nitrogen, sulfur, or oxygen) in addition to carbon atoms. In general, heterocyclic compounds are more difficult to degrade than analogous aromatics that contain only carbon. This is probably due to the higher electronegativity of the nitrogen and oxygen atoms compared with the carbon atom, leading to deactivation of the molecule toward electrophilic substitution. Heterocyclic compounds with five-membered rings and one heteroatom are readily biodegradable, probably because five-membered ring compounds exhibit higher reactivity toward electrophilic agents and hence are more readily biologically hydroxylated. The susceptibility of heterocyclic compounds to biodegradation decreases with increasing number of heteroatoms in the molecules.

PESTICIDES

Pesticides are the biggest nonpoint source of chemicals added to the environment. The majority of the currently used organic pesticides are subject to extensive mineralization within the time of one growing season or less. Synthetic pesticides show a bewildering variety of chemical structures, but most can be traced to relatively simple aliphatic, alicyclic, and aromatic base structures already discussed. These base structures bear a variety of halogen, amino, nitro, hydroxyl, carboxyl, and phosphorus substituents. For example, the chlorophenoxyacetates, such as 2,4-D and 2,4,5-T, have been released into the environment as herbicides over the past 50 years. Half-life is a term used to express the time it takes for 50% of the compound to be degraded. Generally, five half-lives are believed sufficient for the compound to be completely degraded. Finally, in the fourth set, methoxychlor is more degradable than DDT. In this case, the half-lives are even longer, 1 year for methoxychlor and 15.6 years for DDT.

POLYAROMATIC HYDROCARBONS

PAHs are a class of organic compounds made up of two or more benzene rings fused in either linear, angular, or cluster arrangement. Basically, they contain only C and H atoms, although S, N, and O may substitute some carbon atoms in the aromatic rings to form the so-called heterocyclic subclass of polyaromatics. Examples of PAHs are naphthalene, fluorene, acenaphthene, acenaphthylene, phenanthrene, anthracene, fluoranthene, pyrene, benz[a]anthracene, dibenz[a,h]anthracene chrysene, benzo[a] pyrene, coronene, dibenzofuran, and dibenzothiophene.

TOXICITY, SOURCES OF CONTAMINATION, AND LEGAL DISPOSITIONS

PAHs are formed whenever organic matter is burnt but, as environmental pollutants, mainly issue from the processing, disposal, and combustion of fossil fuels. They are constituents of fractionated oil products such as diesel and jet fuels, petroleum, and lubricating oils. PAH contamination at industrial sites is commonly associated with spills from storage tanks, transport, processing, use and disposal of PAH-containing fuels. Heavily polluted industrial sites are often former gas plants, coking plants, petrochemical industries, and wood-preserving products manufactures.

As an indication, creosote and anthracene oil, which were widely used as wood preservatives, contain up to 85% of PAHs by weight. It is now well established that exposure to PAHs constitutes a risk for people living in industrialized areas. Different carcinogenic, mutagenic, and genotoxic activities were found associated with PAHs, ranging from inactive to highly potent. Most non-substituted PAHs with two or three rings (e.g., naphthalene, fluorene, anthracene, phenanthrene, and pyrene) are not carcinogenic in experimental animals. Benz[a]anthracene and chrysene are weak, whereas benzo[a]pyrene is a strong carcinogenic compound. The carcinogenic effects of PAHs on mammalian cells are a consequence of the metabolic activation to diol epoxides, which are highly reactive molecules that covalently bind to DNA. This activation occurs mainly in the microsomes of the endoplasmic reticulum and is catalyzed by monooxygenase enzymes associated to cytochrome P-450. PAHs were also shown to affect the immune system of mammals.

POLYAROMATIC HYDROCARBONS AS LONG-TERM CONTAMINANTS

The persistence of PAHs in the environment is basically due to their low water solubility. PAHs accidentally introduced in soils rapidly escape from the water phase and become associated with sediments and soil particles to which they strongly sorb. From the microbial point of view, this dual segregation leads to a strongly reduced accessibility – i.e., a low bioavailability – of the PAHs as potential carbon sources. As for genotoxicity, hydrophobicity and environmental persistence increase with the molecular size and the number of fused benzene rings. The age of the contamination is also an important factor since the mean sorption distance of PAHs into soil particles is a time-dependent factor. Consequently, the reversal sorption of PAHs to the more accessible fraction of the soil is delayed in soils with long-term contamination history. Besides natural abiotic removal such as volatilization, hydrolysis, and leaching, it is widely accepted that the major natural process through which PAHs are removed from the environment is microbial degradation. In multiphase soil systems, biodegradation is accomplished by natural surface and subsurface microorganisms that can either mineralize PAHs into CO_2 and H_2O or partially transform them. The biodegradation of two- and three-ring PAHs has been shown to be extensive, whereas that of four and more ring PAHs is considerably less significant.

PAHs-Degrading Microorganisms

A wide range of different microorganisms is able to partially metabolize PAHs. Fungi utilize cytochrome P-450 monooxygenases to form arene oxides that are further transformed to *trans-dihydrodiols* but might also rearrange into phenol derivatives that are carcinogenic. Detoxification occurs via conjugation of the phenols with either sulfate or sugar derivatives. White-rot fungi are ligninolytic organisms that produce extracellular peroxidases with no or little substrate specificity that can convert PAHs to quinone derivatives. A well-studied example is *P. chrysosporium,* whose degrading capacities have been reviewed elsewhere. PAH degradation by algae and cyanobacteria has also been reported.

A large number of bacteria with PAH-degrading capabilities have been reported that were able to either completely assimilate a defined range of compounds, or to exhibit just partial metabolism and did so either as isolated organisms, or as part of consortia made up of several different organisms.

Bacterial PAH Degradation Pathways

Unlike fungal and mammalian cells, bacteria characteristically produce dioxygenases which incorporate two oxygen atoms into the substrate to form dioxetanes that are further oxidized to cis-dihydrodiols and dihydroxy products. However, occasional monooxygenation or angular dioxygenation have been reported, for example in the bacterial metabolism of fluorene and dibenzofuran. Catechol (1,2-dihydroxybenzene), gentisic acid (2,5-dihydroxybenzoic acid), and protocatechuic acid (3,4-dihydroxybenzoic acid) are the most common central intermediates in the bacterial PAH degradation pathways. These compounds are, in turn, metabolized to succinate, pyruvate, fumarate, acetate, and acetaldehyde following different pathways that support protein synthesis and energy production and lead to the release of carbon dioxide and water. The intermediate compound produced depends on the position of the hydroxyl groups (*ortho* or *para*) in the initial diphenolic compounds. A number of different metabolic pathways have been established for the bacterial degradation of PAHs. These pathways are essentially based on the chemical analysis of metabolites. PAHs are microbially degraded either as the sole carbon and energy source or by co-metabolism. As unique growth substrates, PAHs can be partially or completely mineralized, i.e., converted to carbon dioxide and water. As co-metabolic substrates, they cannot support growth but are modified by enzymes produced in the course of the degradation of a growth-supporting, usually structurally related compound. Though little is known about the regulatory events controlling co-metabolism, it has been observed to contribute for a big extent to the degradation of high-molecular weight PAHs and is believed to substantially initiate the removal of those recalcitrant compounds. Anaerobic degradation of PAHs has also been observed.

Naphthalene is probably the simplest PAH. Naphthalene-degrading bacteria were used as illustrative examples to model the general principles of PAH metabolic pathways, enzymatic mechanisms, and genetic regulation and have hence been abundantly studied. The choice of naphthalene degradation as a model was explained by the relative ease of the handling and the genetic manipulation of the naphthalene-degrading bacteria, which are mainly pseudomonads. However, it is more and more accepted nowadays that data retrieved from these studies are far from being representative and in many aspects do not reflect the general characteristics of other PAH degraders. Unlike most PAHs, naphthalene is fairly soluble in water (32 mg/L at 25°C, compared to 1.9 mg/L for fluorene, 1.0 mg/L for phenanthrene, and much less than 1 mg/L for all the other PAHs). In addition, DNA hybridization experiments have demonstrated that except for some phenanthrene degradation systems, no significant homology between PAH degradation genes can be observed with the naphthalene system of *Pseudomonas* when bacteria from different genera are considered. Three-ring PAHs such as acenaphthene, acenaphthylene, fluorene, and phenanthrene can be metabolized by a variety of different bacteria, including Gram-negative and Gram-positive representatives. Anthracene, although identical to phenanthrene in the number of aromatic rings, seems much more difficult to degrade probably as a consequence of its low (0.07 mg/L) solubility in water. By contrast, data concerning the isolation and physiological description of microorganisms able to metabolize more complex PAHs are much scarcer.

Early reports in the 1970s described the microbial oxidation of benzo[a]pyrene and benz[a]anthracene by a *Beijerinckia* sp. that had been chemically mutagenized. Co-metabolic biodegradation of fluoranthene and benzo[a]pyrene was also reported. The first bacterium able to extensively metabolize a PAH with four aromatic rings was isolated in the late 1980s from oil field sediments. It turned out later to be a member of the *Mycobacterium* genus. It was described in 1988 as a Gram-positive rod that co-metabolically degraded a number of PAHs including fluoranthene, pyrene, and benzo[a]pyrene. *Sphingomonas yanoikuyae* strain Bl (formerly described as a *Beijerinckia* sp.) was shown the same year to oxidize dibenz[a]anthracene. In 1989, a bacterial community consisting of seven different strains was isolated from a soil contaminated with creosote that could use fluoranthene as sole source of carbon and energy. In 1990, the first organisms able to utilize a four-ring PAH as sole source of carbon and energy were isolated: *Alcaligenes denitrificans* WWI and *Sphingomonas paucimobilis* EPA505 were shown to grow on a mineral medium supplemented with fluoranthene. These organisms were also shown to co-metabolically transform other PAHs composed of four fused aromatic rings. Additional fluoranthene degraders of the *Sphingomonas* genus were isolated in the last decade. *Mycobacterium* sp. strain PYR-1 was isolated during the same period and displayed a remarkable fluoranthene mineralization rate. This bacterium was also shown to metabolize other PAHs to various degrees with the exception of chrysene. Since then, a number of interesting bacteria able to grow on three- and four-ring PAHs have been isolated. Some relevant examples are *Mycobacterium* sp. KR2 which grows on pyrene and *Mycobacterium* sp. BBl which degrades fluoranthene and pyrene; *Rhodococcus* sp. UW1 which degrades the same PAHs plus chrysene; *Mycobacterium* sp. RJGIJ-135, another pyrene-degrader and *Mycobacterium* sp. CH1 which can grow on either pyrene or phenanthrene.

AROMATIC HYDROCARBONS

Spills of aromatic products, such as aromatic naphtha and leaks from underground fuel tanks, contribute significantly to the contamination of groundwater by aromatic compounds. Non-oxygenated monoaromatic hydrocarbons, such as BTEX, are of particular concern. The high-water solubility of the BTEX species enables them to migrate in the subsurface and contaminate drinking water.

Two thermophilic aerobic bacteria (*Thermus aquaticus* and *Thermits* sp.) have been reported to degrade BTEX fractions co-metabolically. However, only small fractions of benzene and toluene were metabolized to carbon dioxide, and biodegradation was inhibited by higher BTEX concentrations, but was enhanced if strains were pre-grown on catechol and o-cresol, indicating that the pre-conditioning can enhance the performance of microbes.

Two anaerobic consortia, consisting of unidentified bacterial cocci, could grow on all BTEX compounds as sole carbon and energy sources at 45°C–75°C (93°F–167°F), with 50°C (122°F) being the optimal temperature. Only a small fraction of toluene was mineralized to carbon dioxide. Biodegradation was coupled by both consortia to sulfate reduction as well as to generation of hydrogen sulfide. No growth or BTEX metabolism occurred when sulfate was omitted. Thus, sulfate-reducing bacteria are most likely the principal species that carry out the biodegradation, while other thermophilic species may use the early water-soluble BTEX metabolites.

POLYNUCLEAR AROMATIC HYDROCARBONS

Polynuclear aromatic hydrocarbons, in the current context, are persistent organic compounds with two or more aromatic rings in various structural configurations. Polynuclear aromatic hydrocarbons constitute a large and diverse class of organic compounds. However, derivatives, such as tetralin (1,2,3,4-tetrahydronaphthalene) and decalin (decahydronaphthalene, bicyclo[4.4.0]decane), are not included in this group, but are included in the alkane group because of the saturated ring.

The chemical properties, and hence the environmental fate, of polynuclear aromatic hydrocarbons are dependent in part upon molecular size (i.e., the number of aromatic rings and the pattern of ring linkage). Ring linkage patterns (also known as molecular topology) in polynuclear aromatic hydrocarbons may occur such that the tertiary carbon atoms are centers of two or three interlinked rings, as in the linear kata-condensed polynuclear aromatic hydrocarbon anthracene or the peri-condensed polynuclear aromatic hydrocarbon pyrene. However, most polynuclear aromatic hydrocarbons occur as hybrids encompassing various structural components, such as in the polynuclear aromatic hydrocarbon benzo[a]pyrene.

Polynuclear aromatic hydrocarbon molecule stability and hydrophobicity are two primary factors which contribute to their persistence in the environment. Polynuclear aromatic hydrocarbons are present as natural constituents in fossil fuels, are formed during the incomplete combustion of organic material, and are therefore present in relatively high concentrations in products of fossil fuel refining. Polynuclear aromatic hydrocarbons released into the environment may originate from petroleum products including gasoline, diesel fuel, and fuel oil. The concentration of polynuclear aromatic hydrocarbons in the environment varies widely, depending on the proximity of the contaminated site to the production source, the level of industrial development, and the mode(s) of polynuclear aromatic hydrocarbon transport.

The toxic, mutagenic, and carcinogenic properties of polynuclear aromatic hydrocarbons have resulted in some of these compounds (including naphthalene, phenanthrene, and anthracene) to be designated as priority pollutants. In addition, the solubility of polynuclear aromatic hydrocarbons in aqueous media is very low, which affects degradation of these compounds and can lead to biomagnification within an ecosystem. Interest in the biodegradation mechanisms and environmental fate of polycyclic aromatic hydrocarbons (polynuclear aromatic hydrocarbons) is prompted by their ubiquitous distribution and their potentially deleterious effects on human health. The biodegradation of polynuclear aromatic hydrocarbons by microorganisms is the subject of many excellent reviews, and the biodegradation of polynuclear aromatic hydrocarbons composed of three rings is well documented.

Active bioremediation strategies (such as biostimulation) for application to polynuclear aromatic-contaminated soils can be used to supply nutrients, oxygen, and other amendments to the subsurface to enhance indigenous microbial activity and contaminant biodegradation. The benefits of adding oxygen and/or nutrients on the biodegradation of polynuclear aromatics hydrocarbons have been reported for contaminated soils from various sites. However, only a few studies have focused on the direct effects of biostimulation on the indigenous microbial community and polynuclear aromatic hydrocarbon-degrading bacteria.

Due to their lipophilic nature, polynuclear aromatic hydrocarbons have a high potential for bioconcentration. In addition to increases in environmental persistence with increasing polynuclear aromatic hydrocarbon molecular size, evidence suggests that in some cases, polynuclear aromatic hydrocarbon toxicity also increases with size up to at least four or five fused benzene rings. The relationship between polynuclear aromatic hydrocarbon environmental persistence and increasing numbers of benzene rings is consistent with the results of various studies correlating environmental biodegradation rates and polynuclear aromatic hydrocarbon molecule size.

Biodegradation of Hydrocarbons

The contamination of soils and aquifers by spilled petroleum is a persistent and widespread pollution problem that causes ecological disturbances and the associated health implications. Once petroleum is released and comes into contact with water, air, the necessary salts, and microorganisms present in the environment, the natural process of petroleum biodegradation begins.

The recognized mechanical and chemical methods for remediation of hydrocarbon-polluted environment are often expensive, technologically complex, and lack public acceptance. Thus, bioremediation is often the method of choice for effective removal of hydrocarbon pollutants from a variety of ecosystems. In fact, petroleum and petroleum products are a rich source of carbon, and the reaction of the hydrocarbons contained therein with aerial oxygen (with the release of carbon dioxide) is promoted by a variety of microorganisms.

However, the rate of microbial degradation of hydrocarbons in soils is affected by several physicochemical and biological parameters, including the number and species of microorganisms present, the conditions for microbial degradation activity (e.g., presence of nutrient, oxygen, pH, and temperature), the quality, quantity, and bioavailability or bioaccessibility of the contaminants, and the soil characteristics, such as particle size distribution.

Hydrocarbon-degrading bacteria and fungi are mainly responsible for the mineralization (conversion of hydrocarbons to carbon dioxide and water) of petroleum-related pollutants, and are distributed in diverse ecosystem. Furthermore, the population of microorganisms found in a polluted environment will degrade petroleum-related constituents differently and at a different rate than microorganisms in a relatively clean environment. It is uncommon to find organisms that could effectively degrade both aliphatic constituents and aromatic constituents, possibly due to differences in metabolic routes and pathways for the degradation of the two classes of hydrocarbons. There are indications of the existence of bacterial species with propensities for simultaneous degradation of aliphatic hydrocarbons and aromatic hydrocarbons. This rare ability may be as a result of long exposure of the organisms to different hydrocarbon pollutants, resulting in genetic alteration and acquisition of the appropriate degradative genes. It is essential to recognize that the environmental impact of petroleum spills is dependent on previous hydrocarbon exposures and the adaptive status of the local microbiota. The different structural and functional response of microbial sub-groups to different hydrocarbons confirms that the overall response of biota is sensitive to petroleum composition. This suggests that the preferred response to anticipated contaminants may be engineered by pre-exposure to representative substrates. The controlled adaption of microbes to a threatening contaminant is the basis of proactive bioremediation technology, including the augmentation of newly contaminated sites with locally remediated soil in which the biota had already been adapted.

Petroleum and petroleum products are mixtures of differing molecular species hydrocarbons, and the constituents of these molecular categories are present in varied proportions, resulting in high variability in petroleum and petroleum products. In terms of bulk fractions, the resin constituents and the asphaltene constituents are of

particular interest (or notoriety) because these constituents generally resist degradation. After a spill, the constituents of petroleum and petroleum products are subjected
to physical and chemical processes, such as evaporation or photochemical oxidation,
which produce changes in the composition of the spilled material.

ALKANES

Alkanes are major constituents of conventional petroleum and petroleum products,
and can be degraded by indigent or added bacteria. Conventional (light) petroleum
contains 10%–40% w/w normal alkanes, but weathered and heavier oils may have
only a fraction of a percent. Higher molecular weight alkanes constitute 5%–20%
w/w of light oils, and up to 60% w/w of the more viscous oils and tar sand bitumen.
Aromatic hydrocarbons are those characterized by the presence of at least one benzene (or substituted benzene) ring. The low-molecular weight aromatic hydrocarbons
are subject to evaporation, and although toxic to much marine life, are also relatively
easily degraded. Conventional (light) petroleum typically contains between 2% and
20% w/w low-boiling aromatic compounds, whereas heavy oils contain less than 2%
w/w aromatic compounds. As molecular weight and complexity increase, aromatics
are less readily degraded. Thus, the degradation rate of polyaromatics is slower than
that of monoaromatics.

Of these, the normal alkane series (straight-chain alkane series) is the most abundant and the most quickly degraded. Compounds with chains of up to 44 carbon
atoms can be metabolized by microorganisms, but those having 10–24 carbon atoms
(C_{10} to C_{24}) are usually the easiest to metabolize. Shorter chains (up to approximately
C_8) also evaporate relatively easily. Only a few species can use C_1 to C_4 alkanes, and
C_5 to C_9 alkanes are degradable by some microorganisms, but toxic to others.

Branched alkanes are usually more resistant to biodegradation than normal
alkanes, but less resistant than cycloalkanes (naphthenes): those alkanes having carbon atoms in ring-like central structures. Branched alkanes are increasingly resistant to microbial attack as the number of branches increases. At low concentrations,
cycloalkanes may be degraded at moderate rates, but some highly condensed cycloalkanes can persist for long periods after a spill.

Generally, with respect to the molecular composition of the aliphatic constituents of petroleum and petroleum-related products, microbial biodegradation attacks
n-alkanes and isoprenoid alkanes. The polycyclic alkanes of sterane and triterpane
type tend to be somewhat resistant to biodegradation. Since this is the case even for
naphthenic-type petroleum (which is originally depleted in n-alkanes), it has been
concluded that the biodegradation of petroleum type pollutants, under natural conditions, will be restricted to n-alkanes and isoprenoids.

In aqueous systems, addition of acclimatized naturally occurring microorganisms
(bioaugmentation) enhances the biodegradation of hydrocarbons. Since dissolved
hydrocarbons are more available for microbiological degradation, application of dispersants and surfactants increase the bioavailability significantly and enhance oil
degradation. Other factors (such as salinity and pH) have considerable effects on
biodegradation of petroleum hydrocarbons in the marine environments as well. For

example, different concentrations of sodium chloride (0%–5% w/w) exert considerable influence on the biodegradation of petroleum and polynuclear aromatic hydrocarbons from the heavy crude oil-contaminated soil. Not surprisingly, increasing the concentration of sodium chloride in soil has a decreasing effect on petroleum biodegradation and the removal of polynuclear aromatic hydrocarbons. The biodegradation of total crude oil was higher in the absence of sodium chloride (41%), while the reduction in the biodegradation of polynuclear aromatic hydrocarbons was observed in the presence of 1% w/w sodium chloride (35%). A lower reduction of petroleum and polynuclear aromatic hydrocarbons was observed in the presence of 5% w/w sodium chloride (12% and 8% respectively). The reduction of phenanthrene, anthracene, and pyrene reduction was higher in the presence of 1% w/w sodium chloride, while fluoranthene and chrysene reduction were higher in the absence of sodium chloride.

FURTHER READING

Ahn Y, Sanseverino J, Sayler G. Analyses of polycyclic aromatic hydrocarbon degrading bacteria isolated from contaminated soils. *Biodegradation*. 1999;10:149–157.

Alexander M. How toxic are toxic chemicals in soil? *Environ. Sci. Technol.* 1995;29: 2713–2717.

Bhatt P, Kumar MS, Mudliar S, Chakrabarti T. Biodegradation of chlorinated compounds—a review. *Crit. Rev. Environ. Sci. Technol.* 2007;37:165–198.

Britton LN. Microbial degradation of aliphatic hydrocarbons. In: Gibson DT (ed.) *Microbial Degradation of Organic Compounds*. New York, NY: Marcel Dekker Inc.; 1984: 89–129.

Das S. *Microbial Biodegradation and Bioremediation*, 1st ed. Elsevier; 2014.

Fennel DE, Nijenhuis I, Wilson SF, Zinder SH, Haggblom MM. *Dehalococcoides ethenogenes* strain 195 reductively dechlorinates diverse chlorinated aromatic pollutants. *Environ. Sci. Technol.* 2004;38:2075–2081.

Gentry TJ, Rensing C, Pepper IL. New approaches for bioaugmentation as a remediation technology. *Crit. Rev. Environ. Sci. Technol.* 2004;34:447–494.

Ghosal D, You IS, Chatterjee DK, Chakrabarty AM. Plasmids in the degradation of chlorinated aromatic compounds. In: Helinski DR, Cohen SN, Clewell DB, Jackson DA, Hollaender A (eds.) *Plasmids in Bacteria*. New York: Plenum Press; 2015:667–686.

Gilliom RJ, Barbash JE, Crawford CG, et al. *Pesticides in the Nation's Streams and Ground Water, 1992–2001*. U.S. Geological Survey Circular 1291; 2006.

Herman DC, Lenhard RJ, Miller RM. Formation and removal of hydrocarbon residual in porous media: effects of bacterial biomass and biosurfactants. *Environ. Sci. Technol.* 1997;31:1290–1294.

Janssen DB, Oldenhuis R, van den Wijngarrd AJ. Hydrolytic and oxidative degradation of chlorinated aliphatic compounds by aerobic microorganisms. In: Kamely D, Chakrabarty A, Omenn GS (eds.) *Biotechnology and Biodegradation*. Houston, TX: Gulf Publishing Company; 1990.

Johnsen AR, Wick LY, Harms H. Principles of microbial PAH-degradation in soil. *Environ. Pollut.* 2005;133:71–84.

Jutras EM, Smart CM, Rupert R, Pepper IL, Miller RM. Field scale biofiltration of gasoline vapors extracted from beneath a leaking underground storage tank. *Biodegradation*. 1997;8:31–42.

Kenawy ER, Worley SD, Broughton R. The chemistry and applications of antimicrobial polymers: a state-of-the-art review. *Biomacromolecules*. 2007;8:1359–1384.

Leahy JG, Colwell RR. Microbial degradation of hydrocarbons in the environment. *Microbiol. Rev.* 1990;54:305–315.

Maier RM. Bioavailability and its importance to bioremediation. In: Valdes JJ (ed.) *Bioremediation.* The Netherlands: Kluwer Academic Publishers; 2000:59–78.

Maier RM, Gentry TJ. Microorganisms and organic pollutants. In: *Environmental Microbiology*, 3rd ed. Elsevier Inc.; 2014:415–439.

Maier RM, Soberon-Chavez G. *Pseudomonas aeruginosa* rhamnolipids: biosynthesis and potential environmental applications. *Appl. Microbiol. Biotechnol.* 2000;54:625–633.

Martienssen M, Schirmer M. Use of surfactants to improve the biological degradation of petroleum hydrocarbons in a field site study. *Environ. Technol.* 2007;28:573–582.

Miller RM, Herman DH. Biotransformation of organic compounds—remediation and ecotoxicological implications. In: Tarradellas J, Bitton G, Rossel D (eds.) *Soil Ecotoxicology.* Boca Raton, FL: Lewis Publishers; 1997:53–84.

Morkin M, Devlin JF, Barker JF, Butler BJ. In situ sequential treatment of a mixed contaminant plume. *J. Contam. Hydrol.* 2000;45:283–302.

Novak JM, Jayachandran K, Moorman TB, Weber JB. Sorption and binding of organic compounds in soils and their relation to bioavailability. In: Skipper H, Turco RF (eds.) *Bioremediation—Science & Applications*, Soil Science Society of America Special Publication Number 43. Madison, WI: Soil Science Society of America; 1995:13–32.

NRC (National Research Council). *In Situ Bioremediation, When Does it Work?* Washington, DC: National Academy Press; 1993.

Palmisano AC, Schwab BS, Maruscik DA, Ventullo RM. Seasonal changes in mineralization of xenobiotics by stream microbial communities. *Can. J. Microbiol.* 1991;37:939–948.

Pavan M, Worth AP. *Review of QSAR Models for Ready Biodegradation. EUR Scientific and Technical Research Series Report EUR 22355.* EN-DG Joint Research Centre, Institute for Health and Consumer Protection; 2006.

Pepper IL, Gerba CP, Brusseau ML. *Environmental and Pollution Science*, 2nd ed. San Diego, CA: Academic Press;2006.

Pepper IL, Gerba CP, Gentry TJ. *Environmental Microbiology.* 3rd ed. Elsevier Academic Press;2015.

Perry JJ. Microbial metabolism of cyclic alkanes. In: Atlas R (ed.) *Petroleum Microbiology.* New York: Macmillan; 1984:61–97.

Pitter P, Chudoba J. *Biodegradability of Organic Substances in the Aquatic Environment.* Ann Arbor, MI: CRC Press; 1990.

Sikkema J, de Bont JAM, Poolman B. Mechanisms of membrane toxicity of hydrocarbons. *Microbiol. Rev.* 1995;59:201–222.

Snyder SA, Westerhoff P, Yoon Y, Sedlak DL. Pharmaceuticals, personal care products, and endocrine disruptors in water: Implications for the water industry. *Environ. Eng. Sci.* 2003;20:449–469.

Stroo HF, Leeson A, Ward CH (eds.) *SERDP ESTCP Environmental Remediation Technology.* Vol 5, New York: Springer; 2013.

Suttinun O, Luepromchai E, Muller R. Cometabolism of trichloroethylene: concepts, limitations, and available strategies for sustained biodegradation. *Rev. Environ. Sci. Biotechnol.* 2013;12:99–114.

Trudgill PW. Microbial degradation of the alicyclic ring. In: Gibson DT (ed.) *Microbial Degradation of Organic Compounds.* New York: Marcel Dekker, Inc.; 1984:131–180.

Tunkel J, Mayo K, Austin C, Hickerson A, Howard P. Practical considerations on the use of predictive models for regulatory purposes. *Environ. Sci. Technol.* 2005;39:2188–2199.

Ulrich AC, Guigard SE, Foght JM, et al. Effect of salt on aerobic biodegradation of petroleum hydrocarbons in contaminated groundwater. *Biodegradation.* 2009;20:27–38.

USEPA (U.S. Environmental Protection Agency). *Use of Bioremediation at Superfund Sites.* Washington, DC: EPA 542-R01-019; 2001.

Uyttebroek M, Vermeir S, Wattiau P, Ryngaert A, Springael D. Characterization of cultures enriched from acidic polycyclic aromatic hydrocarbon-contaminated soil for growth on pyrene at low pH. *Appl. Environ. Microbiol.* 2007;73:3159–3164.

van der Meer JR. Environmental pollution promotes selection of microbial degradation pathways. *Front. Ecol. Environ.* 2006;4:35–42.

Wang X, Bartha R. Effects of bioremediation on residues, activity and toxicity in soil contaminated by fuel spills. *Soil Biol. Biochem.* 1990;22:501–505.

Widdel F, Rabus R. Anaerobic biodegradation of saturated and aromatic hydrocarbons. *Curr. Opin. Biotechnol.* 2001;12:259–276.

10 Biodegradation of Petroleum and Petroleum Products

BIODEGRADATION OF PETROLEUM

The biodegradability of any petroleum constituent is a measure of the ability of that constituent to be metabolized (or co-metabolized) by bacteria or other microorganisms through a series of biological process, which include ingestion by organisms as well as microbial degradation. The chemical characteristics of the contaminants influence biodegradability; in addition, the location and distribution of petroleum contamination in the subsurface can significantly influence the likelihood of success for bioremediation.

The biodegradability of petroleum and petroleum product is inherently influenced by the composition of the substrate upon which the bacteria are acting. For example, petroleum is quantitatively biodegradable, and kerosene, which consists almost exclusively of medium-chain length alkanes, is completely biodegradable under suitable conditions, but for heavy asphaltic crudes, approximately only 6%–10% of the material oil may be biodegradable within a reasonable time period, even when the conditions are favorable for biodegradation (Bartha, 1986; Okoh et al., 2002; Okoh, 2003, 2006). In addition, biodegradation of petroleum-related constituents can be enhanced by use of a consortium of different bacteria compared to the activity of single bacterium species (Ghazali et al., 2004; Milic et al., 2009).

In addition to the composition of the petroleum-related substrate, petroleum and petroleum products introduced to the environment are immediately subject to a variety of changes caused by physical, chemical, and biological effects, usually referred to collectively as weathering. Physical and chemical processes include (1) evaporation, (2) dissolution of petroleum constituents in a water system (or aquifer), (3) dispersion, (4) photochemical oxidation, (5) formation of water-oil emulsions, and (6) adsorption onto suspended particulate material. These processes are not sequential, and typically occur simultaneously and cause important changes in the composition and properties of the original pollutant, which in turn may affect the rate or effectiveness of biodegradation.

Specifically, the biodegradation of petroleum typically (1) raises the viscosity and decreases the API gravity, which adversely reduces the ability of the degraded product to flow, (2) decreases the hydrocarbon content, thereby increasing the residuum content, (3) increases the concentration of certain metals, (4) increases the sulfur content, and (5) increases oil acidity and adds compounds, such as carboxylic acids and phenols. All of these changes are seen in the product relative to the unchanged (non-biodegraded) petroleum.

DOI: 10.1201/9781003272618-12

The commercial practice of bioremediation focuses primarily on the cleaning up of petroleum hydrocarbons (Del'Arco and de Franqa, 1999). Thus, successful application of bioremediation technology to a contaminated ecosystem requires knowledge of the characteristics of the site and the parameters that affect the microbial biodegradation of pollutants (Sabate et al., 2004).

Despite the difficulty of degrading certain fractions, some hydrocarbons are among the most easily biodegradable naturally occurring compounds. Biodegradation gradually destroys petroleum-related spills by the sequential metabolism of various classes of compounds present in the oil (Bence et al., 1996). When biodegradation occurs in an oil reservoir, the process dramatically affects the fluid properties, and hence, the value and producibility of an oil accumulation. Specifically, petroleum biodegradation typically raises viscosity of the residual material (which reduces oil producibility) and reduces the API gravity (which reduces the value of the produced oil). It increases the asphaltene content (relative to the saturated and aromatic hydrocarbon content and the starting material), the concentration of certain metals, the sulfur content, and oil acidity.

There are indications that petroleum biodegradation involves more biological components than just the microorganisms that directly attack petroleum constituents (the primary degraders) and shows that the primary degraders interact with these components (Head et al., 2006). In addition, primary degraders need to compete with other microorganisms for limiting nutrients, and the non-petroleum-degrading microorganisms can be affected by metabolites and other compounds that are released by oil-degrading bacteria and vice versa. The environment, having been exposed to petroleum inputs for thousands of years, can assimilate the hydrocarbons under the proper conditions. However, areas of particular concern are low energy environments common to estuarine systems. These environments, such as marshes, mud flats, and subtidal areas, are vital to marine fisheries and estuarine productivity, and are especially sensitive to contaminant impacts. These systems are particularly vulnerable to impacts of petroleum where research has shown petroleum can persist in these systems for years. The removal processes for petroleum in wetlands are (1) evaporation, (2) photooxidation, (3) dissolution of specific constituents, (4) microbial degradation, and (5) physical flushing. However, once incorporated into the sediment, biodegradation and dissolution are the primary removal mechanisms. Petroleum biodegradation in wetland environments can be limited by anoxia and nutrient availability. Consequently, estuarine wetlands are the most vulnerable of the low-energy intertidal areas to petroleum spills.

EFFECTS OF BIODEGRADATION

During biodegradation, the properties of the petroleum fluid change because different classes of compounds in petroleum have different susceptibilities to biodegradation. The early stages of biodegradation (in addition to any evaporation effects) are characterized by the loss of n-paraffins (n-alkanes or branched alkanes) followed by loss of acyclic isoprenoids (e.g., norpristane, pristane, and phytane). Compared with those compound groups, other compound classes (such as highly branched and cyclic saturated hydrocarbons, as well as aromatic compounds) are more resistant to biodegradation. However, even the more resistant compound classes are eventually destroyed as biodegradation proceeds.

CONDITIONS FOR BIODEGRADATION

The composition and inherent biodegradability of the petroleum hydrocarbon pollutant are perhaps the first and most important considerations when the suitability of a cleanup approach is to be evaluated. Heavier crude oil is generally much more difficult to biodegrade than lighter ones, just as heavier crude oils could be suitable for inducing increased selection pressure for the isolation of petroleum hydrocarbon degraders with enhanced efficiency. Also, the amount of heavy crude oil metabolized by some bacterial species increases with increasing concentration of the contaminant, while degradation rates may appear to be more pronounced within a specific concentration range (Okoh et al., 2002; Rahman et al., 2002).

Important aspects of the conditions for biodegradation at a spill site are the activity of microorganisms and the ability of the organisms to produce enzymes to catalyze metabolic reactions, which is governed by the genetic composition of the organism(s). Enzymes produced by microorganisms in the presence of carbon sources cause initial attack on the hydrocarbon constituents, while other enzymes are utilized to complete the breakdown of the hydrocarbon. Thus, lack of an appropriate enzyme either prevents attack or is a barrier to complete hydrocarbon degradation.

Biodegradation of petroleum-related constituents by bacteria can occur under both aerobic (oxic) and anaerobic (anoxic) conditions (Zengler et al., 1999), usually by the action of different consortia of microorganisms. In the subsurface, biodegradation occurs primarily under anaerobic conditions, mediated by sulfate-reducing bacteria in cases where dissolved sulfate is present or methanogenic bacteria in cases where dissolved sulfate is low. Although subsurface oil biodegradation does not require oxygen, there is a requirement for the presence of essential nutrients (such as nitrogen, phosphorus, and potassium), which can be provided by dissolution/alteration of minerals in the water layer. In the absence of nutrients, the potential for hydrocarbon degradation in anoxic sediments is markedly reduced.

In situ groundwater can be an effective medium for biodegradation of petroleum hydrocarbons. While there are some notable exceptions (such as MTBE, which is not a hydrocarbon), the short-chain, low molecular weight, more water-soluble constituents are degraded more rapidly, and to lower residual levels, than are long-chain, high molecular weight, less soluble constituents. However, as with all bioremediation efforts, petroleum and petroleum products (such as residual fuel oil and asphalt) typically have a residual high-boiling fraction composed of resin and asphaltene constituents, which is composed of complex, polynuclear aromatic systems (Speight, 2007).

Microbial utilization of hydrocarbons (being fully reduced substrates) requires an exogenous electron sink. In the initial attack, this electron sink has to be molecular oxygen. In the subsequent steps too, oxygen is the most common electron sink. In the absence of molecular oxygen, further biodegradation of partially oxygenated intermediates may be supported by nitrate or sulfate reduction.

Uptake and utilization of water-insoluble substrates, such as petroleum alkanes, require specific physiological adaptations of the microorganisms. The synthesis of specific amphiphilic molecules (i.e., biosurfactants) is often surmised to be a prerequisite for either specific adhesion mechanisms to large oil drops or emulsification of oil, followed by uptake of submicron oil droplets. In fact, various species of bacteria have been observed to adopt the requisite strategy to deal with water-insoluble substrates, such as hydrocarbons. Hence, to facilitate hydrocarbon uptake through the hydrophilic outer membrane, many hydrocarbon-utilizing microorganisms produce cell wall associated or extracellular surface active agents. This includes low-molecular weight compounds, such as fatty acids, triacyl-glycerol derivatives, and phospholipids, as well as the heavier glycolipids, examples of which include emulsan and liposan.

Studies with bacteriophages, antibodies, and emulsan-deficient mutants have demonstrated that: (1) as the cells approach stationary phase, emulsan accumulates on the cell surface before release into the medium (Goldman et al., 1982); (2) cell-bound emulsan serves as a specific receptor and acts as stabilizer for the oil-water interface (Pines and Gutnick, 1981); (3) this indicates that the cell-bound form of emulsan is required for growth on crude oil: species without cell-bound emulsan no longer grow well on petroleum-related substrate (Pines and Gutnick, 1984); and (4) the affinity of emulsan for the oil-water interface suggests that it might affect microbial degradation of emulsified oils (Gutnick and Minas, 1987)

FACTORS AFFECTING BIODEGRADATION OF PETROLEUM

Effect of Nutrients

Different types of nutrients (primarily nitrogen and phosphorus) have been applied to improve petroleum hydrocarbon degradation, including classic (water soluble) nutrients and oleophilic and slow-release fertilizers.

Bioavailability is one main factor that influences the extent of biodegradation of hydrocarbons. Generally, hydrocarbons have low-to-poor solubility in water, and, as a result, are adsorbed on to clay or humus fractions, so they pass very slowly to the aqueous phase, where they are metabolized by microorganisms. Cyclodextrins are natural compounds that form soluble inclusion complexes with hydrophobic molecules and increase degradation rate of hydrocarbons in vitro. In the perspective of an in situ application, β-cyclodextrin does not increase eluviation (the lateral or downward movement of the suspended material in soil through the percolation of water) of hydrocarbons through the soil, and, consequently, does not increase the risk of groundwater pollution. Furthermore, the combination of bioaugmentation and enhanced bioavailability due to β-cyclodextrin was effective for a full degradation. Thus, in situ bioremediation of polynuclear aromatic hydrocarbon-polluted soil can be improved by the augmentation of degrading microbial populations and by the increase of hydrocarbon bioavailability (Bardi et al., 2007).

Inadequate mineral nutrients, especially nitrogen and phosphorus, often limit the growth of hydrocarbon utilizers in water and soils. Iron has been reported to be limiting only in clean, offshore seawater. Sulfur, in the form of sulfate ions, is plentiful in seawater, but could be limiting in some freshwater environments. The slight alkaline pH of seawater seems to be quite favorable for petroleum hydrocarbon degradation, but in acidic soils limiting to pH 7.8–8.0 had a definite stimulatory effect. Nutrients are very important ingredients for successful biodegradation of hydrocarbon pollutants, especially nitrogen, phosphorus, and in some cases, iron. Depending on the nature of the impacted environment, some of these nutrients could become limiting, thus affecting the biodegradation processes. When a major oil spill occurs in freshwater and/or marine ecosystems, the supply of carbon is dramatically increased, and the availability of nitrogen and phosphorus generally becomes the limiting factor for oil degradation. This is more pronounced in marine environments, due to the low background levels of nitrogen and phosphorus in seawater, unlike in freshwater systems that regularly fluctuate in nutrient status as result of perturbations and receipt of industrial and domestic effluents and agricultural runoff. Freshwater wetlands are typically considered to be nutrient limited, due to heavy demand for nutrients by the plants, which can be considered to be nutrient traps, since a substantial amount of nutrients is often found in the indigenous biomass (Mitsch and Gosselink, 1993).

Furthermore, it is not surprising that the chemical form of those nutrients is also important, the soluble forms (i.e., iron or nitrogen in the form of phosphate, nitrate, and ammonium) being the most frequent and efficient, due to their higher availability for microorganisms. Depending on the microbial community and their abundance, another factor that may improve polynuclear aromatic hydrocarbon degradation is the addition of readily assimilated carbon sources, such as glucose (Zaidi and Imam, 1999).

EFFECT OF TEMPERATURE

Temperature plays an important role in the biodegradation of petroleum-related hydrocarbons, not only because of the direct effect on the chemistry of the pollutants but also because of the effect on the physiology and diversity of the microbial surroundings. In short, temperature can play the role of increasing a microbial reaction or inhibiting a microbial reaction in a similar manner to the general rules for the influence of temperature on chemical reactions.

Typically, biodegradation of petroleum and petroleum products occurs at temperatures less than 80°C. At higher temperatures (unless the microbes are of a specific thermophilic type), many of the microorganisms involved in subsurface oil biodegradation cannot exist. The ambient temperature of an environment affects both the properties of spilled petroleum or petroleum product and the activity or population of microorganisms. At low temperatures, the viscosity of the oil increases, while the volatility of toxic low-molecular weight hydrocarbons is reduced, delaying the onset of biodegradation. Temperature also variously affects the solubility of hydrocarbons (Foght et al., 1996).

Although hydrocarbon biodegradation can occur over a wide range of temperatures, the rate of biodegradation generally decreases with decreasing temperature. Highest degradation rates generally occur in the range of 30°C–40°C (86°F–104°F) in soil environments, 20°C–30°C (68°F–86°F) in some freshwater environments, and 15°C–20°C (59°F–68°F) in marine environments. In fact, the biodegradability of petroleum is highly dependent not only on composition but also on microbial incubation temperature (Atlas, 1975): at 20°C (68°F), conventional petroleum has higher abiotic losses and is more susceptible to biodegradation than heavy oil. As expected from petroleum chemistry and composition, the rate of mineralization for the heavy oil is significantly lower at 20°C (68°F) than for conventional oil.

During biodegradation, some preference is shown for removal of the paraffin constituents over the aromatic and asphaltic constituents, especially at low temperatures. Branched paraffins, such as pristane, are degraded at both 10°C and 20°C (50°F and 68°F). This was confirmed by showing that the residual material (after an incubation period of 42 days) had a lower relative percentage of paraffins and higher percentage of asphaltic constituents (usually resin and asphaltene constituents) than fresh or weathered oil.

Effect of Dispersants

It has been suggested that dispersants tend to increase oil biodegradation by increasing the surface area for microbial attack and encouraging migration of the droplets through the water column making oxygen and nutrients more readily available. However, dispersants can have a detrimental (toxic) effect on microbial processes, thereby retarding the rate of petroleum degradation. It would appear that the dual capability of dispersants (increasing the surface area of dispersed oil and affecting the growth of hydrocarbon degraders) is related to the chemistry of the dispersant, which influences the effectiveness of dispersants for bioremediation.

There are indications that the use of surfactants in situations of petroleum-related contamination may have a stimulatory, inhibitory, or neutral effect on the bacterial degradation of the petroleum constituents. It is clear that the introduction of external non-ionic surfactants (the main components of oil spill dispersants) will influence the alkane degradation rate.

Rates of Oil Biodegradation

Different results for microbial activity measurements may be obtained in laboratory studies, depending on pretreatment and size of the sample, even when the environmental conditions are mimicked (Bjorklof et al., 2008). These differences may be related to, among other factors, differences in the bioavailability of the contaminant in different analyses.

The most important critical stage of petroleum degradation during the first 48 hours of a spill is usually evaporation, the process by which the lower molecular weight (lower-boiling constituents) constituents of petroleum volatilize into the atmosphere. Evaporation can be responsible for the loss of one- to two-thirds of the mass of spilled material (assuming that the spilled material is conventional crude oil or a distillate product) during this period, with the loss rate decreasing rapidly over time. The constituents of heavy oil, tar sand bitumen, and asphalt do not evaporate to the same extent, the lower boiling constituents being generally absent from these materials. Evaporative loss is controlled by the composition and physical properties of the petroleum or petroleum derivative, the surface of the spill, wind velocity, and temperature.

Weathered petroleum (i.e., petroleum and petroleum-related products) that has been exposed to air and oxidized and to other influences, such as evaporation, offers a different challenge to bioremediation efforts.

Oxygen is often the limiting factor in aerobic bioremediation at many sites. The degradation of petroleum hydrocarbons occurs much faster under aerobic conditions compared to anaerobic conditions. Therefore, the addition of oxygen can significantly increase the remediation rates. Oxygen addition is most frequently used to address dissolved phase contamination, such as total petroleum hydrocarbons and BTEX, as well as contamination in the capillary fringe zone. Oxygen can only be effective if the hydrocarbons are bioavailable and there is no nutrient limitation.

BIODEGRADATION OF NAPHTHA AND GASOLINE

Naphtha and gasoline spills are common, and can cause water contamination issues as hydrocarbons dissolve in groundwater and travel offsite in the aquifer. Hydrocarbons naturally degrade in the subsurface due to microbial-mediated reactions. However, the reaction rates are slow, because electron acceptors, like oxygen, are quickly depleted in contaminated ground water and are slowly recharged. Contaminated groundwater has significant hydrocarbon concentrations, but depleted electron acceptors, whereas the overlying unsaturated zone contains oxygen, but lower hydrocarbon concentrations. Early response and intervention are the keys to minimizing extent and costs of remedial action for gasoline and its components, and are essential to protecting public health and the environment.

The term naphtha refers to any one of several low-boiling flammable liquid mixtures of hydrocarbons that is a distillation product from petroleum; naphtha is also produced by the destructive distillation of coal and/or oil shale.

Gasoline is not a direct product of petroleum refining, but is a composite of several product streams and additives (Speight, 2007). Thus, the gasoline produced by a refinery is a blend of all the appropriate available streams, including, but not limited to (1) straight-run naphtha, (2) naphtha, from the fluid catalytic cracking unit, (3) hydrocracked naphtha, and (4) reformate.

The aromatic hydrocarbons in naphtha and gasoline are generally more toxic than aliphatic compounds with a similar number of carbon atoms, and have more mobility in water due to their solubility being 3–5 times higher. By virtue of its properties, gasoline contains a high proportion of volatile organic compounds.

REMEDIATION

When attempting to clean up naphtha or gasoline spills, timely and comprehensive source control and associated hydrogeological investigations are needed once a release is detected. These include (1) immediate control and cessation of the release, (2) repair or removal of the release source, such as a tank or pipe, (3) removal/recovery of free product in both the saturated/unsaturated zones, and (4) removal of residual free product from the subsurface soils. Any remedial action initiated before the source is controlled is ineffective, and has the potential of expanding the scope of the remedial action as uncontrolled sources continue to migrate in the subsurface.

SITE REMEDIATION

Two major objectives of site remediation include destruction of residual or dissolved naphtha or gasoline constituents or their removal from the impacted area. Destruction can range from total mineralization/oxidation or reduction to inorganic components or transformation to some unlisted form.

For in situ treatment, recent demonstrations have confirmed the effectiveness of both chemical and biological oxidation processes for the destruction of gasoline. Refinements in the formulation of chemical oxidants (e.g., stabilized hydrogen peroxide, chelated catalysts) for in situ treatments provide greater control, extended longevity, and more thorough treatment than earlier, uncontrolled formulations. This, together with uniform delivery of the oxidant using an approach called deep remediation injection systems (DRIS), has avoided many of the bounce-back problems observed with well-point injection programs. For in situ bioremediation, providing higher concentrations of oxygen than available from simple aeration accelerates biological processes. Oxygen concentrators, pure oxygen sources, and oxygen release compounds now deliver abundant electron acceptors to the biologically active zone.

For ex situ destruction, ventilation blowers or pumping systems transfer gasoline constituents directly from the subsurface to the surface. Indirect phase-transfer processes may require several steps to move the gasoline constituents to the surface, where they can subsequently be removed from the air or water matrix before discharge.

In the subsurface ventilation evacuation method (vacuum or pressurization), gas exchange is increased in the subsurface and is an effective method for moving vapor-phase gasoline hydrocarbons to the surface for destruction. Another form of the subsurface ventilation evacuation method is performed in situ in the unsaturated (or vadose) zone. This type of soil-remediation technology, in which a vacuum is applied to the soil, pulls gas-phase volatiles and some semi-volatile contaminants from the soil through extraction wells to the surface. The subsurface ventilation evacuation method is also known as in situ soil venting, in situ volatilization, enhanced volatilization, or soil vacuum extraction. The gas leaving the soil may be treated to recover or destroy the contaminants, depending on local and state air discharge regulations.

Vertical extraction vents are typically used at depths of 1.5 m (5 ft) or greater, and they have been successfully applied as deep as 91 m (300 ft). Horizontal extraction vents (installed in trenches or horizontal borings) can be used as warranted by contaminant-zone geometry, drill-rig access, or other site-specific factors.

Bioventing is a process of stimulating the natural, in situ biodegradation of contaminants in soil by providing air or oxygen to existing soil microorganisms. Bioventing uses low air-flow rates to provide only enough oxygen to sustain microbial activity in the vadose zone. Oxygen is most commonly supplied through direct air injection into residual contamination in soil.

In addition to degradation of adsorbed fuel residuals, volatile compounds move slowly through biologically active soil or shallow sediments where the vapors are biodegraded. The primary mechanism of volatile removal by the subsurface ventilation evacuation method is mass transfer. However, circulating air in the vadose zone stimulates microbial activity to degrade volatile and some semi-volatile compounds directly in the soil or sediment matrix. The major difference between the subsurface ventilation evacuation method and the bioventing method is the volume of air moved through the subsurface. Low-flow ventilation and gas exchange favor biological activity, while higher flows and extraction favor volatile removal. Bioventing can reduce vapor-treatment costs and can also result in the remediation of semi-volatiles that cannot be removed by direct volatilization alone. Bioventing uses the same blowers as systems used in the subsurface ventilation evacuation method to provide specific distribution and flux of air through the contaminated vadose zone to stimulate the indigenous microorganism to degrade hydrocarbons. The thermal enhancement and direct heating of subsurface soils and sediments by radio frequency heating or by resistance heating with electrode pairs or indirect heating through the injection of steam or hot air through the injection wells enhance the subsurface ventilation evacuation method performance dramatically, if the system is designed and operated properly.

Heated air or steam helps to "loosen" some fewer volatile compounds from the soil or sediments in a process similar to steam distillation. When sufficient power is available, direct heating of soils with electrode pairs has been very effective for the focused remediation of small areas at several sites.

Air sparging injection of compressed air at controlled pressures and volumes into the water-saturated soils or sediments remediates the sediments and groundwater by three processes:

1. In situ air stripping (ISAS) of dissolved volatile organic compounds;
2. volatilization of trapped- and adsorbed-phase contamination present below the water table and in the capillary fringe; and
3. aerobic biodegradation of both dissolved and adsorbed phase contaminants.

Stripping and volatilization are the dominant removal mechanisms, with biodegradation becoming more significant over the long term. Air sparging is a low-cost technique applicable to a wide range of contaminant concentrations, and it is flexible enough to accommodate diverse geological and hydrogeological limitations. For air sparging to be successful, the soil or sediment in the saturated zone must have sufficient permeability to allow the injected air to readily escape up into the unsaturated zone. Air sparging, therefore, will work fastest at sites with coarse-grained soil or sediment, like sand and gravel.

In Situ Air Stripping

In situ air stripping (ISAS) combines three technologies, air sparging, horizontal wells, and soil vapor extraction. ISAS uses horizontal wells to inject (sparge) air into the groundwater. The horizontal wells provide more effective access to a horizontal groundwater plume. As the air comes into contact with contaminants, they volatize and rise through the soil sediment. The volatile organic compounds are then extracted from overlying soils by standard soil vapor extraction. The air sparging process eliminates the need for surface groundwater treatment systems, such as air strippers.

The subsurface ventilation method and air sparging (replacing the air with the ozone) in the subsurface lead to the in situ destruction of many gasoline constituents. Ozone, the third strongest direct oxidant after fluorine and hydroxyl radical, can oxidize organic contaminants by direct oxidation, in addition to the oxidation by free hydroxyl radical. In situ destruction of volatile organic compounds reduces surface treatment costs and often pretreats adsorbed semi-volatile compounds to improve their susceptibility to biodegradation. Subsequent oxygen release also stimulates in situ biodegradation.

Biosparging, like its sister technique bioventing, uses lower airflow ($0.5-3\,ft^3$/ minute per injection point) than needed for air stripping and volatilization. The objective is to provide enough oxygen and gas exchange to drive in situ aerobic biodegradation in the saturated zone without stripping the hydrocarbons.

Pump-and-Treat Systems

Pump-and-treat systems operate by pumping groundwater containing dissolved or non-aqueous phase liquid hydrocarbon to the surface. Free-phase hydrocarbons are separated out, dissolved constituents are removed, and the water is either reinjected or discharged to a surface water body or municipal sewage plant. Contaminant recovery is limited by the behavior of the contaminant in the subsurface (primarily solubility and the adsorption/partition coefficients), site geology and hydrogeology, and the extraction system design. It is further complicated by residual contamination in the saturated zone and adsorbed contamination in the capillary fringe, as well as insoluble non-aqueous phase liquids (NAPLS) floating on the water table.

Pump-and-treat systems are also used to contain contaminated groundwater, provide hydraulic control, and recover contaminant mass in either gas or liquid phase by creating a capture zone around the pumping well. The natural hydrogeological property of the site and the rate at which ground water is extracted limit the capture zone of the recovery well. As refinements to the pump-and-treat system, pulse pumping/ reverse flow pumping addresses low-permeability formation, channeling, capillary fringe, cost, and the delivery of nutrients to stimulate in situ bioremediation.

Vacuum-Enhanced Recovery

The application of a vacuum to the extraction point (vacuum-enhanced recovery, VER) provides a method to further enhance the capture zone. A high-vacuum or negative pressure applied to a recovery well and to the formation enhances liquid recovery of the well by increasing the net effective drawdown. VER increases the mass removal of the volatile and semi-volatile contaminants by maximizing dewatering and facilitating volatilization from previously saturated sediments via increased air movement. Physical removal of significant hydrocarbon mass increases subsurface oxygen levels for aerobic biodegradation of residual contaminant. VER is cost effective to enhance the overall recovery of contaminants, especially under low-permeability conditions.

SINGLE-PHASE VACUUM EXTRACTION

In the single-phase vacuum extraction (SPVE) method, a single pump removes fluid, via a drop tube, and applies a vacuum to the well and formation. The well produces both liquid- and vapor-phase material. Compared to pumping alone, single-phase vacuum extraction increases the capture zone, and therefore, reduces the number of recovery wells needed, and accelerates the recovery of both liquid and residual contaminants. Although limited to depths of less than 25 ft, this technique is one of few enhancements for mass removal from low-permeability sites.

DUAL-PHASE VACUUM EXTRACTION

Similar in its advantages to the single-phase system, the dual-phase vacuum extraction (also called the multiphase-phase vacuum extraction) uses a submersible pump for liquid recovery at greater depths (>25 ft), and a diaphragm or liquid-ring pump for evacuation of the formation. A single vacuum pump can be used for multiple wells under a variety of design strategies. When coupled with surfactants, a pump-and-treat system can be used as an in situ soil washing system.

SURFACTANT-ENHANCED AQUIFER REDEMPTION

Surfactant-enhanced aquifer redemption techniques have been particularly effective at removing NAPLS and the residual material (from NAPLS) from highly permeable subsurface systems. Surfactant selection must be paired with the target constituents and must exhibit some short-term resistance to biodegradation by the indigenous microflora.

IN SITU-ENHANCED BIOREMEDIATION

Bioremediation uses indigenous or introduced (augmented) microorganisms (primarily bacteria) to degrade organic contaminants into harmless substances, biomass, or carbon dioxide and water. The kinetics of biodegradation are limited by substrate availability, electron donors, nutrients, pH, moisture content, other carbon and energy sources, and other factors beyond the scope of this summary of techniques employed in remediation, often involving some form of bioremediation.

The end products of the biodegradation of petroleum hydrocarbon degradation are carbon dioxide and water, which are also measures of microbial respiration and activity in soils. Higher rates of biodegradation by soil microbial activity as shown from the carbon dioxide production occurred during the first 4 weeks; the increase in carbon dioxide production was progressive initially, but started decreasing after week 4. This is also collaborated by the rates of biodegradation, which was highest in the first 4 weeks (Obire and Nwaubeta, 2001).

The amount of carbon dioxide produced in the control and in the hydrocarbon-contaminated soil is in the following order:

Gasoline > kerosene > diesel oil.

Respiration of microbes occurred very rapidly during the initial period of incubation, when the lighter and more readily degraded fractions were degraded but slowed down as the residue became more difficult to degrade on account of the increase of the heavier fractions. Carbon dioxide production, which is a measure of microbial activity and respiration, occurred in all soils. This confirmed the fact that the microorganisms metabolized the petroleum hydrocarbons, which resulted in a progressive increase in carbon dioxide evolved, which decreased later, following exhaustion of metabolizable fractions.

The indigenous microbial flora of the soil showed hydrocarbon degradation potentials, as was revealed by total hydrocarbon data obtained and carbon dioxide production in the soils. There was significant difference between control soil and gasoline-contaminated soil at both 5% and 1% level of probability.

While it is known that n-alkanes become more readily depleted by biodegradation than branched and cyclic alkanes, details are limited about the relative susceptibilities of branched alkanes, alkylcyclohexane derivatives, alkyl cyclopentane derivatives, and aromatic hydrocarbons to biodegradation and about how the accompanying carbon isotope distribution is affected.

Activated sludge represents a natural consortium of broadly diverse active bacteria, which are capable of degrading organic waste compounds. Encapsulated activated sludge beads have been applied to the biodegradation of a number of environmental pollutants, including 2,4,6-trichlorophenol in groundwater, phenol in wastewater, and mixtures of acetaldehyde and propionaldehyde in waste gas. Non-encapsulated activated sludge cultures have been previously used for the biodegradation of gasoline hydrocarbons.

Gasoline is a major groundwater contaminant as the result of release from leaking underground storage tanks, accidental spills, leaking process equipment, and other industrial activities (Day et al., 2001). The use of encapsulated sludge microbeads for the biodegradation of gasoline hydrocarbons may be feasible as suspension in a pump-and-treat surface bioreactor system (ex situ bioremediation) or as slurry, filling a subsurface permeable reactive barrier (in situ bioremediation). Alternatively, the encapsulated sludge microbeads may have the potential to be used for the start-up of bioreactors or the maintenance of high biodegradation activities in bioreactors (bioaugmentation).

BIODEGRADATION OF KEROSENE AND DIESEL

Kerosene (kerosine), also called paraffin or paraffin oil, is a flammable, pale yellow or colorless, oily liquid with a characteristic odor. Also called paraffin oil, it should not be confused with the much more viscous paraffin oil used as a laxative: use of this type of paraffin oil was used by miners to lubricate the alimentary tract that would contain coal dust residues inhaled during working hours in the mine. Chemically, kerosene is a mixture of hydrocarbons; the chemical composition depends on its source, but it usually consists of about ten different hydrocarbons, each containing 10–16 carbon atoms per molecule; the constituents include n-dodecane ($C_{12}H_{26}$) and alkyl benzenes, as well as naphthalene and its derivatives.

Diesel fuel is a complex fuel mixture primarily consisting of paraffin, olefin, and aromatic hydrocarbons and smaller quantities of substances containing sulfur, nitrogen, metals, and oxygen. The hydrocarbon molecules contain from 8 to 40 atoms of carbon, and are generally heavier than those found in gasoline.

Diesel oil leakages from underground storage tanks, distribution facilities, and various industrial operations represent an important source of soil and aquifer contamination. This fuel is a complex mixture of normal, branched and cyclic alkanes, and aromatic compounds obtained from the middle-distillate fraction during petroleum separation (Speight, 2007).

Diesel oil consists mostly of linear and branched alkanes with different chain lengths, and it contains a variety of aromatic compounds. Many of these compounds, especially linear alkanes, are known to be easily biodegradable. However, due to their low water solubility, the biodegradation of these compounds is often limited by slow rates of dissolution, desorption, or transport. In general, the bioavailability of hydrophobic compounds is determined by their sorption characteristics and dissolution or partitioning rates, and by transport process to microbial cell (Sticher et al., 1997).

Kerosene and diesel fuel are middle distillate petroleum hydrocarbon products of intermediate volatility and mobility. As intermediate products, kerosene has a combination of (mostly) lighter, less persistent, and more mobile compounds, as well as (some) heavier, more persistent, and less mobile compounds. These two different groups are associated with two distinctly different patterns of fate/pathway concerns.

The relatively lighter, more volatile, mobile, and water-soluble compounds in kerosene will tend to fairly quickly evaporate into the atmosphere or migrate to groundwater. When exposed to oxygen and sunlight, most of these compounds will tend to break down relatively quickly. In groundwater, many of these compounds tend to be more persistent than in surface water, and readily partition on an equilibria basis back and forth between water and solids (soil and sediment) media. Cleaning up groundwater without cleaning up soil contamination will usually result in a rebound of higher concentrations of these compounds partitioning from contaminated soils into groundwater.

BIOREMEDIATION

Generally, sources of kerosene pollution include accidental spills, pipe leakage, deliberate disposal of oily wastes, corrosion of pipes, kerosene seeps, and other operational deficiencies. Apart from the esthetic and economic damage caused by kerosene spills, plant, vertebrates, invertebrates, and microorganisms in both the terrestrial and aquatic environments are adversely affected.

In general, kerosene/jet fuel components biodegrade significantly under aerobic conditions, provided sufficient nutrients are present for conversion of the hydrocarbons to microbial biomass. There is a complex interplay between partitioning, and thus bioavailability, and biodegradation in the various media in the environment. Lower molecular weight, normal hydrocarbons are most readily biodegraded, but tend to partition to air rather than water, while more complex, higher molecular weight polynuclear aromatics and substituted aromatics tend to sorb to soil or sediment; both processes limit bioavailability and can slow biodegradation. The hydrocarbons in kerosene are generally not inhibitory to microbial activity, though changes in microbial community composition may occur in spill or impacted areas, due to the proliferation of species that can biodegrade the compounds.

DIESEL FUEL

Kerosene-range and diesel-range hydrocarbons are degraded primarily by bacteria and fungi, and adaptation by prior exposure of microbial communities to hydrocarbons increases hydrocarbon degradation rates.

Diesel oil, which is one of the major products of crude oil, constitutes a major source of environmental pollution. With the combined dependence on diesel oil by some vehicles and generators, greater quantities are being transported over long distances. Therefore, diesel oil can enter into the environment through wrecks of oil tankers carrying diesel oil, cleaning of diesel tanks by merchants, and war ships carrying diesel oil. Diesel oil spills on agricultural land generally reduce plant growth. Suggested reasons for the reduced plant growth in diesel oil contaminated soils range from direct toxic effect on plants and reduced germination to unsatisfactory soil condition, due to insufficient aeration of the soil because of the displacement of air from the space between the soil particles by diesel oil.

The biodegradation rate of hydrocarbons is an important consideration in determining the time scale for bioremediation in oil-contaminated environments. Two naturally occurring bacterial cultures, *Exiguobacterium aurantiacum* and *Burkholderia cepacia*, were capable of utilizing diesel oil as the sole source of carbon and energy by induction of hydrophobic cell surfaces (Mohanty and Mukerji, 2008). The cultures demonstrated good degradation characteristics for diesel range n-alkanes (C_9 to C_{26}), and were also capable of degrading pristane. Biodegradation altered the relative abundance of n-alkanes in diesel and resulted in a loss of symmetry in n-alkane distribution: C_9, C_{17} to C_{19}, and C_{26} were completely degraded by both of the cultures. In *B. cepacia*, the residual diesel was enriched in the higher carbon number n-alkanes C_{20} to C_{25}.

Bioremediation of diesel oil in soil can occur by natural attenuation, or treated by biostimulation or bioaugmentation. In this study, all three technologies were evaluated for the degradation of total petroleum hydrocarbons in soil.

BIODEGRADATION OF FUEL OIL

Fuel oil is a fraction obtained from petroleum distillation, either as a distillate or a residue. The term fuel oil is also used in a stricter sense to refer only to the heaviest commercial fuel that can be obtained from crude oil, but it has higher boiling point and higher density than naphtha and gasoline.

Upon release into the environment, heavy fuel oil will break into small masses and will not spread as rapidly as less viscous oil (Mehrasbi et al., 2003). The density of some heavy fuel oils means that they may sink on release to water, rather than float on the surface like other petroleum fuels. Loss of the lower molecular weight components due to volatility and dissolution will increase the density of the floating oil causing it to sink. This heavy fraction will assume a tar-like consistency and stick to exposed substrates, or become adsorbed to particulates. Weather conditions and temperature during the period after the spill will significantly influence the rate of dispersion; wind and wave action will tend to disperse oil into the water column, while higher temperatures will increase the rate of evaporation of lighter hydrocarbons. Water temperature is a major factor in determining the extent of the environmental impact following a heavy fuel oil spill, since higher temperatures will enhance loss of lighter constituents by evaporation as well as degradation processes.

For fuel oil-contaminated soil, landfarming was more effective than bioventing during the initial phase of treatment and this effect was more pronounced when nutrients were added. The lag period that occurs prior to rapid microbial growth and increased activity appeared to have a greater effect in the heavy-oil-contaminated soil, compared to the diesel contaminated soil. It is possible that the measured increase in the concentration of total petroleum hydrocarbons is an analytical artifact of microbial fatty acids and related microbial biomass that are produced during adaptation of the soil microbiota to new conditions and carbon sources.

BIODEGRADATION OF LUBRICATING OIL

The economic importance of crude oil is to be found in the numerous possible products obtainable from crude oil through refining: the uses of these products justify the resources committed to its exploration and production. Motor fuels, domestic fuels, industrial fuels for heating and power generation, and lubricants are among the products derivable from crude oil. Of particular interest to this is lubricating oil, which may also be related (but not always in the true chemical sense, because of the additives that are included in lubricating oil) to atmospheric and vacuum gas oil.

The uses of various kinds of lubricating oils (also called mineral oils) in various industrial situations have made them an indispensable ingredient of the industrialization and development that has characterized the past century. The class of petroleum products known as mineral oils can be generally understood to include a variety of products that go by different names, such as white oils, lubricating oils, light fuel oils, residual fuel oils, as well as transformer and cable oils (Gary and Handwerk, 2001; Speight, 2007).

The materials that fall into the lubricating oil category are complex petroleum mixtures, composed primarily of saturated hydrocarbons with carbon numbers ranging from C_{15} to C_{50} (some ranges give C_{15} to C_{40}). At ambient temperatures, lubricating base oils are liquids of varying viscosities with negligible vapor pressures. Base oils are produced by first distilling crude oil at atmospheric pressure to remove lighter components (e.g., gasoline and distillate fuel components), leaving a residue (residuum) that contains base oil precursors. This atmospheric residuum is then distilled under vacuum to yield a range of distillate fractions (unrefined distillate base oils) and a vacuum residuum. Removal of the asphalt components of the vacuum residuum results in unrefined residual base oils. These distillate and residual base oil fractions may then undergo a series of extractive or transforming processes that improve the base oils' performance characteristics and reduce or eliminate undesirable components.

Lubricating oils are generally recognized to cause pollution of soil and water when spilled accidentally or when disposed affecting plant and animal life. They cannot be regarded as readily biodegradable, and hence, their harmful effects often persist in the environment. In fact, petroleum-based lubricating oils, greases, and hydraulic fluids are usually toxic and not readily biodegradable; because of these characteristics, if these materials escape to the environment, the impacts tend to be cumulative and consequently harmful to plants, fish, and wildlife.

In spite of the large number of works on petroleum, petroleum derivatives, and hydrocarbon biodegradation, little information is available on petroleum lubricant oils biodegradation by microbial inoculation (Nocentini et al., 2000). In practice, an effective bacterial inoculum should be able to tolerate high levels of petroleum lubricant while maintaining a level of activity to provide efficient biodegradation. Understanding the physiological and biochemical properties of lubricant oil-degrading bacteria is required before optimum use of bacteria in environmental applications. Intensive investigation of microbial degradation of petroleum lubricants is thus imperative, and of important practical significance.

BIODEGRADATION OF RESIDUA AND ASPHALT

The residuum or asphaltic fraction of petroleum is the fraction that is either (1) the residue remaining after the completion of atmospheric distillation or (2) the residue remaining after the completion of vacuum distillation. On the other hand, asphalt is the product produced from residua that is used, among other uses, for road and highway surfacing (Speight, 2007).

The main hazard associated with asphalt is from the polynuclear aromatic hydrocarbon derivatives that can move into the ecosystem from the breakdown of asphalt. Since asphalt contains so many toxic and carcinogenic compounds, and since leaching of harmful polynuclear aromatic hydrocarbon derivatives has been documented even in water pipe use, asphalt should be kept out of rivers, streams, and other natural waters to the extent possible.

Bioremediation of petroleum residua and asphalt has become increasingly important due to the damage caused by the spills of such materials. To lay a solid foundation of bioremediation processes, it is a necessary (as with all petroleum products) that there is an understanding to the composition and properties of residua and asphalt. The biodegradation of residua requires a complex metabolic pathway, which usually can be observed in a microbial community. Many studies have been carried out on the biodegradation of petroleum hydrocarbons using a consortium of microorganisms. In literature, it is revealed that in soils that are permeated with the heavy hydrocarbons, bacteria, including indigenous ones, would survive and function after contaminations seeped through the soil. Selection of bacterial communities for petroleum substances occurs rapidly after even short-term exposures of soil to petroleum hydrocarbons. During adaptation of microbial communities to hydrocarbon components, particularly complex ones, genes for hydrocarbon-degrading enzymes that are carried on plasmids or transposons may be exchanged between species, and new catabolic pathways eventually may be assembled and modified for efficient regulation. Other cell adaptations leading to new ecotypes may include modifications of the cell envelope to tolerate solvents and development of community level interactions that facilitate cooperation within consortia.

Biodegradation of asphalt occurs mostly on the surface of the asphalt exposed to oxygen, where the primary nutrients (asphalt and other trace minerals) are all present in adequate levels. Surface degradation is limited by oxygen diffusion and diffusion or presence of needed minerals/ions (i.e., phosphorous and nitrogen). As time progresses, degradation should diminish as all of these become limited. In addition, the mechanism for biodegradation may be such that a residue that is resistant to biodegradation remains on the surface, protecting the underlying asphalt from further biodegradation. As the saturated and naphthene aromatic hydrocarbons are emulsified, a residue remains on the surface that is resistant to biodegradation and protects the underlying asphalt from biodegradation. The most potent asphalt-degrading bacterium, *Acinetobacter calcoaceticus*, excretes an emulsifier that is capable of emulsifying the saturated and naphthene aromatic fractions of asphalt cement-20.

FURTHER READING

Agarry SE, Owabor CN, Yusuf RO. Studies on biodegradation of kerosene in soil under different bioremediation strategies. *Bioremed. J.* 2010;14(3):135–141.

Atlas RM. Effects of temperature and crude oil composition on petroleum biodegradation. *Appl. Microbiol.* 1975;30(3):396–403.

Atlas RM. Microbial degradation of petroleum hydrocarbons: an environmental perspective. *Microbiol. Rev.* 1981;45:180–209.

Bardi L, Martini C, Opsi E, Bertolone E, Belviso S, Masoero G, Marzona M, Ajmone Marsan F. Cyclodextrin-enhanced in situ bioremediation of polyaromatic hydrocarbons-contaminated soils and plant uptake. *Incl. Phenom. Macrocycl. Chem.* 2007;57:439–444.

Bartha R. *Microbial Ecology: Fundamentals and Applications.* Reading, MA: Addison-Wesley Publishers; 1986.

Bence AE, Kvenvolden KA, Kennicutt MC. Organic geochemistry applied to environmental assessments of Prince William sound, Alaska, after the Exxon Valdez oil spill – a review. *Organic Geochem.* 1996;24:7–42.

Bento FM, Camargo FAO, Okeke BC, Frankenberger WT. Comparative bioremediation of soils contaminated with diesel oil by natural attenuation, biostimulation and bioaugmentation. *Bioresource Technol.* 96:1049–1055.

Bjorklof K, Salminen J, Sainio P, Jorgensen K. Degradation rates of aged petroleum hydrocarbons are likely to be mass transfer dependent in the field. *Environ. Geochem. Health.* 2008:30:101–107.

Day MJ, Reinke RF, Thomson JAM. Fate and transport of fuel components below slightly leaking underground storage tanks. *Environ. Foren.* 2001;2:21–28.

Del'Arco JP and de Franga FP. Biodegradation of crude oil in a sandy sediment. *Int. Biodeter. Biodegrad.* 44:87–92.

Foght JM, Westlake D, Johnson WM, Ridgway HF. Environmental gasoline-utilizing isolates and clinical isolates of pseudomonas aeruginosa are taxonomically indistinguishable by chemotaxonomic and molecular techniques. *Microbiology.* 1996;142:1333–1338.

Gary JH, Handwerk GE. *Petroleum Refining Technology and Economics*, 4th ed. New York: Marcel Dekker Inc.; 2001.

Ghazali FM, Rahman RNZA, Salleh AB, Basri M. Biodegradation of hydrocarbons in soil by microbial consortium. *Ml. Biodeter. Biodegrad.* 2004;54:61–67.

Goldman S, Shabtai Y, Rubinovitz C, Rosenberg E, Gutnick DL. Emulsan in Acinetobacter calcoaceticus RAG-1: distribution of cell-free and cell-associated cross-reacting material. *Appl. Environ. Microbiol.* 1982;44:165–170.

Gutnick DL, Minas W. Perspectives on microbial surfactants. *Biochem. Soc. Trans.* 1987;15:22S–35S.

Head IM, Jones DM, Roling WEM. Marine microorganisms make a meal of oil. *Nat. Rev. Microbiol.* 4:173–182.

Mehrasbi MR, Haghighi, B, Shariat, M, Naseri, S, Naddafi, K. Biodegradation of petroleum hydrocarbons in soil. *Iranian J. Publ. Health.* 2003;32(3):28–32.

Milic JS, Beskoski VP, Ilic MV, Ali SAM, Gojgic-Cvijovic GD, Vrvic MM. 2009. Bioremediation of soil heavily contaminated with crude oil and its products: composition of the microbial consortium. *Serb. Chem. Soc.* 74(4):455–460.

Mitsch WJ, Gosselink JG. *Wetlands.* 2nd ed. New York: John Wiley & Sons Inc.; 1993.

Mohanty G, Mukherji S. Biodegradation rate of diesel range n-alkanes by bacterial cultures *Exiguobacterium aurantiacum* and *Burkholderia cepacia. Int. Biodeter. Biodegrad.* 2008;61:240–250.

Nocentini M, Pinelli D, Fava F. Bioremediation of a soil contaminated by hydrocarbon mixtures: the residual concentration problem. *Chemosphere.* 2000;41:1115–1123.

Obire O, Nwaubeta O. 2001. Biodegradation of refined petroleum hydrocarbons in soil. *J. Appl. Sci. Environ. Manag.* 5(1):43–46.

Okoh AI. Biodegradation of bonny light crude oil in soil microcosm by some bacterial strains isolated from crude oil flow stations saver pits in Nigeria. *Afr. J. Biotech.* 2003;2(5):104–108.

Okoh AI. Biodegradation alternative in the cleanup of petroleum hydrocarbon pollutants. *Biotech. Mol. Biol. Rev.* 2006;1(2):38–50.

Okoh AI, Ajisebutu S, Babalola CO, Trejo-Hernandez MR. Biodegradation of Mexican heavy crude oil (Maya) by *Pseudomonas aeruginosa. J. Trop. Biosci.* 2002;2(1):12–24.

Pines O, Gutnick DL. Relationship between phage resistance and emulsan production. Interaction of phages with the cell-surface of Acinetobacter alcoaceticus RAG-1. *Arch. Microbiol.* 1981;130:129–133.

Pines O, Gutnick DL. Alternate hydrophobic sites on the cell surface of Acinetobacter calcoaceticuls RAG-1. *FEMS Microb. Lett.* 1984;22:307–311.

Rahman KSM, Thahira-Rahman J, Lakshmanaperumalsamy P, Banat IM. 2002. Towards efficient crude oil degradation by a mixed bacterial consortium. *Biores. Tech.* 85:257–261.

Sabate J, Vinas M, Solanas AM. Laboratory-scale bioremediation experiments on hydrocarbon-contaminated soils. *Intern. Biodeter. Biodegrad.* 2004;54:19–25.

Solano-Serena F, Marchal R, Huet T, Lebeault JM, Vandecasteele JP. Biodegradability of volatile hydrocarbons of gasoline. *Appl. Microbiol. Biotechnol.* 2000;54:121–125.

Speight JG. *The Chemistry and Technology of Petroleum*, 4th ed. Boca Raton, FL: CRC Press, Taylor & Francis Group; 2007a.

Speight JG. *The Chemistry and Technology of Petroleum*, 5th ed. Boca Raton, FL: CRC Press, Taylor & Francis Group; 2007b.

Speight JG, Arjoon KK. *Bioremediation of Petroleum and Petroleum Products.* Salem, MA, USA : Scrivener Publishing; 2012.

Speight JG, Lee S. *Environmental Technology Handbook.* 2nd ed. Boca Raton, FL: CRC Press, Taylor & Francis Group; 2000.

Sticher P, Jaspers MC, Stemmler K, Harms H, Zehnder AJ, van der Meer JR. Development and characterization of a wholecell bioluminescent sensor for bioavailable middle-chain alkanes in contaminated groundwater samples. *Appl. Environ. Microbiol.* 1997;63:4053–4060.

Zaidi BR, Imam SH. Factors affecting microbial degradation of polycyclic aromatic hydrocarbon phenanthrene in Caribbean coastal water. *Mar. Pollut. Bullet.* 1999;38:738–774.

Zengler K, Richnow HH, Rossello-Mora R, Michaelis W, Widdel F. Methane formation from long-chain alkanes by anaerobic microorganisms. *Nature.* 1999;401:266–269.

11 Microbial Degradation of Chlorinated Aromatic Compounds

Chlorinated aromatic compounds represent an important class of environmental pollutants. Most of these compounds can be biologically degraded aerobically and/or anaerobically. Besides being subject to co-metabolic turnover, a process which is unproductive for microorganisms, because it is not coupled to energy conservation and biomass production, chloroaromatics can serve for growth. While a variety of microbial populations that mediate dechlorination of chloroaromatics under anaerobic conditions have been reported, only few studies on dechlorination with pure cultures of anaerobic bacteria are known. These organisms use chlorinated compounds such as chlorobenzoates, chlorophenols, or chlorohydroxyphenylacetate as an electron acceptor in a so-called dehalorespiration process leading to dechlorination. In contrast, a larger number of aerobic bacteria are known which use chloroaromatics as sole carbon and energy source. Dechlorination reactions are essential for mineralization which finally results in the formation of chloride, carbon dioxide, and biomass. Aerobic organisms growing with chloroaromatics can be differentiated on the basis of the catabolic pathways used for removal of the chlorine substituents. Chlorine substituents can be cleaved off prior to ring cleavage by oxygenolytic, reductive or hydrolytic reactions. Further conversion of the chlorine-free metabolites can then occur via classical pathways for the metabolism of aromatic compounds such as the 3-oxoadipate, the protocatechuate, or the meta-cleavage pathway. However, the majority of the organisms that are able to mineralize chlorinated aromatics do not possess enzymes capable of initial dechlorination prior to ring cleavage. They use oxygenases to convert chlorinated aromatics to chlorocatechols, and dichlorination takes place after ring cleavage. The enzymatic reactions leading to the formation of chlorocatechols are similar to those used for the degradation of non-chlorinated aromatic compounds.

DEGRADATION OF CHLOROCATECHOLS VIA THE MODIFIED ORTHO-CLEAVAGE PATHWAY

It is well established that in most organisms that degrade chloroaromatics, the chlorocatechol degradative pathway starts with *ortho-cleavage* by chlorocatechol 1,2-dioxygenases which introduce molecular oxygen and produce the corresponding chloro-cis, cis-muconate. The pathways used for the degradation of 3-chloro-, 4-chloro-, and 3,5-dichlorocatechol have mostly been studied with the 3-chlorobenzoate-degrading *Pseudomonas* sp. strain B13 and the 2,4-dichlorophen oxyacetate-degrading *Ralstonia eutropha* strain JMP134. Elimination of a chlorine substituent occurs spontaneously when 3-chloro- and 2,4-dichloro-cis,cis-muconate are subsequently converted by chloromuconate cycloisomerases to 4-chloro- and 2,4-dichloromuconolactone respectively. The dienelactones are formed by *anti*-elimination of hydrogen chloride and formation of an exocyclic double bond (Schmidt and Knackmuss 1980). For example, cis-dienelactone results from 3-chloro-cis,cis-muconate. 2,4-Dichloro-cis,cis-muconate, the intermediate in the 3,5-dichlorocatechol degradation, is converted to 2-chlorodienelactone probably in the cis-configuration. Only recently, it was realized that the dechlorination of 2-chloro-cis,cis-muconate, the intermediate in the degradation of 3-chlorocatechol, is catalyzed by the chloromuconate cycloisomerase instead of being a spontaneous step. Both cycloisomerization, leading to (+)5-chloromuconolactone, and dechlorination are carried out by the same enzyme to give *trans-dienelactone* (Vollmer and Schlomann 1995; Vollmer et at. 1994). The dienelactones are further converted to the respective maleylacetates by dienelactone hydrolases (Figure 11.1).

An additional dechlorination takes place with 2-chloromaleylacetate, the intermediate in the degradation of 3,5-dichlorocatechol, by a reductive reaction. The

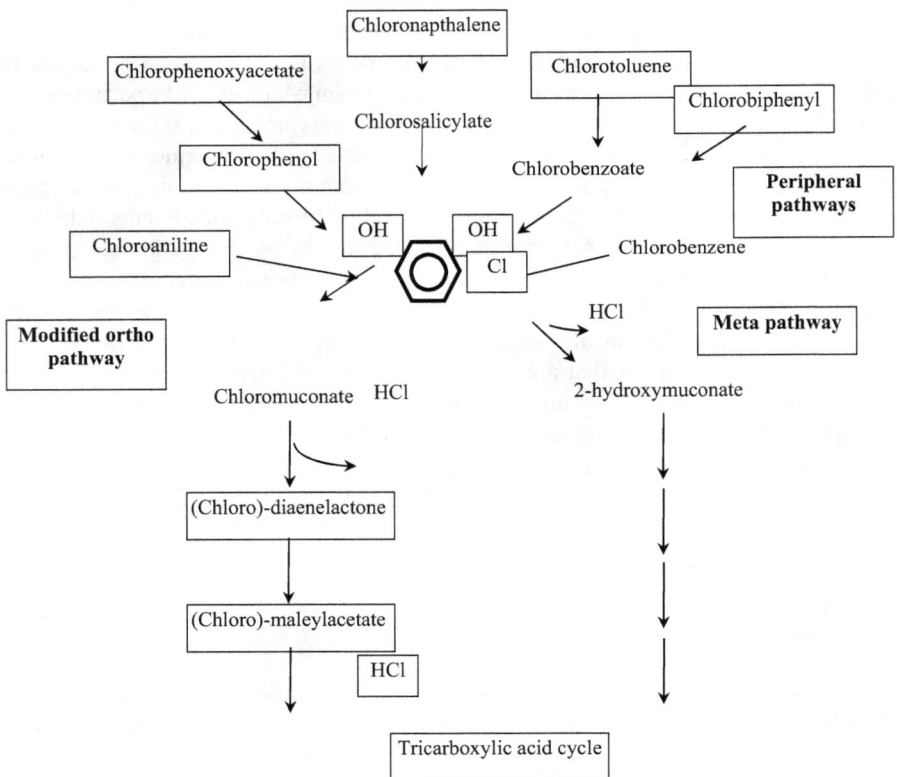

FIGURE 11.1 Schematic presentation of the mineralization of chloroaromatics with chlorocatechols as key metabolites.

elimination is thought to occur spontaneously during the maleylacetate reductase reaction. The modified *ortho-cleavage* pathway of *Pseudomonas chlororaphis* strain RW71 tolerates substitution at the aromatic ring up to four chlorine atoms in the degradation of 1,2,3,4-tetrachlorobenzene via tetrachlorocatechol (Potrawtke et al. 1998).

DEGRADATION OF CHLOROCATECHOLS VIA THE META-CLEAVAGE PATHWAY

Problems with the *Meta-Cleavage* Pathway in the Case of Chloroaromatics

It was assumed for a long time that organisms cannot use the *meta-cleavage* pathway for the degradation of chlorinated aromatic compounds. The reasons for this statement were the following:

- Studies on the degradation of chlorocatechols with strains harboring both the modified *ortho-* and *meta-cleavage* pathway indicated that channeling of chlorocatechols into the *meta-cleavage* pathway was unproductive or even counterproductive. The cells exclusively used the modified *ortho-cleavage* pathway for the degradation of 3-chloro-, 4-chloro-, and 3,5-dichlorocatechol (Reineke et al. 1982).
- 3-Chlorocatechol has often been observed to act as a strong inhibitor of extradiol dioxygenases. The inactivation of catechol 2,3-dioxygenase of *Pseudomonas putida* PaW1 (XylE) by 3-chlorocatechol is an irreversible process that requires oxygen. It was proposed that the enzyme cleaves 3-chlorocatechol to an acyl chloride that acts as a suicide product by reacting with nucleophilic groups in the enzyme. The proposed acyl chloride, which has not yet been identified, can also react with water, yielding 2-hydroxymuconate. Small amounts of this product were identified during the conversion of 3-chlorocatechol by the catechol 2,3-dioxygenase of strain PaW1. In contrast, the catechol 2,3-dioxygenase of *P. putida* strain F1 was inactivated by 3-chlorocatechol in a reversible way. This was explained by chelation of the ferrous iron cofactor of the enzyme by 3-chlorocatechol. The absence of oxygen uptake by whole cells when confronted with 3-chlorocatechol might suggest that the catechol 2,3-dioxygenase from strain F1 does not convert any 3-chlorocatechol. The absence of a suicide inactivation mechanism was also indicated by experiments with 14C-labeled 3-chlorocatechol. Labeling was found at various sites when the catechol 2,3-dioxygenase from strain F1 was proteolytically separated into peptides after the incubation with 14C-labeled 3-chlorocatechol. With a suicide inactivation mechanism, labeling at a single site would be expected. Preliminary experiments with the catechol 2,3-dioxygenase from strain PaW1 showed that the homodimeric protein band of native enzyme migrated slightly faster on SDS-polyacrylamide gels than that of enzyme that was inactivated with 3-chlorocatechol, provided that the samples were not boiled prior to application on the gel (Mars et al., 1999). Modification of the catechol 2,3-dioxygenase by 3-chlorocatechol also made the enzyme

less susceptible to digestion with proteinase Lyse. Complete digestion of dialyzed, inactivated XylE was only possible in the presence of 1 M urea, while the native enzyme could be digested in the absence of urea. These results indicate that inactivation by 3-chlorocatechol involves a labile covalent modification which influences the multimeric form of XylE.

- In contrast to the situation of catechol 2,3-dioxygenases, when exposed to 3-chlorocatechol, various extradiol dioxygenases can convert catechols that are chlorinated at the 4-position (Moon et al., 1996). However, growth on substrates via 4-chlorinated catechols and a *meta-cleavage* pathway was usually absent.

Productive Use of the *Meta-Cleavage* Pathway for Chlorinated Catechols Substituted in *Para-Position*

In recent years, some bacterial strains were described that utilize chlorinated compounds like 4-chlorobenzoate, 4-chlorobiphenyl, 3-chloro-2-methylbenzoate, and 4-chlorophenol through a *meta-cleavage* pathway (Seo et al., 1997). The chlorinated substrates are converted to 4-chlorinated catechols, which are cleaved by catechol 2,3-dioxygenases. The chlorine substituents are removed after cleavage of the aromatic ring, although it is unknown where the dechlorination occurs.

Productive Degradation of 3-Chlorocatechol via the *Meta-Cleavage* Pathway

An exception to the general rule that the *meta-cleavage* pathway is insufficient for the degradation of 3-chlorocatechol and that productive simultaneous degradation of methyl- and chloroaromatics is not possible by natural isolates was reported about 10 years ago. *P. putida* strain GJ31, which was obtained by conventional batch enrichment, i.e., without prolonged adaptation time, from sediment samples taken from the river Rhine using chlorobenzene as sole carbon source, was described to utilize toluene and chlorobenzene simultaneously (Oldenhuis et al., 1989). More recently, it was found that *P. putida* GB1 rapidly degrades chlorobenzene via 3-chlorocatechol. However, instead of using the modified *ortho-cleavage* pathway as do other chlorobenzene degraders described in the literature, strain GJ31 uses a *meta-cleavage* pathway. The pathway is initiated by proximal cleavage between C-2 and C-3.

In contrast to other catechol 2,3-dioxygenases, which are subject to inactivation, the 2,3-dioxygenase of strain GB1 productively converts 3-chlorocatechol. Stoichiometric displacement of chlorine subsequently leads to the production of 2-hydroxymuconate; because of that property, the enzyme is termed chlorocatechol 2,3-dioxygenase. Since 2-hydroxymuconate is a common metabolite in the degradation of catechol through the *meta-cleavage* pathway, further degradation runs accordingly via 4-oxalocrotonate, 2-oxopent-4-enoate, 4-hydroxy-2-oxovalerate, and finally, pyruvate and acetaldehyde.

POLYCHLORINATED BIPHENYLS (PCBs)

Polychlorinated biphenyls (PCBs) are a family of compounds produced commercially by direct chlorination of biphenyls. The molecule of biphenyls is made up of two connected rings of six carbon atoms, and it contains one or more chlorine atom attached to the biphenyl nucleus. Carbon has a high selectivity for those PCBs which can assume a planar conformation; these PCBs contain no ortho-chlorines. As the degree of ortho-substitution increases (up to four chlorines in 0,0' positions), the retention time of the compound decreases. Thus, a PCB with four ortho-chlorines would elute from a carbon column before other PCBs. This selectivity is useful for fractionation of PCBs. In the toxicity of PCBs, the degree of ortho-substitution affects the toxicity. Therefore, isolation and characterization of different fraction of the congeners from commercial mixtures has been of interest. Although the non-ortho PCBs are often described as "the coplanar congeners," all PCBs regardless of substitution pattern are twisted.

The PCB molecule consists of two phenyl molecules joined together with two or more hydrogen atoms replaced by chlorine atoms. PCB comprises a group of 209 structurally different congeners with the empirical formula $C_{12}H_{10-n}CL_n$. The congeners of PCBs are created by replacing the chlorine atoms at various corners of the carbon ring.

Today, PCBs can be found in all environmental compartments from the bottom of the oceans to the aerial Polar Regions. PCBs are found in air, water, soil, and food. Majority of the PCBs in the environment find its way during their manufacturing, usage as well as during disposal. This can be in the form of spillages and leakages during production, transportation, and other exposure units. Other sources of PCB emission include treatments, storage, disposal facilities, and landfills; hazardous waste sites; steel and iron reclamation facilities like auto scrap burning as well as an accidental release of PCB to the atmosphere. They spread into the environment from dumps, landfills, combustion process, and from their use in various open and close systems. PCBs are lipophilic and are enriched in adipose tissues of predators, mainly through consumption of contaminated food. They have also been found to cause a multitude of toxic responses in wildlife and humans.

Recorded effects of its toxicity include dermal toxicity, immune toxicity, reproductive effects and tera toxicity, endocrine disruption, and carcinogenicity. The first step in toxicity mechanism is mediated by the binding of PCB to the Aryl hydrocarbon (Ah) cellular receptor. Toxicity of PCBs is said to range from low-moderate levels. Absorption of PCBs by human and animals is through the skin, the lungs, and the gastrointestinal tract. Once inside the body, PCBs are transported through the bloodstream to liver and to various muscles and adipose tissue where they accumulate.

Biological Transformation of PCBs

The ability of PCBs to be degraded or be transformed in the environment depends on the degree of chlorination of the biphenyl molecule as well as isomeric substitution pattern. At present, employing the biochemical abilities of microorganisms is the most popular strategy for the biological treatment of contaminated soils. The relatively recent development of bioremediation has added to existing cleanup strategies currently available for the restoration and rehabilitation of contaminated sites and can be conducted either in situ or ex situ. This biological strategy is dependent on the catabolic activities of the indigenous microflora, optimizing the conditions in situ for growth and biodegradation.

Biodegradation is done in two ways: mineralization and co-metabolism. Mineralization is a process whereby the organic pollutant is used as a source of carbon and energy by the organism resulting in the reduction of the pollutant to its constituent elements. Co-metabolism on the other hand requires a second substance as its source of carbon and energy for the microorganisms but the target pollutant is transformed at the same time. When the products of co-metabolism are ready for further degradation, they can be mineralized; otherwise, incomplete degradation occurs. This may then result in the formation and accumulation of metabolites that are more toxic than the present molecule requiring a consortium of microorganisms, which can utilize the new substance as source of nutrients. The effectiveness of biodegradation depends on many environmental factors. Rates vary depending on the conditions present in the environment. These factors include the structure of the compound, the presence of exotic substituent and their position in the molecule, solubility of the compound, and concentration of the pollutant. In the case of aromatic halogenated compounds, a high degree of halogenations requires high energy by the microorganisms to break the stable carbon–hydrogen bonds. Chlorine also acts as the substituent that alters the resonant properties of the aromatic substance as well as the electron density of specific sites. This may result in deactivation of the primary oxidation of the compound by microorganisms. There are also stereochemical effects on the affinity between enzymes and their substrate molecules on the positions occupied by substituent chlorines.

Degradation of PCB by Microorganisms

Recalcitrance of PCBs to biodegradation by microbes was as a result of its chemical stability. Just as higher chlorine constitution increases chemical stability and lowers water solubility, it makes higher chlorinated congeners more resistant to remediation. Metabolism of PCBs is usually unfavorable energetically, thus requiring additional source of carbon to aid its co-metabolism. PCBs are regarded as persistent organic pollutants (POPs), however; its degradation by microbes has been well reported. There are two known metabolic pathways of microbes in PCB: aerobic and anaerobic, and these depend on the degree of chlorination of the congener, the types of microbes involved as well as the redox conditions.

Anaerobic PCB Dechlorination

PCB congeners that contain four or more chlorine substituents undergo anaerobic reductive dechlorination. This is an energy yielding process in which PCBs serve as the electron acceptor for the oxidation of organic substrates. Anaerobic bacteria possess characteristics that are suited for high carbon-concentration pollutants because of the limitation in oxygen diffusion in a high concentration. A predominant anaerobic environment is conducive for the reductive transformation resulting in the displacement of chlorine by hydrogen. The dechlorinated compound is suitable for the oxidative attack of the aerobic bacteria. Aerobic bacteria grow faster than anaerobes and can sustain high degradation rate resulting in mineralization of the compound. Theoretically, the biological degradation of PCBs should give carbon dioxide, chlorine, and water. This process involves the removal of chlorine from the biphenyl ring followed by cleavage and oxidation of the resulting compound. Transformation of chlorinated organic compounds anaerobically involves reductive dehalogenation where the halogenated organic compounds serve as the electron acceptor; the halogen substituent is replaced with hydrogen. Here, chlorine atoms are preferentially taken out from the meta- and para-positions on the biphenyl structure, thereby leaving lower chlorinated ortho-substituted congeners.

The current paradigm is that aerobic biodegradation is generally limited to PCB mixtures having an average mass percentage of 42% chlorine (e.g., Aroclor 1242). Although metabolism of higher chlorinated congeners (e.g., hexachlorobiphenyls) has been demonstrated in cultures grown with biphenyl, the process has, so far, not been implemented in soils and sediments contaminated by Aroclors 1254 (54% Cl) or 1260 (60% Cl). These more highly chlorinated congeners, nevertheless, undergo slow reductive dehalogenation in flooded soils and sediments to less chlorinated congeners, which would be susceptible to aerobic biodegradation. Coupled anaerobic and aerobic metabolism in sediments is probably how biodegradation of highly chlorinated congeners occurs in sediments. Studies have shown that the addition of brominated analogs or ferrous sulfate enhances dehalogenation of PCBs.

Dechlorination in the absence of oxygen can attack a large array of chlorinated aliphatic and aromatic hydrocarbons. Several bacteria involved in this reaction have been isolated; they include *Desulfomanile tiedjel*, *Disulfiro bacterium*, *Dehalobacter restrictus*, *Dehalococcoides ethenogenes* and the facultative anaerobes *Enterobacter* strain MS1 and *Enterobacter agglomerans*. Others are *Dehalospirillum multivorans* and *Desulfuromanas chloroethenica*. Most of these bacteria reductively dechlorinate the chlorinated compounds in a co-metabolism reaction; others however utilize the chlorinated compounds as electron acceptors in their energy metabolism. Examples of phenomena that are common to the dehalogenators include the following:

a. Aryl reductive dehalogenators function in a syntrophic communities and may be dependent on such a community.
b. This aryl reductive dehalogenation is catalyzed by enzymes that are inducible.
c. There is exhibition of distinct substrate specificity by this enzyme.
d. Aryl dehalogenators obtain their metabolic energy from reductive dehalogenation. Hence, microorganisms with these distinctive dehalogenating enzymes each exhibit a unique pattern of congener activity.

Reductive dechlorination of PCBs occurs in soil and sediments under anaerobic condition and it is these microorganisms with the dehalogenating enzymes that are responsible. The route, extent, and even the rate of these activities depend on the makeup of the active microbial community which tends to be influenced by the factors of the environment like the presence of carbon source, hydrogen or other electron donors, the presence or absence of electron acceptors other than PCBs, temperature, and pH.

For every anaerobically mediated dechlorination of PCB, the significant evidence was dependent on the observed modification of the substance in the sediments devoid of oxygen. When the distribution patterns of PCB in both the anaerobic sediments and commercial mixtures introduced to the river were compared, it showed that the sediments have a high proportion of the mono- and di-congeners and a reduction of the higher congeners. These inferences however were consistent with reductive dechlorination through meta- and para-chlorine removal. Confirmation of these findings was later done at the laboratory and the evidence was obtained that microbial numbers in the sediment could reductively dechlorinate most of the congeners of Aroclor 1242 at the meta- and para-positions, and proportions of mono- and di-chlorobiphenyls increased considerably.

Aerobic Biodegradation of PCB

Sparsely chlorinated PCB congeners which form as a result of dechlorination of the higher congeners are substrates for aerobic bacteria. Those PCB congeners undergo co-metabolic aerobic oxidation which is mediated by an enzyme deoxygenases, bringing about a ring opening hence completing mineralization of the molecule. A lot of bacterial strains are implicated in oxidative degradation of PCBs, among them are *Pseudomonas* spp., *Burkholderia* spp., *Comamonas* spp., *Rhodococcus* spp., as well as *Bacillus* spp. Chlorine numbers per molecule and its placement are important factors in aerobic biodegradation. PCB congeners with three or less chlorine atoms per molecule are easily degraded, but ones with more than three chlorine atoms are recalcitrant, therefore requiring reductive dechlorination prior to oxidative mineralization. PCB destruction in the presence of oxygen involves two gene clusters. The first one enables transformation of PCB congeners to chlorobenzoic acid and the second involves degradation of the chlorobenzoic acid. A common growth substrate for PCB-degrading bacteria is biphenyl or monochlorobiphenyls. During utilization of biphenyls, a yellow meta-ring cleavage product is formed as observed in most studied bacteria, for example, *Pseudomonas* spp. and *Micrococcus* spp. Through 1,2-dioxygenative ring cleavage, benzoate results as a common byproduct of biphenyl degradation. Some other bacterial species seem to produce benzoate through PCB metabolism and further breakdown differs among microbes but their byproducts are less toxic compounds. Since PCBs persist more at increasing chlorination of the congeners, aerobic biodegradation involving ring cleavage is restricted to the lightly chlorinated congeners.

While biphenyls and monochlorobiphenyls can serve as growth substrates, the degradation of PCB congeners with more than one chlorine atom proceeds by a co-metabolic process in which biphenyl is used as carbon and energy source while oxidizing PCBs. Biphenyls therefore serve as an indicator of degrading enzymes. Earlier study reported that two species of *Achromobacter* are capable of growing on biphenyls and 4-chlorobiphenyl. The degradation of PCB by *Myocardial* spp. and *Pseudomonas* spp. increased upon addition of biphenyls. This was reported to enhance co-metabolism of Aroclor 1242 in the presence of acetate using mixed cultures of *Alcaligenes odorans, Alcaligenes denitrificans*, and an unidentified bacterium. Increased mineralization of Aroclor 1242 by *Acinetobacter* spp. strain P6 by addition of biphenyls and 4-chlorobiphenyl was also observed. Furthermore, these microorganisms co-metabolize Aroclor 1254 in the presence of biphenyl.

The relationship between chlorine substitution and the microbial breakdown of PCBs is as follows:

1. The rate of degradation of PCBs is inversely proportional to the increase in chlorine substitution.
2. PCBs containing two chlorine in the ortho-position of a single ring (i.e., 2, 3, 6) and each ring (i.e., 2, 2′) shows a striking resistance to degradation.
3. PCBs which have all of its chlorine on its single ring degrade much faster than those with same number on double rings.
4. PCBs having two chlorines at the 2,3-position of one ring such as 2,3,2′,3′-, 2,3,2′,5′-, and 2,4,5,2′,3′-chlorobiphenyls are susceptible to microbial attack compared with other tetra- and penta-chlorobiphenyls, though this series of PCBs is metabolized through the alternative pathway.
5. Initial deoxygenation followed by ring cleavage of the biphenyl molecule occurs with a non-chlorinated or less chlorinated ring.

BIODEGRADATION OF PESTICIDES

A pesticide is any agent used to kill or control undesired insects, weeds, rodents, fungi, bacteria, or other organisms. Thus, the term "pesticides" includes insecticides, herbicides, rodenticides, fungicides, nematicides, and acaricides as well as disinfectants, fumigants, wood preservatives, and plant growth regulators. Pesticides play a vital role in controlling agricultural, industrial, home/garden, and public health. The era of pesticides began in the nineteenth century when sulfur compounds were developed as fungicides. In the late nineteenth century, arsenic compounds were introduced to control insects that attack fruit and vegetable crops, for example, lead arsenate was used widely on apples and grapes. These substances were acutely toxic. In the 1940s, the chlorinated hydrocarbon pesticides, most notably DDT (dichloro diphenyltrichloroethane), were introduced. DDT and similar chemicals were used extensively in agriculture and in the control of malaria and other insect-borne diseases. Because they had little or no immediate toxicity, they were widely hailed and initially believed to be safe.

Before the development of synthetic pesticides, farmers used naturally occurring substances such as arsenic and pyrethrum. Following World War II, pesticides were a component of what was predicted to be a "green revolution" of abundant food for the world. Over the past 50 years, agricultural production in many areas of the world has increased dramatically, partly because of the use of herbicides and insecticides. Health benefits, such as those related to the eradication of malaria-carrying mosquitoes, were also foreseen and, in many cases, attained.

THREE MAJOR GROUPS OF CONVENTIONAL PESTICIDES

The first group consists of chlorinated hydrocarbons, also known as organochlorines. These pesticides generally break down very slowly and can remain in the environment for long periods of time. Dieldrin, chlordane, aldrin, DDT, and heptachlor are pesticides of this type. The second group is known as organic phosphates or organophosphates. These pesticides are often highly toxic to humans, but generally do not remain in the environment for long. Diazinon, malathion, dimethoate, and chlorpyrifos are pesticides of this classification. The last group is the carbamates. They are generally less toxic to humans, but concerns persist about the potential effects of some carbamates on immune and central nervous systems. Carbaryl, carbofuran, and methomyl are examples of carbamates.

FATE OF PESTICIDES IN THE ENVIRONMENT

Ideally, a pesticide stays in the treated area long enough to produce the desired effect and then breaks down into harmless materials. Three primary modes of degradation occur in soils:

- Biological—breakdown by microorganisms.
- Chemical—breakdown by chemical reactions, such as hydrolysis (soluble decomposition) and oxidation.
- Photochemical—breakdown by ultraviolet or visible light.

The rate at which a chemical degrades is expressed as the half-life, which is the amount of time it takes for half of the pesticide to be converted into something else or until its concentration is half of its initial level. The half-life of a pesticide depends on the soil type, its formulation, and environmental conditions such as temperature and moisture levels. Other processes that influence the fate of the chemical include plant absorption, soil adhesion, leaching, and vaporization. If pesticides migrate from their targets due to wind drift, runoff, or leaching, they are considered to be pollutants. The potential for pesticides to move depends on the chemical properties and formulation of the pesticide, soil properties, the rate and method of application, pesticide persistence, frequency and timing of rainfall, irrigation, and depth to groundwater.

PESTICIDE TOXICITY

Toxicity is the inherent ability of a pesticide to cause injury or death, indicating how poisonous the chemical is. Acute toxicity is the ability of a substance to cause harm as a result of a single dose or exposure to a chemical. Chronic toxicity is the ability of a substance to cause injury as a result of repeated doses or exposures over time. Any chemical substance is toxic if it is ingested or absorbed in excessive amounts. Table salt, for example, if consumed in excess, can be toxic. The degree of danger or hazard when using a pesticide is determined by multiplying toxicity times exposure. The designation given to a pesticide indicating its relative level of toxicity is called the lethal dose or LD_{50} value. This value identifies the dosage necessary to kill 50% of a test population. The lethal dose is expressed in milligrams of chemical per kilogram of body weight of the test population. The lower the LD number, the more toxic the material. The toxicity rating is important as an indicator, but the length of exposure, type of exposure, and other factors also impact the relative hazard of any pesticide. The toxicity of pesticides is often measured using an LD_{50} (lethal dose) or an LC_{50}

(lethal concentration). Both LD_{50} and LC_{50} measure only acute effects and therefore provide no information about a chemical's connection to long-term health issues.

Agricultural pesticides have prevented pest damage of between 5% and 30% of potential production in many crops. Pesticides, however, have posed a number of problems for agriculture, including the killing of beneficial insects, secondary pest outbreaks, and the development of pesticide-resistant pests. Consequently, it has become necessary to use larger doses and more frequent applications of pesticides. Combining pesticides, or substituting more expensive, toxic, or ecologically hazardous pesticides, occurs more frequently.

Pesticides not absorbed by plants and soils or broken down by sunlight, soil organisms, or chemical reactions may ultimately reach groundwater sources of drinking water. This depends on the nature of the soil, depth to groundwater, chemical properties of the pesticide, and the amount and timing of precipitation or irrigation in an area. Usually, the faster a pesticide moves through the ground, as with sandy soils and heavy rainfall or irrigation, the lesser the filtration or breakdown. Heavier soils, combined with lower moisture levels and warmer temperatures, provide a greater opportunity for pesticides to break down before reaching groundwater. The amount of a pesticide detected in well samples also relates to the kind of pesticide and the amount originally applied. Contamination problems can result from using high concentrations of water-soluble pesticides for a specific crop in a vulnerable area. Pesticides are, of course, designed to be toxic for certain insects, animals, plants, or fungi. But when used without regard to site characteristics, such as adsorption capacity of the soil (adhesion), solubility, climatic conditions, and irrigation patterns, a given pesticide can create greater environmental problems than the damage the target pest could cause. Once in groundwater, pesticides continue to break down, but usually much slower than in surface layers of soil. Groundwater carrying pesticides away from the original point of application can lead to contaminated well samples years later in a different location.

When pesticides are found in water supplies, they normally are not present in high enough concentrations to cause acute health effects such as chemical burns, nausea, or convulsion. Instead, they typically occur in trace levels, and the concern is primarily for their potential to cause chronic health problems. To estimate chronic toxicity, laboratory animals are exposed to lower than lethal concentrations for extended periods of time. Measurements are made of the incidence of cancer, birth defects, genetic mutations, or other problems such as damage to the liver or the central nervous system. Although we may encounter many toxic substances in our daily lives, in low enough concentrations, they do not impair our health. Caffeine, for example, is regularly consumed in coffee, tea, chocolate, and soft drinks. Although the amount of caffeine consumed in a normal diet does not cause illness, just 50 times this amount is sufficient to kill a human. Similarly, the oxalic acid found in rhubarb and spinach is harmless at low concentrations found in these foods, but will lead to kidney damage or death at higher doses.

The pathways of human exposure to pesticides are numerous. Pesticide residues are found virtually everywhere: in the office and home, on food, in drinking water, and in the air. Throughout more than a half century of pesticide use, most pesticides have never been systematically reviewed to evaluate their full range of long-term health effects on humans, such as potential damage to the nervous, endocrine, or immune systems. The Environmental Protection Agency (EPA) considers only cancers in determining the potential threat of pesticides to human health.

Although pesticides do offer certain benefits for farmers and others, new scientific research is revealing some important health-related issues associated with their usage. For example, some scientists have become convinced that there is a relationship between pesticides that mimic the estrogen hormone and the disruption of the endocrine systems in humans and wildlife. This potentially could contribute to serious health problems, including breast and other types of cancer in humans, and reproductive disorders.

ENZYMES INVOLVED IN PESTICIDE BIODEGRADATION

Fungi and bacteria are considered as the extracellular enzyme-producing microorganisms for excellence. White-rot fungi have been proposed as promising bioremediation agents, especially for compounds not readily degraded by bacteria. This ability arises from the production of extracellular enzymes that act on a broad array of organic compounds. Some of these extracellular enzymes are involved in lignin degradation, such as lignin peroxidase, manganese peroxidase, laccase, and oxidases. Several bacteria that degrade pesticide have been isolated and the list is expanding rapidly. The three main enzyme families implicated in degradation are esterases, glutathione S-transferases (GSTs), and cytochrome P_{450}.

For pesticides degradation, there are mainly three enzyme systems involved: hydrolases, esterases (also hydrolases), the mixed function oxidases (MFOs), these systems in the first metabolism stage, and the GST system, in the second phase. Several enzymes catalyze metabolic reactions including hydrolysis, oxidation, addition of an oxygen to a double bond, oxidation of an amino group (NH_2) to a nitro group, addition of a hydroxyl group to a benzene ring, dehalogenation, reduction of a nitro group (NO_2) to an amino group, replacement of a sulfur with an oxygen, metabolism of side chains, ring cleavage. The process of biodegradation depends on the metabolic potential of microorganisms to detoxify or transform the pollutant molecule, which is dependent on both accessibility and bioavailability.

Metabolism of pesticides may involve a three-phase process. In Phase I metabolism, the initial properties of a parent compound are transformed through oxidation, reduction, or hydrolysis to generally produce a more water-soluble and usually a less toxic product than the parent. The second phase involves conjugation of a pesticide or pesticide metabolite to a sugar or amino acid, which increases the water solubility and reduces toxicity compared with the parent pesticide. The third phase involves conversion of Phase II metabolites into secondary conjugates, which are also non-toxic. In these processes, fungi and bacteria are involved producing intracellular or extracellular enzymes including hydrolytic enzymes, peroxidases, and oxygenases.

HYDROLASES

Hydrolases are a broad group of enzymes involved in pesticide biodegradation. Hydrolases catalyze the hydrolysis of several major biochemical classes of pesticide (esters, peptide bonds, carbon-halide bonds, ureas, thioesters, etc.) and generally operate in the absence of redox cofactors, making them ideal candidates for all of the current bioremediation strategies. The degradation pathway of carbofuran is an example of the catalytic activity of enzymes hydrolases. This pesticide can be transformed in the environment, and different metabolites are generated and accumulated in potentially contaminated sites (soil, water, and sediments, mainly).

PHOSPHOTRIESTERASES (PTEs)

Among the most studied pesticide degrading enzymes, the PTEs are one of the most important groups. These enzymes have been isolated from different microorganisms that hydrolyze and detoxify organophosphate pesticides (OPs). The first isolated phosphotriesterase belongs to the *Pseudomonas diminuta* MG species; this enzyme shows a high catalytic activity toward OPs. The phosphotriesterases are encoded by a gene called *opd* (organophosphate degrading). *Flavobacterium* ATCC 27551 presents the *opd* gene encoding to a PTE. The gene was cloned and sequenced. These enzymes specifically hydrolyze phosphoester bonds, such as P–O, P–F, P–NC, and P–S, and the hydrolysis mechanism involves a water molecule at the phosphorus center. Different microbial enzymes with the capacity to hydrolyze MP have been identified, such as organophosphorus hydrolase (OPH; encoded by the *opd* gene), methyl-parathion hydrolase (MPH; encoded by the *mpd* gene), and hydrolysis of coroxon (HOCA; encoded by the *hocA* gene), which were isolated from *Flavobacterium* sp., *Plesiomonas* sp. strain M6, and *Pseudomonas moteilli* respectively.

The phosphotriesterase enzyme is a homodimeric protein with a monomeric molecular weight of 36 kDa. As a first step in the PTE organophosphorus pesticide hydrolysis mechanism, the enzymatic active site removes a proton from water, activating this molecule; then, the activated water directly attacks the central phosphorus of the pesticide molecule producing an inversion in its configuration. The oxygen is polarized by the active site, with the participation of a zinc atom. This enzyme has potential use for the cleaning of organophosphorus pesticides contaminated environments.

ESTERASES

Esterases are enzymes that catalyze hydrolysis reactions over carboxylic esters (carboxylesterases), amides (amidases), phosphate esters (phosphatases), etc. Many insecticides (organophosphates, carbamates, and pyrethroids) have associated a carboxylic ester, and the enzymes capable of hydrolyzing such ester bond are known with the name of carboxylesterases.

The esterases are a group of enzymes highly variable, which has been recognized as one of the most important in the metabolism of xenobiotics and its mechanism is associated with the mass production of multifunctional hydrolytic enzymes. Organophosphate pesticides can be hydrolyzed by such enzymes. There are different types of esterases and with very different distribution in tissues and organisms. The Carboxylesterases (type B esterases) are a group that hydrolyze, in addition to endogenous compounds, xenobiotics with ester, amide, thioester, phosphate esters (parathion, paraoxon), and acid anhydrides (DIPFP=DFP) in mammals.

Esterases A contain a Cys residue in the active center and esterases B contain a Ser residue. In esterases A, the organophosphates interact with the functional group –SH forming a bond between P=S, which is easily hydrolyzed by H_2O. In the esterase B, organophosphates interact with the SER-OH forming a P=O bond that is not hydrolyzed by H_2O. Organophosphates that bind to the esterase B stoichiometrically inhibit its enzymatic activity.

Esterases are a diverse group that protects the target site (acetylcholinesterase) by catalyzing the hydrolysis of insecticides or acting as an alternative blank. Esterases in general have a wide range of substrate specificities; they are capable of binding to phosphate triesters, esters, thioesters, amides, and peptides.

OXIDOREDUCTASES

Oxidoreductases are a broad group of enzymes that catalyze the transfer of electrons from one molecule (the reductant or electron donor) to another (the oxidant or electron acceptor). Many of these enzymes require additional cofactors to act as either electron donors, electron acceptors, or both. Oxidoreductases have been further subclassified into 22 subclasses. Several of these have applications in bioremediation, albeit their need for cofactors complicates their use in some applications. There are enzymes that catalyze an oxidation/reduction reaction by including the molecular oxygen (O_2) as electron acceptor. In these reactions, oxygen is reduced to water (H_2O) or hydrogen peroxide (H_2O_2). The oxidases are a subclass of the oxidoreductases.

As an example of many functions of these enzymes in the degradation of pesticides, endosulfan degradation pathway is one example. In this process, not only oxidoreductase enzymes are involved but also different microorganisms and catalytic activities, in combination, can lead to complete mineralization of a pesticide. Endosulfan (*1,2,5,6,7,7-hexachloro-5-norbornene-2,3-dimethanolcyclic sulfite*) is an organochlorine insecticide of the cyclodiene family of pesticides. It is highly toxic and endocrine disruptor, and it is banned in European Union and several countries. Because it has been extensively applied directly to fields, it can detect a considerable distance away from the original site of application. Contamination of drinking water and food and detrimental effects to wildlife are important concerns. The molecular structure has two stereochemical isomers α- and β-endosulfan. The end-use product of endosulfan is a mixture of two isomers, typically in a 2:1 ratio.

Microorganisms play a key role in the removal of xenobiotics like endosulfan from the contaminated sites because of their dynamic, complex, and complicated enzymatic systems which degrade these chemicals by eliminating their functional groups of the parent compound. This pesticide can undergo either oxidation or hydrolysis reactions. Several intensive studies on the degradation of endosulfan have been conducted showing the primary metabolites to normally be endosulfan sulfate and endosulfan diol (endodiol). Endosulfan sulfate will be present in the environment as a result of the use of endosulfan as an insecticide. If endosulfan sulfate is released to water, it is expected to absorb to the sediment and may bioconcentrate in aquatic organism. This metabolite has a similar toxicity as endosulfan and has a much longer half-life in the soil compared to endosulfan. Therefore, production of endosulfan sulfate by biological systems possesses an ecological hazard in that it contributes to long persistence of endosulfan in soil. Endodiol is much less toxic to fish and other organisms than the parent compound.

Thus, it is important to note that some microbial enzymes are specific to one isomer or catalyze at different rates for each isomer. For example, a *Mycobacterium tuberculosis* ESD enzyme degrades β-endosulfan to the monoaldehyde and hydroxyether (depending on the reducing equivalent stoichiometry), but transforms α-endosulfan to the more toxic endosulfan sulfate. However, oxidation of endosulfan or endosulfan sulfate by the monooxygenase encoded by *ese* in *Arthrobacter* sp. KW yields endosulfan monoalcohol. Both *ese* and *esd* proteins are part of the unique Two-Component Flavin-Dependent Monooxygenase Family, which require reduced flavin. They are conditionally expressed when no or very little sulfate or sulfite is available, and endosulfan is available to provide sulfur in these starved conditions.

Alternatively, hydrolysis of endosulfan in some bacteria (*Pseudomonas aeruginosa*, *Burkholderia cepacia*) yields the less toxic metabolite endosulfan diol. Endosulfan can spontaneously hydrolyze to the diol in alkaline conditions, so it is difficult to separate bacterial from abiotic hydrolysis. The diol can be converted to endosulfan ether or endosulfan hydroxyether and then endosulfan lactone. Hydrolysis of endosulfan lactone yields endosulfan hydroxycarboxylate. These various branches of endosulfan degradation all result in desulfurization while leaving the chlorines intact, exhibiting the recalcitrance to bioremediation found in many organohalogen aromatics.

MIXED FUNCTION OXIDASES (MFOs)

In the reaction catalyzed by the MFO, an atom of one molecule of oxygen is incorporated into the substrate, while the other is reduced to water. For this reason, the MFO requires Nicotinamide-adenine dinucleotide phosphate (NADPH) and O_2 for its operation. It is an enzyme system comprising two enzymes, cytochrome P_{450} and NADPH-cytochrome P_{450} reductase, both membrane proteins. They are also known as dependent cytochrome P_{450} monooxygenases or P_{450} system. The genes encoding different isozymes comprise a superfamily of over 200 genes grouped into 36 families based on their sequence similarity. Cytochrome P_{450} enzymes are active in the metabolism of wide variety of xenobiotics.

The cytochrome P_{450} family is a large, well-characterized group of monooxygenase enzymes that have long been recognized for their potential in many industrial processes, particularly due to their ability to oxidize or hydroxylate substrates in an enantiospecific manner using molecular oxygen. Many cytochrome P450 enzymes have a broad substrate range and have been shown to catalyze biochemically recalcitrant reactions such as the oxidation or hydroxylation of non-activated carbon atoms. These properties are ideal for the remediation of environmentally persistent pesticide residues. Over 200 subfamilies of P_{450} enzymes have been found across various prokaryotes and eukaryotes. All contain a catalytic iron-containing porphyrin group that absorbs at 450 nm upon binding of carbon monoxide. In common with many of the other oxidoreductases described before, P_{450} enzymes require a non-covalently bound cofactor to recycle their redox center (most frequently NAD(P)H is used), which limits their potential for pesticide bioremediation to strategies that employ live organisms.

In insects, MFOs are found in the endoplasmic reticulum and mitochondria, and they are involved in a large number of processes such as growth, development, reproduction, and detoxification. MFOs are involved in the metabolism of both endogenous and exogenous substances; for this reason, these compounds promote their induction. Due to its high inspecificity, the MFOs metabolize a wide range of compounds such as organophosphates, carbamates, pyrethroids, DDT, inhibitors of the chitin synthesis, and juvenile hormone mimics.

Glutathione S-Transferase (GST)

The GSTs are a group of enzymes that catalyze the conjugation of hydrophobic components with the tripeptide glutathione. In this reaction, the thiol group of glutathione reacts with an electrophilic place in the target compound to form a conjugate which can be metabolized or excreted, and they are involved in many cellular physiological activities, such as detoxification of endogenous and xenobiotic compounds, intracellular transport, biosynthesis of hormones, and protection against oxidative stress.

BIODEGRADATION OF PESTICIDES

In 1950, Audus isolated the first soil bacteria able to degrade 2,4-D. This was the ultimate proof that pesticides could be subject to biological degradation in the soil environment. Since this time, 2,4-D has been considered as a model xenobiotic. As a consequence, its degradation pathway(s), the microorganisms involved as well as the genetic factors supporting its catabolism were extensively studied.

Phenoxyacetic Acid: 2,4-D

Microorganisms have the ability to interact, both chemically and physically, with substances leading to structural changes or complete degradation of the target molecule. Among the microbial communities, bacteria, fungi, and actinomycetes are the main transformers and pesticide degraders. Fungi generally biotransform pesticides and other xenobiotics by introducing minor structural changes to the molecule, rendering it non-toxic. The biotransformed pesticide is released into the environment, where it is susceptible to further degradation by bacteria.

Ralstonia eutropha (frequently reported as *Alcaligenes eutrophus*) JMP134 (*pJP4*) is able to use 2,4-D as sole carbon and energy source. Its metabolic pathway encoded by the *tfd* genes carried by the plasmid pJP4 is the most extensively studied. The three first steps lead to the opening of the aromatic ring. 2,4-D is firstly transformed to 2,4-dichlorophenol by a α-ketoglutarate-dependent dioxygenase, *TfdA* (Fukumori and Hausinger, 1993). The aromatic ring is then hydroxylated in position 2 by the phenol hydroxylase *TfdB* to form 3,5-dichlorocatechol. A single operon with the *tfdCDEF* genes encodes for the formation of 2,4-dichloromuconate, the first aliphatic metabolite, and further metabolism to β-ketoadipate which is then metabolized by chromosome-encoded enzymes (van der Meer et al., 1992). The uptake of 2,4-D has been shown to be an active process, the uptake system being inducible and encoded by another gene located on plasmid *pJP4*, *tfdK* (Leveau et al., 1998).

Different genera (*Sphingomonas*, *Pseudomonas*, or *Alcaligenes*) of microorganisms able to degrade 2,4-D have been isolated. The similarity of the degrading genes was found to be linked to the phylogeny of the degrading microorganisms (Fulthorpe et al., 1995; Ka et al., 1994). *Sphingomonas* is reported to carry the most dissimilar genes with only in some cases some similarity to the *tfdB* gene but not to *tfdA* and *tfdC*, whereas *Pseudomonas* and *Alcaligenes* species generally carry genes similar to *tfdA*, *tfdB*, and *tfdC*. *Burkholderia* strains generally exhibited genes highly similar to *tfdB* and *tfdC* but only weakly similar to *tfdA*. *Rhodoferax fermentans* strains exhibited as a whole very good similarity with *tfdA* but weak similarities to *tfdB* and *tfdC*.

In the 80-kb plasmid pJP4 of *Ralstonia eutrophus* JMPI34, the *tfd* genes encoding the 2,4-D catabolic pathway are contiguous and in the order *ACDEFB*. Two regulatory genes (*tfdR* and *tfdS′*), a gene coding for an uptake system of the herbicide (*tfdK*), as well as copies of *tfdA* and *tfdC* are also carried by this plasmid. Inoculation of *R. eutrophus* JMP134 in a nonsterile soil amended with 2,4-D resulted in the spreading of pJP4 throughout the indigenous soil microflora.

A S-TRIAZINE: ATRAZINE

For a long time, atrazine has been considered as a moderately persistent compound. The removal of the chloro-substituent was exclusively attributed to abiotic processes, whereas the biotic degradation was mainly attributed to fungi and limited to N-dealkylation. The biodegradation of its s-triazine ring was considered to be very slow and bound residue formation an important way of dissipation (Winkelmann and Klaine, 1991). It is only 40 years after its introduction on the market that rapid atrazine degradation by soil microorganisms was observed. This rapid biodegradation appeared to be coupled with a new catabolic pathway leading to extensive mineralization of the s-triazine ring (Vanderheyden et al., 1997).

The N-Dealkylation Dead-End Cytochrome P$_{450}$ Pathway

Fungal degradation of atrazine appears to be limited to the side chains of the s-triazine ring. In liquid culture, *Phanerochaete chrysosporium* only partially biotransformed atrazine, producing mainly Deethylatrazine (CIAT) but Deisopropylatrazine (CEAT) and Hydroxyatrazine (OEIT). This activity was attributed to a P450 monooxygenase. It must be pointed out that fungal biodegradation studies of atrazine were always carried out with species that were not isolated or enriched for this purpose so that their action is to be attributed to their wide-spectrum enzymes that they produce "naturally." Different actinomycetes were also reported to be able to catabolize the substituents of the atrazine ring. Atrazine metabolism by *Rhodoccocus* strains that was linked to the 77-kb plasmid that was required for S-ethyl dipropylcarbamothioate (EPTC) degradation. Another member of the *Rhodococcus* group (*Rhodococcus corallinus*) was shown to be able to carry out dechlorination and deamination on both deisopropylatrazine and deethylatrazine due to an s-triazine hydrolase.

THE MINERALIZATION PATHWAY

The three first steps of atrazine mineralization by *Pseudomonas* ADP were identified (Sadowsky et al., 1998). AtzA, the first enzyme of the pathway, converts atrazine to hydroxyatrazine via hydrolytic dechlorination. Hydroxyatrazine transformation is then carried out by two amidases (AtzB and AtzC) to successively produce N-isopropylammelide and cyanuric acid, the central metabolite to all *s-triazine* mineralization pathways reported. Surprisingly, this pathway differed from the catabolic reactions previously reported. First, N-dealkylation was previously shown to be a prerequisite to dechlorination. Second, the two N-alkyl substituents undergo deamidation and not N-dealkylation. Third, the three first reactions are here of hydrolytic nature, whereas oxidative process was generally reported. It is also remarkable to quote that this is the most direct way to cyanuric acid, the metabolite allowing the mineralization of the *s*-triazine ring. This new pathway is followed by diverse atrazine mineralizing microorganisms belonging to different genera (*Pseudomonas*, *Ralstonia*, *Alcaligenes*, *Agrobacterium*, and *Clavibacter*) and that shared highly conserved degradative genes. A protein similar to AtzA was also reported from a *Rhizobium* sp. isolate capable of atrazine dichlorination (Bouquard et al., 1997). Cyanuric acid is transformed into biuret, an easily metabolizable compound that is degraded in urea and ultimately to carbon dioxide and ammonia. The s-triazine ring is thus essentially a N source since the C atoms of the cycle are already at their maximum oxidation stage.

A CARBAMATE PESTICIDE: CARBOFURAN

Carbofuran is a soil-applied pesticide that was used extensively to control insects and nematodes in potato, sugar beet, com, rice, and other crops. Two different hydrolytic enzymes degrading carbofuran to carbofuran-phenol and methylamine have been reported. The first hydrolase was purified from an *Achromobacter* sp. strain WM 111 (Derbyshire et al., 1987) and the mcd gene encoding this enzyme is localized on the >100-kb plasmid pDLl1 (Tomasek and Karns, 1989). The second carbamate hydrolase was produced by the *Pseudomonas* strain CRL-OK (Mulbry and Eaton, 1991). Both enzymes also transformed carbaryl and aldicarb, but the *Pseudomonas* enzyme exhibited a somewhat narrower substrate range. The metabolite methylamine appears to be the primary substrate that can be used as nitrogen or carbon source by a number of degrading *Pseudomonas* or *Flavobacterium* isolates. A methylotrophic bacterium was found to contain a 120-kb plasmid that was very similar to pOLl 1 and that carried the mcd gene (Top et al., 1993).

Carbofuran phenol is more recalcitrant to biodegradation and its catabolism is less often reported. *Arthrobacter* sp. strain catabolizes carbofuran via carbofuran phenol (Ramanand et al., 1988). This metabolite was transiently detected during carbofuran degradation and the strain was also able to metabolize carbofuran phenol as main substrate.

Rhodococcus TEl was found to co-metabolize carbofuran as well as two other N-methylcarbamates: propoxur and carbaryl (Behki et al., 1994). The end product of carbofuran transformation was 5-hydroxycarbofuran. The gene encoding this transformation was located on the same 77-kb plasmid that is involved in the biodegradation of EPTC (a thiocarbamate) and atrazine (an s-triazine). Since substitution of aromatic rings by hydroxy groups generally ease mineralization, the production of 5-hydroxy-7-phenol carbofuran might allow to go further in the metabolism of the carbofuran ring.

BIODEGRADATION OF EXPLOSIVES

Characteristically, organic explosives have nitro ($-NO_2$) groups attached to ring structures. Historically, the most widely used explosive is 2,4,6-trinitrotoluene (TNT), which is found as a contaminant in soils where ammunitions have been stored. Other explosives used include hexahydro-1,3,5-trinitro-1,3,4-triazine (RDX) and octhydro-1,3,5,7-tetranitro-1,3,5,7-triazine (HMX), and these are soil contaminates as well. Nitroaromatic respiration by *Desulfobacterium* and *Desulfovibrio* occurs when TNT and 2,6-dinitrotoluene are used as final electron acceptors (Boonpathy et al., 2007). Under nitrogen-limited conditions, cultures of sulfate-reducing bacteria and *Veillonella alkalescens* reduce TNT to triaminotoluene before deamination is proposed to release ammonium. In an analogous reaction, *Desulfobacterium aniline* will reductively metabolize aniline to produce benzoic acid. The reduction of the nitro group of TNT to amino moieties may be attributed to nitrite reductase, which is commonly found in anaerobic bacteria. When TNT is the carbon and energy source for *Desulfovibrio,* acetic acid is proposed as the product of catabolism. Anaerobic metabolism of TNT by *Clostridium bifermentans* has also been reported, and it is tentatively attributed to the fermentative capabilities of this obligate anaerobe (Staley et al., 2007).

Bioremediation is replacing incineration of soil and lagoon sediments at military instillations contaminated with TNT, RDX, and HMX. A composting process has been useful in remediation of explosives-contaminated soil, and field demonstrations have been conducted at several sites (Keehan and Sisk, 1996). The compost pile consisted of contaminated material, straw, alfalfa hay, fertilizer, and animal manure. After several months with continuous aeration at 55°C, the mixed microbial system reduced the level of explosives by over 99% and no toxicity was associated with metabolic end products.

FURTHER READING

Agarwal SK. *Pesticide Pollution*. APH Publishing; 2009.

Behki RM, Topp EE, Blackwell BA. Ring hydroxylation of N-methylcarbamate insecticides by *Rhodococcus* TEl. 1. *Agric. Food Chem*. 1994;42:1375–1378.

Boonpathy R. Anaerobic metabolism of nitroaromatic compounds and bioremediation of explosives by sulphate-reducing bacteria. In: Barton LL, Hamilton WA (eds.) *Sulphate-Reducing Bacteria—Environmental and Engineered Systems*. Cambridge: Cambridge University Press; 2007:503–524.

Borja J, Taleon DM, Auresenia J, Gallardo S. Polychlorinated biphenyls and their biodegradation. *Process Biochem*. 2005;40:1999–2013.

Bouquard C, Ouazzani J, Prome JC, Michel-Briand Y, Plesiat P. Dechlorination of atrazine by a *Rhizobium* sp. isolate. *Appl. Environ. Microbiol*. 1997;63:862–866.

Chapman PJ. Degradation mechanisms. In: Bourquin AW, Pritchard PH (eds.) *Microbial Degradation of Pollutants in Marine Environments*. EPA-600/9-79-0 12. Gulf Breeze, FL: Environmental Protection Agency; 1979: 28–66.

Derbyshire MK, Karns S, Kearney PC, Nelson O. Purification and characterization of an N-methylcarbamate pesticide hydrolizing enzyme. *J. Agric. Food Chern.* 1987;35:871–877.

Focht DD, McCullar MV, Searles DB, Koh SC. Mechanisms involving the aerobic biodegradation of PCB in the environment. In: Agathos SN, Reineke W (eds). *Biotechnology for the Environment: Strategy and Fundamentals. Focus on Biotechnology*, Vol 3A. Dordrecht: Springer; 2002.

Fukumori F, Hausinger RP. 1993. *Alcaligenes eutrophus* JMP134 "2,4-dichlorophenoxyacetate monooxygenase" is an a-ketoglutarate-dependent dioxygenase. *J. Bacteriol.* 175: 2083–2086.

Fulthorpe RR, McGowan C, Maltseva OV, Holben WE, Tiedje JM. 2,4-Dichlorophenoxyacetic acid-degrading bacteria contain mosaics of catabolic genes. *Appl. Environ. Microbiol.* 1995;61:3274–3281.

Hill IR, Wright SJL. *Pesticide Microbiology: Microbiological Aspects of Pesticide Behaviour in the Environment*. Academic Press; 1978.

Hirose J, Kimura N, Suyama A, Kobayashi A, Hayashida S, Furukawa K. Functional and structural relationship of various extradiol aromatic ring-cleavage dioxygenases of *Pseudomonas* origin. *FEMS Microbiol. Lett.* 1994;118:273–278.

Ka JO, Holben WE, Tiedje JM. Genetic and phenotypic diversity of 2,4-dichlorophenoxyacetic acid (2,4-D)-degrading bacteria isolated from 2,4-D-treated field soils. *Appl. Environ. Microbiol.* 1994a;60:1106–1115.

Ka JO, Holben WE, Tiedje JM. Use of gene probes to aid in recovery and identification of functionally dominant 2,4-dichlorophenoxyacetic acid-degrading populations in soil. *Appl. Environ. Microbiol.* 1994b;60:1116–1120.

Keehan KR, Sisk WE. The development of composting for the remediation of explosives contaminated soils. In: Hickey RF, Smith G (eds.) *Biotechnology in Industrial Waste Treatment and Bioremediation*, Boca Raton, FL: Lewis Publishers; 1996:69–79.

Kim KP, Seo DI, Min KH, Ka JO, Park YK, Kim C-K. Characterization of catechol 2,3-dioxygenase produced by 4-chlorobenzoate-degrading *Pseudomonas* sp. S-47. *J. Microbiol.* 1997;35:295–299.

Leveau JH, Zehnder AJ, van der Meer JR. The *tfdK* gene product facilitates uptake of 2,4-dichlorophenoxyacetate by *Ralstonia eutropha* JMPI34(pJP4). *J. Bacteriol.* 1998;180:2237–2243.

Mars AE, Kingma J, Kaschabek SR, Reineke Wand Janssen DB. Conversion of 3-chlorocatechol by various catechol 2,3-dioxygenases and sequence analysis of the chlorocatechol dioxygenase region of *Pseudomonas putida* GJ31. *J. Bacteriol.* 1999;18:1309–1318.

Moon J, Min KR, Kim C-K, Min K-H, Kim Y. Characterization of the gene encoding catechol 2,3-dioxygenase of *Alcaligenes* sp. KF711: overexpression, enzyme purification, and nucleotide sequencing. *Arch. Biochem. Biophys.* 1996;332:248–254.

Mulbry WW, Eaton RW. Purification and characterization of the N-methylcarbamate hydrolase from *Pseudomonas* strain CRL-OK. *Appl. Environ. Microbiol.* 1991;57:3679–3682.

Oldenhuis R, Kuijk K, Lammers A, Janssen DB, Witholt B. Degradation of chlorinated and non-chlorinated aromatic solvents in soil suspensions by pure bacterial cultures. *Appl. Microbiol. Biotechnol.* 1989;30:211–217.

Potrawfke T, Timmis KN, Wittich RM. Degradation of I, 2,3,4-tetrachlorobenzene by *Pseudomonas chlororaphis* RW71. *Appl. Environ. Microbiol.* 1998;64:3798–3806.

Racke KD, Coats JR. *Enhanced Biodegradation of Pesticides in the Environment*. American Chemical Society; 1990.

Ramanand K, Sharmila M, Sethunathan N. 1988. Mineralization of carbofuran by a soil bacterium. *Appl. Environ. Microbiol.* 54:2129–2133

Rasul Chaudhry G. *Biological Degradation and Bioremediation of Toxic Chemicals.* Dioscorides Press; 1994.

Rathore HS, Nollet LML. *Pesticides: Evaluation of Environmental Pollution.* CRC Press; 2012.

Reineke W, Jeenes DJ, Williams PA, Knackmuss H-J. TOL plasmid pWWO in constructed halobenzoate-degrading *Pseudomonas* strains: Prevention of meta pathway. *J. Bacteriol.* 1982;150:195–201.

Rochkin D. *Microbiological Decomposition of Chlorinated Aromatic Compounds.* CRC Press; 1986.

Sadowsky MJ, Tong Z, de Souza M, Wackett LP. AtzC is a new member of the amidohydrolase protein superfamily and is homologous to other atrazine-metabolizing enzymes. *J. Bacteriol.* 1998;180:152–158.

Safe S, Safe L, Mullin M. In: Safe S (eds.) *Polychlorinated Biphenyls (PCBs): Mammalian and Environmental Toxicology.* Springer-Verlag Berlin Heidelberg; 1987.

Schmidt E, Knackmuss H-J. Chemical structure and biodegradability of halogenated aromatic compounds. Conversion of chlorinated muconic acids into maleoylacetic acid. *Biochem. J.* 1980;192:339–347.

Seo D-I, Lim JY, Kim YC, Min KH, Kim C-K. Isolation of *Pseudomonas* sp. S-47 and its degradation of 4-chlorobenzoic acid. *J. Microbiol.* 1997;35:188–192.

Singh SN (eds.) *Microbe-Induced Degradation of Pesticides.* Springer International Publishing; 2017.

Staley JT, Gunsalus RP, Lory S, Perry JJ. *Microbial Life.* 2nd ed. Sunderland, MA: Sinauer Associates; 2007.

Stucki G, Thiier M. Experiences of a large-scale application of 1,2-dichloroethane degrading microorganisms for groundwater treatment. *Environ. Sci. Technol.* 1995;29:2339–2345.

Talley J. *Bioremediation of Recalcitrant Compounds.* CRC Press; 2005.

Tomasek PH, Karns JS. Cloning of a carbofuran hydrolase gene from *Aehromobaeter* sp. strain WM 111 and its expression in gram-negative bacteria. *J. Bacteriol.* 1989;171:4038–4044.

Topp E, Hanson RS, Ringelberg DB, White DC, Wheatcroft R. Isolation and characterization of an N-methylcarbamate insecticide-degrading methylotrophic bacterium. *Appl. Environ. Microbiol.* 1993;59:3339–3349.

U. S. Department of Health and Human Services. *Toxicological profiles - Polychlorinated Biphenyls*; 2000.

van der Meer JR, de Vos WM, Harayama S, Zehnder AJB. Molecular mechanisms of genetic adaptation to xenobiotic compounds. *Microbiol. Rev.* 1992;56:677–694.

Vanderheyden V, Debongnie P, Pussemier L. Accelerated degradation and mineralization of atrazine in surface and subsurface soil materials. *Pestic. Sci.* 1997;49:237–242.

Vollmer MD, Fischer P, Knackmuss H-J, Schltimann M. Inability of muconate cycloisomerases to cause dehalogenation during conversion of 2-chloro-cis, cis-muconate. *J. Bacteriol.* 1994;176:4366–4375.

Vollmer MD, Schltimann M. Conversion of 2-chloro-cis, cis-muconate and its metabolites 2-chloro-and 5-chloromuconolactone by chloromuconate cycloisomerases of pJP4 and pAC27. *J. Bacteriol.* 1995;177:2938–2941.

Winkelmann DA, Klaine SJ. Degradation and bound residue formation of four atrazine metabolites, deethylatrazine, deisopropylatrazine, dealkylatrazine and hydroxyatrazine, in a western Tennessee soil. *Environ. Toxicol. Chern.* 1991;10:347–354.

Young LY, Cerniglia CE. *Microbial Transformation and Degradation of Toxic Organic Chemicals.* Wiley; 1995.

Zhang H, Jiang X, Lu L, Xiao W. Biodegradation of polychlorinated biphenyls (PCBs) by the novel identified cyanobacterium Anabaena PD-1. *PLoS One.* 2015;10(7):e0131450.

12 Biodegradation of Azo Dyes

The recognition that environmental pollution is a worldwide threat to public health has given rise to new initiatives for environmental restoration for both economic and ecological reasons. The industrial effluents contain toxic and hazardous pollutants. One particular class of synthetic chemicals which is of major concern is synthetic dyes and dye intermediates. The dyes are extensively used in textile, paper printing, color photography, cosmetic, pharmaceutical, and leather industries.

Azo dyes, which are aromatic compounds with one or more $-N=N-$ groups, constitute the largest class of synthetic dyes in commercial applications. The azo groups are in general bound to a benzene or naphthalene ring, but they can also be attached to heterocyclic aromatic molecules or to enolizable aliphatic groups. On the basis of the characteristics of the processes in which they are applied, the molecule of the dye is modified to reach the best performances; so they can be acid dyes, direct dyes, reactive dyes, disperse dyes, or others.

ACID DYES

Acid dyes constitute a large group of water-soluble anionic colorants with relatively low molecular weights, typically characterized by the presence of strong water-solubilizing substituents, especially sulfonate groups. They are mainly composed of aromatic monoazo compounds, but they also include bisazo, nitro, 1-aminoanthraquinone triphenylmethine, and other groups of dyes. Aromatic sulfonates are not only easily accessible synthetically but also have the advantage of being negatively charged in aqueous solution over an extremely broad pH range. Anionic monoazo dyes and their metal salts are widely used for either dyeing paper and leather or as pigments. Their main application, however, constitutes the dyeing of proteins, that is animal hair fibers (wool, silk) and synthetic fibers (nylon). In this context, the term acid dyes is often used, since the corresponding dyeing process takes place in a weakly acidic solution (pH 2–6). Attachment to the fiber is attributed, at least partly, to the salt formation between anionic groups in the dyes and cationic groups in the fiber: animal protein fibers and nylon fibers contain many cationic sites. A certain amount of dyestuff always remains in water after dyeing.

DIRECT DYES

Direct dyes are attracted to the textile, according to their "substantivity," by intermolecular forces without the need of mordant. They are used to color cotton and paper leather, silk, and nylon, and are also used as pH indicators or as biological stains. The

DOI: 10.1201/9781003272618-14

water solubility is assured by sulfonate groups (usually 2–4), and direct dyeing is normally carried out in a neutral or slightly alkaline dyebath; washing is easy and fast.

REACTIVE DYES

Reactive dyes contain substituent that, when activated, reacts with the $-OH$ groups of cellulose (i.e., cotton) or with $-NH_2$ and $-SH$ groups of protein fibers (i.e., wool) forming covalent bonds, making them among the most permanent of dyes.

DISPERSE DYES

Disperse dyes are almost insoluble in water; they do not contain any basic or acidic group in the molecule. They are finely ground mixed to a dispersing agent and disposed as powder or paste, and then used as aqueous suspensions. They are usually used to dye cellulose acetate, nylon, triacetate, and polyester fibers; also, acrylics can be dyed with disperse dyes, but with poor intensity. High temperature and pressure of dyebath is required in some cases, and dyeing rate is influenced by the particle size and the chosen dispersing agent.

TOXICITY CAUSED BY AZO DYES

Dyes can be toxic and mutagenic, and if they are discharged directly into the environment, they persist as environmental pollutant as well as traverse through the entire food chains, leading to biomagnifications. Azo dyes represent the largest group of organic dyes synthesized and account for about 70% of all textile dyes produced. During the dying process, most reactive dyes are hydrolyzed and later released into waterways. Although these dyes are not toxic by themselves, after release into the aquatic environment, they may be converted into potentially carcinogenic amines (Chung and Stevens, 1993; Spadaro et al., 1992) that impacted the ecosystem downstream from the mill.

The discharge of azo dyes into the environment is a concern due to coloration of natural waters and their absorption and reflection of sunlight falling in the water bodies. This interferes with the growth of bacteria and plants, causing an annoyance to the ecology of the receiving water body due to the toxicity, mutagenicity, and carcinogenicity of the dyes and their biotransformation products. Therefore, substantial attention has been given to evaluate the fate of azo dyes during wastewater treatment and in the natural environment.

Azo dyes and their degradation intermediates vary in their recalcitrance to biodegradation due to their complex structures and xenobiotic nature, and in some cases are both mutagenic and carcinogenic (Chung and Cerniglia, 1992; Pinheiro et al., 2004). Treatment of dye-contaminated wastewater discharged from the textile and other dye-stuff industries is necessary to prevent contamination of soil and surface and groundwater. Currently, there are several physicochemical and biological methods for the removal of dyes from effluents.

It is generally observed that dyes resist biodegradation in conventional activated sludge treatment units. Bioaugmentation is a process in which various microorganisms including indigenous, wild type, or genetically engineered are introduced to the bio-reactor or the polluted sites/matrices to accelerate the desired biological processes and achieve more consistent results (Ritman and Whiteman, 1994; van Limbergen, 1998). It is now known that several microorganisms including yeasts, algae, bacteria, and fungi or their enzymes can decolorize and even completely mineralize many azo dyes under certain environmental conditions.

FUNGAL DEGRADATION OF DYES

Fungus has proved to be a suitable organism for the treatment of textile effluent and dye removal. Based on the mechanism involved, these studies can be grouped into bioac-cumulation, biodegradation, and biosorption. Biodegradation is an energy-dependent process and involves the breakdown of dye into various byproducts through the action of various enzymes. Fungi can produce the lignin-modifying enzymes, such as laccase, lignin peroxidase (LiP), and manganese peroxidase (MnP), to mineralize and/or to decolorize azo dyes. Biosorption is defined as binding of solutes to the bio-mass by processes that do not involve metabolic energy or transport, although such processes may occur simultaneously where live biomass is used. Therefore, it can occur in either living or dead biomass. Many genera of fungi have been employed for the dye decolorization either in living or dead form.

Many bacteria and fungi are used for the development of biological processes for the treatment of textile effluents. Containing various substituents such as nitro and sulfonyl groups, synthetic dyes are not uniformly susceptible to decomposition by activated sludge in a conventional aerobic process. Attempts to develop aerobic bacte-rial strains for dye decolorization often resulted in a specific strain, which showed a strict ability on a specific dye structure. The use of lignin-degrading white-rot fungi (WRF) has attracted increasing scientific attention, as these organisms are able to degrade a wide range of recalcitrant organic compounds. Their lignin modifying enzymes (LMEs), that is MnP, LiP, and laccases, are directly involved in the deg-radation of not only lignin in their natural lignocellulosic substrates but also various xenobiotic compounds including dyes. Peroxidases and laccases of WRF are oxidative enzymes, which do not need any other cellular components to work. They have broad substrate specificity and are able to transform a wide range of toxic compounds. These enzymes, which are widely distributed in nature, have been studied for many years because of their potential use as biocatalysts in pulp and paper bleaching, wastewater treatment, soil remediation, on-site waste destruction, and medical diagnostics.

ENZYMES OF WHITE-ROT FUNGI INVOLVED
IN AZO DYE DECOLORIZATION

WRF are key regulators of the global C-cycle. Some WRF produce all three LMEs, while others produce only one or two of them (Hatakka, 1994). The main LMEs are oxidoreductases, that is two types of peroxidases, LiP and MnP, and a phenoloxidase Laccase. LMEs are produced by WRF during their secondary metabolism. Synthesis and secretion of these enzymes are often induced by limited nutrient (C or N) levels (Wesenberg et al., 2003).

The proposed mechanism for the functionality of MnP involves the oxidation of manganous ions Mn^{2+} to Mn^{3+}, which is then chelated with organic acids. The chelated Mn^{3+} diffuses freely from the active site of the enzyme and can oxidize secondary substrates (Kariminiaae-Hamedaani et al., 2007).

LiP catalyzes several oxidations in the side chains of lignin and related compounds (Tien and Kirk, 1983) by one-electron abstraction to form reactive radicals. Also, the cleavage of aromatic ring structures has been reported (Umezawa and Higuchi, 1987). The role of LiP in ligninolysis could be the further transformation of lignin fragments, which are initially released by MnP.

Fungal laccases as part of the ligninolytic enzyme system are produced by almost all wood-rotting basidiomycetes. This group of N-glycosylated extracellular blue oxidases with molecular masses of 60–390 kDa (Call and Mucke, 1997) contains four copper atoms in the active site (as Cu^{2+} in the resting enzyme). Laccases catalyze the oxidation of a variety of aromatic hydrogen donors with the concomitant reduction of oxygen to water. Laccase is an oxidase with a redox potential of 780 mV and can catalyze the oxidation of organic pollutants by reduction of molecular oxygen straightforwardly to water in the absence of hydrogen peroxide or even other secondary metabolites. While anthraquinone was directly oxidized by the laccase, azoic and indigo dyes were not the substrates of laccase, and small molecule metabolites mediated the interaction between the dyes and the enzyme (Levin and Forchiassin, 2001).

DYE DEGRADATION BY IMMOBILIZED FUNGI

Fungal cultures are used as free or immobilized cultures for decolorization processes under static and/or agitated conditions. Free cell cultures could decolorize

White Rot Fungi and Their Enzymes Able to Decolorize Azo Dyes (Khalid et al., 2010)

White-Rot Fungi	Enzyme	Azo Dye
Coriolus versicolor	Laccase	Drimarene blue
Dichomitus squalens	Laccase and MnP	Orange G
Funalia trogii	Laccase	Astrazone blue, drimarene blue
Ganoderma lucidum	Laccase	Reactive black 5
Ischnoderma resinosum	Laccase	Orange G
Lentinula edodes	MnP	Congo red, trypan blue, amido black
Phanerochaete chrysosporium	LiP	Diazo dyes, reactive brilliant red K-2BP
	MnP and β-glucosidase	Amaranth, new coccine, and orange G
Phanerochaete sordida	MnP	Reactive red 120
Pleurotus calyptratus	Laccase	Orange G
Pleurotus ostreatus	Laccase	Drimarene blue
	LiP	Disperse orange 3, Disperse yellow 3
Pleurotus sajorcaju	Laccase	Amaranth, new coccine, orange G, Reactive black 5

Immobilized Fungi Able to Decolorize Azo Dyes

Immobilized Fungi	Azo Dyes
Aspergillus fumigatus	Reactive dye K-2BP
Funalia trogii	Acid black 52, astrazon red dye, drimarene blue, reactive black 5
Phanerochaete chrysosporium	Diazo dye red 533, orange ii, reactive black 5, acid orange, acid red 114, Congo red, direct yellow 12, acid black 1, reactive orange 16, basic blue 41, reactive dye K-2BP.
Trametes pubescens	Reactive red 243, reactive black 5
Tinea versicolor	Acid violet 7, amaranth

the dye and/or textile effluent, but it has some operational problems such as shear force and cell stability in agitated conditions. Immobilized fungal cells offer some advantages over free cells, which enhance decolorization efficiency, cell stability, reuse of biomass easier liquid–solid separation, and minimal clogging in continuous flow systems.

DEGRADATION OF DYES BY YEASTS

In spite of the fact that most investigations of microbial azo dye degradation utilize non-yeast microorganisms, a growing number of research groups have reported on several yeast species capable of decolorizing azo dyes. Yeasts are resilient microorganisms and are able to resist unfavorable environments such as low pH, high salt concentration, and high-strength organic wastewaters such as the case of textile effluents. Bioremediation of colored effluents by yeasts usually mentions non-enzymatic processes as the major mechanism for azo dye decolorization. In a first approximation based on the cellular viability status, these processes can be divided into two different types: bioaccumulation and biosorption. Bioaccumulation usually refers to an active uptake mechanism carried out by living microorganisms (actively growing yeasts). The possibility of further dye biotransformation by redox reactions may also occur due to the involvement of the yeast metabolism. The main advantage of using bioaccumulating yeasts in color removal is avoiding the need for a separate biomass production step. On the other hand, the growth and performance of bioaccumulating yeasts will be mainly constrained by the nutrients' availability, notably carbon and nitrogen sources.

Biosorption is a general phenomenon that can occur in either dead or living biomass. However, this process usually refers to a passive uptake mechanism carried out by nonviable microorganisms (dead yeasts). The biosorption process involves physical–chemical interactions between the yeast surface and the azo dyes, as well as possible passive diffusion inside dead cells.

Using nonviable cellular biomass for azo dye removal has some advantages, namely the ability to function under extreme conditions of temperature and pH, and without addition of growth nutrients. Also, waste yeast biomass, which is a byproduct of industrial fermentations such as beer production, can be used as a relatively cheap source for biosorption of azo dyes. An important setback is the fact that the use of biomass for dye removal leads to an increase in the sludge amount, which requires further removal and treatment.

It was observed that a few ascomycetous yeast species such as *Candida zeylanoides, Candida tropicalis, Debaryomyces polymorphus, Issatchenkia occidentalis, Saccharomyces cerevisiae, Candida oleophila,* and *Candida albicans* perform a putative enzymatic biodegradation and concomitant decolorization of several azo dyes. The unique member of basidiomycetous yeasts allegedly performing a putative enzymatic biodegradation of azo dyes seems to be *Trichosporon* sp. (closely related to the *Trichosporon multisporum–Trichosporon laibachii* complex).

The yeast-mediated enzymatic biodegradation of azo dyes can be accomplished either by reductive reactions or by oxidative reactions. In general, reductive reactions led to cleavage of azo dyes into aromatic amines, which are further mineralized by yeasts. Enzymes putatively involved in this process are NADH-dependent reductases and an azoreductase, which is dependent on the extracellular activity of a component of the plasma membrane redox system, identified as a ferric reductase. Significant increase in the activities of NADH-dependent reductase and azoreductase was observed in the cells of *Trichosporon beigelii* obtained at the end of the decolorization process.

Yeast-mediated color removal by a putative process of biosorption of azo dyes by yeast biomass belonging to *Rhodotorula* sp. is reported. Yeast species such as *Kluyveromyces marxianus* removed the diazo dye remazol black B, *Candida catenulata* and *Candida kefyr* removed more than 90% of amaranth by biosorption. Biosorption uptake of the textile azo dyes remazol blue, reactive black, and reactive red by *S. cerevisiae* and *C. tropicalis* varied according to the selected dye, dye concentration, and exposure time is also reported.

Yeast	Azo Dyes
Candida curvata	Orange II
Geotrichum candidum	Reactive red 33, acid red 73, acid blue 324, reactive black 5
Candida zeylanoides	Methyl orange, orange II
Debaryomyces polymorphus	Reactive black 5, reactive red, reactive yellow, reactive brilliant red
Candida tropicalis	Reactive red
Issatchenkia occidentalis	Methyl orange, orange II
Saccharomyces italicus	Reactive brilliant red
Candida krusei	Reactive brilliant red, acid brilliant red B
Pseudozyma rugulosa	Reactive brilliant red, reactive black KN-B, acid mordant yellow
Candida oleophila	Reactive black 5
Trichosporon multisporum	Reactive red 141
Galactomyces geotrichum	Methyl red, amido black 10B
Candida albicans	Direct violet 51
Trichosporon beigelii (syn. *Trichosporon cutaneum*)	Reactive blue 171, reactive red 141, reactive green 19 a, reactive yellow 17, reactive orange 94

FURTHER READING

Call HP, Mucke I. Minireview: History, overview and applications of mediated ligninolytic systems, especially laccase-mediator-systems (Lignozym-Process). *J. Biotechnol.* 1997;53:163–202.

Chung KT, Cerniglia CE. Mutagenicity of azo dyes: Structure–activity relationships. *Mut. Res.* 1992;277:201–220.

Chung KT, Stevens SE. Decolourisation of azo dyes by environmental microorganisms and helminthes. *Environ. Toxicol. Chem.* 1993;12:2121–2132.

Erkurt HA. *The Handbook of Environmental Chemistry.* Springer-Verlag Berlin Heidelberg; 2010.

Gioia L, Ovsejevi K, Manta C, Míguez D, Menendez P. Biodegradation of acid dyes by an immobilized laccase: an ecotoxicological approach. *Environ. Sci. Water Res. Technol.* 2018;4:2125–2135.

Hatakka A. Lignin-modifying enzymes from selected white-rot fungi: Production and role in lignin degradation. *FEMS Microbiol. Rev.* 1994;13:125–135.

Kariminiaae-Hamedaani HR, Sakurai A, Sakakibara M. Decolorization of synthetic dyes by a new manganese peroxidase-producing white rot fungus. *Dyes Pigm.* 2007;72:157–162.

Khalid A, Arshad M, Crowley D. *Biodegradation of Azo Dyes.* Erkurt HA (eds.) Springer-Verlag Berlin Heidelberg; 2010.

Lee YH, Pavlostathis SG. Reuse of textile reactive azo dyebaths following biological decolorization. *Water Environ. Res.* 2004;76:56–66.

Levin L, Forchiassin F. Ligninolytic enzymes of the white rot basidiomycete *Trametes trogii. Acta Biotechnol.* 2001;21:179–186.

Pinheiro HM, Touraud E, Thomas O. Aromatic amines from azo dye reduction: Status review with emphasis on direct UV spectrophotometric detection in textile industry wastewaters. *Dyes Pigm.* 2004;61:121–139.

Rittman BE, Whiteman R. Bioaugmentation: A coming of age. *Biotechnology.* 1994;1:12–16.

Singh SN. *Microbial Degradation of Synthetic Dyes in Wastewaters.* New York Dordrecht London: Springer International Publishing; 2014.

Spadaro JT, Gold MH, Renganadhan V. Degradation of azo dyes by the lignin degrading fungus *Phanerochaete chrysosporium. Appl. Environ. Microbiol.* 1992;58:2397–2401.

Tien M, Kirk TK. Lignin-degrading enzyme from the *Hymenomycete Phanerochaete chrysosporium* Burds. *Science* 1983;221:661–663.

Umezawa T, Higuchi T. Mechanism of aromatic ring cleavage of h-O-4 lignin substructure models by lignin peroxidase. *FEBS Lett.* 1987;218:255–260.

van Limbergen HV, Top EM, Verstrate W. Bioaugmentation in activated sludge: Current features and future perspectives. *Appl. Microbiol. Biotechnol.* 1998;50:16–23.

Wesenberg D, Kyriakides I, Agathos N. White-rot fungi and their enzymes for the treatment of industrial dye effluents. *Biotechnol. Adv.* 2003;22:161–187.

13 Biosurfactants

There are large number of reports of various chemicals, produced synthetically or occurring naturally, showing the properties of surfactant which show specific and preferential interaction at surfaces and interfaces between fluid phases having different degrees of polarity and hydrogen bonding, e.g., oil and water or air and water interfaces (Desai and Banat, 1997). This is the result of the presence of both hydrophobic and hydrophilic moieties at the surface of these molecules which results in their orientation at the interface and is known as surfactants. Surfactants are very versatile and have found uses in as detergents lowering the interfacial tension, emulsifiers, dispersants, de-emulsifiers, wetting agents, foam retardants, stabilizers, gelling agents, etc. (Lin, 1996; Makkar and Cameotra, 2002; Mukherjee et al., 2006; Singh et al., 2007). Chemical surfactants are generally produced as byproducts of the petrochemical industry and consist primarily of alkylbenzene sulfonates, alkyl phenol ethoxylates, synthetic fatty alcohols and their derivatives. These products are believed to account for 70%–75% of the surfactant consumption in the industrialized countries. Around 60% of the surfactant production is used in household detergents, 30% in industrial and technical applications, 7% in industrial and institutional cleaning, and 6% in personal care (Edsar, 2006). Although chemical surfactants are both inexpensive and efficient, they have adverse effect on the environment causing pollution. The potential advantages of biosurfactant include biodegradability resulting in lower levels of pollution, low toxicity, biocompatibility, and digestibility which allows their application in cosmetics, pharmaceuticals, and as functional food additives, can be produced from cheap raw materials which are available in large quantities. Similarly, they show selectivity and specificity toward hydrocarbon substrates. Their compatibility with chemical product generally leads to novel formulations. Earlier work on biosurfactants mainly focused on the properties, biosynthesis, and chemistry which have been reviewed by many workers (Banat et al., 2000; Maier, 2003).

ENHANCING BIOAVAILABILITY

SYNTHETIC SURFACTANTS AND BIOREMEDIATION

The lack of available pollutant as the microbial substrate is one of the serious barriers to bioremediation because the timescales become protracted due to mass transfer to the aqueous phase becoming the rate-limiting step. This applies to both contaminated soils and groundwater. Some surfactants enhance the solubilization and removal of contaminants.

There are major concerns about the large-scale use of surfactants in this manner. In particular, surfactants vary greatly in their toxicity to humans and ecotoxicity, and their resistance to biodegradation may lead to increased pollution. This is one of the

DOI: 10.1201/9781003272618-15

major barriers to the development of the technique. For soil remediation systems, another technology-inhibiting observation is that the addition of surfactants to soils can form highly viscous emulsions that are difficult to remove. Large quantities of surfactants are also required, and in a soil system, large quantities of aqueous chemicals can ruin soil permeability.

The problem of large volumes and soil saturation may be overcome by the use of surfactant foams. Large volumes of air per unit volume of foam are injected into soil, and the foam contains 70%–90% air. In a laboratory study that compared Triton X-100 liquid and foam injection into soil contaminated with PCP, twice as much PCP was removed by the foam than with liquid surfactant solution.

A POTENTIAL ROLE FOR BIOSURFACTANTS

Many, but not all, biosurfactants are produced as a response to the low water solubility of n-alkanes as growth substrates. Oxygen is more soluble in the oil phase than the aqueous phase, and the low water solubility of alkanes places an additional pressure on microbial growth: that of mass transfer from the oil into the aqueous phase. Biosurfactants may offer several advantages over synthetic surfactants. They are generally less toxic and more biodegradable than synthetic surfactants. They are very effective surfactants, having about a 10- to 40-fold lower critical micelle concentration than synthetic surfactant.

TYPES OF BIOSURFACTANTS

Diverse molecules are produced as biosurfactants by prokaryotic and eukaryotic microorganisms, but they can be divided into groups based on overall structure: glycolipids, lipopeptides, and high-molecular weight biopolymers. Those most commonly associated with oil-degrading bacteria are the glycolipids, and the hydrophilic head group is normally rhamnose or trehalose. The trehalose glycolipids have trehalose, a nonreducing disaccharide, linked by an ester bond to long-chain fatty acids.

These compounds vary in overall chain length, with the largest and most complex being found in the mycobacteria (C_{20} to C_{90}). This variability in structure means that the molecules have different hydrophile–lipophile balances, and therefore, might be useful in solubilizing a wide range of pollutants.

BIOSURFACTANTS AND ALKANE METABOLISM

Three types of alkane uptake response have been postulated, namely, uptake of monodispersed dissolved alkanes in the aqueous phase (minimal for alkanes above C_{10}), direct contact of cells with large oil droplets, and contact with fine oil droplets. An interesting comparison can be drawn between pseudomonads and rhodococci in their biosurfactant physiology. In the rhodococci, the nonionic glycolipids are not excreted from the cell, and they render the cell surface hydrophobic, which may then facilitate the attachment and subsequent passive transport of alkanes into the cell, requiring neither energy nor specialized membrane components. This is consistent with the observation that rhodococci almost always are highly adherent to alkanes and seldom render the aqueous phase of dual-phase culture media turbid. Being cell wall associated, the nonionic trehalose lipids may also impede the wetting of the cell surface. This would increase the direct cell–hydrocarbon contact at the aqueous–hydrocarbon interface. In contrast, the anionic rhamnolipids from some Pseudomonas strains are released into the culture medium and cause turbidity by emulsification. The third mechanism is likely to be quantitatively more important in pseudomonads, since these cells do not have a highly hydrophobic surface.

PROPERTIES OF BIOSURFACTANTS

Biosurfactants are of increasing interest for commercial use because of the continually growing spectrum of available substances. The main distinctive features of biosurfactants and a brief description of their properties are as given below:

Surface and Interfacial Activity

A good surfactant can lower the surface tension of water from 72 to 35 mN/m and the interfacial tension of water/hexadecane from 40 to 1 mN/m. Surfactin from *Bacillus subtilis* can reduce the surface tension of water to 25 mN/m and interfacial tension of water/hexadecane to <1 mN/m. Rhamnolipids from *Pseudomonas aeruginosa* decrease the surface tension of water to 26 mN/m and the interfacial tension of water/hexadecane to <1 mN/m.

Temperature, pH, and Ionic Strength Tolerance

Many biosurfactants and their surface activities are not affected by environmental conditions such as temperature and pH.

Biodegradability

Unlike synthetic surfactants, microbial-produced compounds are easily degraded and particularly suited for environmental applications such as bioremediation and dispersion of oil spills.

Emulsion Forming and Emulsion Breaking

Stable emulsion can be produced with a life span of months and year. Higher molecular mass biosurfactants are in general better emulsifier that the low molecular mass biosurfactants. Sophorolipids from *Torulopsis bombicola* have been shown to reduce surface tension, but are not good emulsifiers. By contrast, liposan does not reduce the surface tension, but has been used successfully to emulsify edible oils. Polymeric surfactants offer additional advantages because they coat droplets of oil, thereby forming the stable emulsions. This property is especially useful for making emulsion for cosmetics and food.

Chemical Diversity

The chemical diversity of naturally produced biosurfactants offer a wide selection of surface-active agents with properties closely related to specific applications.

Low Toxicity

Microbial surfactants are generally considered as the low or non-toxic products and therefore are appropriate for pharmaceutical, cosmetic, and food industries.

A report suggested that a synthetic anionic surfactant (corexit) displayed an LC50 (concentration lethal to 50% of test species) against *Photobacterium phosphoreum* ten times lower than rhamnolipids, demonstrating the higher toxicity of chemical-based surfactants. It was also reported that biosurfactants showed higher EC50 (effective concentration to decrease 50% of test population) value than synthetic surfactants (Poremba et al., 1991).

POTENTIAL APPLICATIONS OF BIOSURFACTANTS

Most important aspect of biosurfactants is their environmental acceptability, because they are readily biodegradable and have low toxicity than synthetic surfactants. These unique properties of biosurfactants allow their use and possible replacement of chemically synthesized surfactants in a great number of industrial applications. Some of the major applications of biosurfactants in pollution and environmental control are microbial enhanced oil recovery (MEOR), hydrocarbon degradation, hexachlorocyclohexane (HCH) degradation, and heavy metal removal from contaminated soil.

MICROBIAL ENHANCED OIL RECOVERY (MEOR)

An area of considerable potential for biosurfactant application is MOER. In MEOR, microorganisms in reservoir are stimulated to produce polymers and surfactants, which aid MEOR by lowering interfacial tension at the oil–rock interface. To produce microbial surfactants in situ, microorganisms in the reservoir are usually provided with low-cost substrates, such as molasses and inorganic nutrients. However, to be useful for MEOR in situ, bacteria must be able to grow under extreme conditions encountered in oil reservoirs such as high temperature, pressure, salinity, and low oxygen level. Several aerobic and anaerobic thermophiles tolerant of pressure and moderate salinity have been isolated which are able to mobilize crude oil in the laboratory (Post and Al-Harjan, 1988).

HYDROCARBON DEGRADATION

Hydrocarbon-utilizing microorganisms excrete a variety of biosurfactants. An important group of such surfactants is mycolic acids which are the α-alkyl, β-hydroxy very long-chain fatty acids contributing to some characteristic properties of a cell such as acid fastness, hydrophobicity, adherability, and pathogenicity. This product has many applications in agrochemistry, mineral flotation, and bitumen production and processing. Further, the product may be used as an emulsifying and dispersing agent while formulating herbicides, pesticides, and growth regulator preparations. The constituent fatty acids of biolipid extract also have anti-phytoviral and antifungal activities, and therefore, can be applied in controlling plant diseases (Voigt et al., 1985).

HYDROCARBON DEGRADATION IN SOIL ENVIRONMENT

Degradation is dependent on the presence in soil of hydrocarbon-degrading species of microorganisms, hydrocarbon composition, oxygen availability, water, temperature, pH, and inorganic nutrients. Addition of synthetic surfactants or microbial surfactants results in increased mobility and solubility of hydrocarbon, which is essential for effective microbial degradation.

HYDROCARBON DEGRADATION IN AQUATIC ENVIRONMENT

When oil is spilled in aquatic environment, the lighter hydrocarbon components volatilize, while the polar hydrocarbon components dissolve in water. However, because of low solubility (<1 ppm) of oil, most of the oil components will remain on the water surface. The primary means of hydrocarbon removal are photooxidation, evaporation, and microbial degradation. Since hydrocarbon-degrading organisms are present in seawater, biodegradation may be one of the most efficient methods of removing pollutants (Atlas, 1981).

Emulsan, a high MW lipopolysaccharide produced by *Acinetobacter calcoaceticus* RAG-1 has been proposed for a number of applications in the petroleum industry such as to clean oil and sludge from barges and tanks, reduce viscosity of heavy oils, enhance oil recovery, and stabilize water-in-oil emulsions in fuels.

BIOSURFACTANT AND HCH DEGRADATION

HCH is still the highest-ranking pesticide used in India and many other countries. Of the eight known isomers of HCH, the alpha-form constitutes more than 70% of the technical product, which is not only known insecticidal but also a suspected carcinogen. The poor solubility is one of the limiting factors in the microbial degradation of alpha-HCH. Presence of six chlorines in the molecule is another factor that renders HCH lipophilic and persistent in the biosphere. It has been reported that addition of biosurfactant from *Pseudomonas* Ptm+ strain facilitated 250-fold increase in dispersion of HCH in water. Addition of either this organism or biosurfactant dislodged surface-borne HCH residues from many types of fruits, seeds, and vegetables as well.

Laboratory-scale studies have revealed that microbial surfactants are very efficient in cleaning the containers where HCH residues were sticking to the wall.

Some more applications of biosurfactants include the following:

i. Binding of heavy metals. A rhamnolipid biosurfactant has been shown to be capable of removing Cd, Pb, and Zn from soil. The mechanism by which rhamnolipid reduces metal toxicity may involve a combination of rhamnolipid complexation of Cd and rhamnolipid interaction with the cell surface to alter Cd uptake.

ii. Food industry. Lecithin and its derivatives, fatty acid esters containing glycerol, sorbitan or ethylene glycol, and ethoxylated derivatives of monoglycerides including a recently synthesized oligopeptide are currently in use as emulsifier in the food industry.

iii. Cosmetic industry. A large number of compounds for cosmetic applications are prepared by enzymatic conversion of hydrophobic molecules by various lipases and whole cells.

The cosmetic industry demands surfactants with a minimum shelf life of 3 years. Therefore, saturated acyl groups are preferred over the unsaturated compounds. Monoglycerides, one of the widely used surfactants in the cosmetic industry, has been reported to be produced from glycerol–tallow (1.5:2) with a 90% yield by using *Pseudomonas fluorescens* lipase treatment.

iv. Medicinal uses. A deficiency of pulmonary surfactant, a phospholipid–protein complex, is responsible for the failure of respiration in prematurely born infants. The isolation of genes for protein molecules of this surfactant and cloning in bacteria has made possible its fermentative production for medical application. 1% Emulsion of rhamnolipids is successfully used for the treatment of *Nicotiana glutinosa* infected with tobacco mosaic virus and for the control of potato virus-x disease.

The properties of surfactant molecules make them the most versatile of process chemicals appearing in wide range of product starting from household usage to medicinal chemistry and then to industries. The last decade has seen the extension of surfactant applications to high-technology areas such as electronic printing, magnetic recording, microelectronic, biotechnology, and diversified medicinal research. In surge of green chemistry, the biologically compatible surfactants are in demand to replace some of the existing chemical surfactants. The reason for the popularity of biosurfactants as high-value microbial products is primarily because of their specific action, low toxicity, higher biodegradability, effectiveness at extreme temperatures, wide spread applicability, and their structure which provide different properties than that of the classical surfactants.

Biological surfactants are highly sought-after biomolecules as fine specialty chemicals; biological control agents; and new generation molecules for pharmaceutical, cosmetic, and health care industries.

FURTHER READING

Atlas RM. Microbial degradation of petroleum hydrocarbons: An environmental. *Microbiol. Rev.* 1981;45(1):180–209.
Banat IM, Makkar RS, Cameotra SS. Potential commercial applications of microbial surfactants. *Appl. Microbiol. Biotechnol.* 2000;53:495–508.
Desai JD, Banat IM. Microbial production of surfactants and their commercial potential. *Microbial. Mol. Biol. Rev.* 1997;61:47–64.
Edsar C. Focus on surfactants. *Latest Market Anal.* 2006;5:1–2.
Kourkoutas Y, Banat IM. Biosurfactant production and application. In: Pandey AP (ed.) *The Concise Encyclopedia of Bioresource Technology*. Philadelphia, PA: Haworth Reference Press; 2004:505–515.
Lang S, Wagner F. Structure and properties of biosurfactants. In: Kosaric N, Cairns WL, Gray NCC (eds.) *Biosurfactants and Biotechnology*. New York: Marcel Dekker;1987:21–45.
Lin SC. Biosurfactants: Recent advances. *J. Chem. Tech. Biotechnol.* 1996;66:109–120.
Maier RM. Biosurfactants: Evolution and diversity in bacteria. *Adv. Appl. Microbiol.* 2003;52:101–121.
Makkar RS, Cameotra SS. An update on the use of unconventional substrates for biosurfactant production and their new applications. *Appl. Microbiol. Biotechnol.* 2002;58:428–434.

Mukherjee S, Das P, Sen R. Towards commercial production of microbial surfactants. *Trends Biotechnol.* 2006;24:509–515.

Mulligan CN, Gibbs BF. Factors influencing the economics of biosurfactants. In: Kosaric N (ed.) *Biosurfactants: Production: Properties: Applications.* New York: Marcel Dekker; 1993:329–371.

Mulligan CN, Sharma SK, Mudhoo A. *Biosurfactants: Research Trends and Applications.* CRC Press; 2014.

Poremba K, Gunkel W, et al. Toxicity testing of synthetic and biogenic surfactants on marine microorganisms. *Environ. Toxicol. Water Qual.* 1991;6:157–163.

Post FJ, Al-Harjan FA. Surface activity of halobacteria and potential use in microbial enhanced oil recovery system. *Appl. Microbiol.* 1988;11:97–101.

Sen R. *Biosurfactants.* Landes Bioscience and Springer Science+Business Media; 2010.

Singh A, Hamme V, Jonathan D et al. Surfactants in microbiology and biotechnology: Part 2. Application aspects. *Biotechnol. Adv.* 2007;25:99–121.

Soberon-Chavez G. *Biosurfactants: From Genes to Applications.* Springer; 2010.

Syldatk C, Wagner F. Production of biosurfactants. In: Kosaric, N, Cairns WL, Gray NC (eds.) *Biosurfactants and Biotechnology,* Vol. 25: Surfactant Science Series. New York: Marcel Dekker;1987:89–120.

Voigt B, Mueller H et al. Antiphytovirale Aktivität von lipophilen Fraktionen aus der Hefe *Lodderomyces elongisporus* IMET H 128. *Acta Biotechnol.* 1985;5:313–317.

14 Biosensors as Environmental Monitors

A biosensor is a two-component analytical device comprised of a biological recognition element that outputs a measurable signal to an interfaced transducer. Biorecognition typically relies on enzymes, whole cells, antibodies, or nucleic acids, whereas signal transduction exploits electrochemical (amperometric, chronoamperometric, potentiometric, field-effect transistors, conductometric, capacitive), optical (absorbance, reflectance, luminescence, chemiluminescence, bioluminescence, fluorescence, refractive index, light scattering), piezoelectric (mass sensitive quartz crystal microbalance), magnetic, or thermal (thermistor, pyroelectric) interfaces.

The detection of specific analytes of importance to environmental monitoring can be achieved with great precision using analytical techniques that center around mass spectrometry (MS), such as gas chromatography (GC)-MS, liquid chromatography (LC)-MS, liquid chromatography coupled to tandem MS (LC-MS2), ion trap (IT)-MS, and quadrupole linear ion trap (QqLIT)-MS. With great precision, however, comes significant time, effort, and expense. Samples must be collected and transported to the obligatory confinements of the laboratory, and requisite preconcentration and cleanup steps must be performed prior to the sample being analyzed on an expensive, high-technology instrument by accompanying trained technical personnel. Considering that some percentage of the samples collected will be negative, either not being contaminated or containing the target analyte at concentrations too low to be detected, the adjusted cost on a per positive sample basis can be extensive.

Although biosensors cannot unequivocally replace the replicate accuracy and reproducibility of conventional analytical instrumentation, they can complement and supplement their operation through ease of sample pre-processing, which is often minimal to none, on-site field portability, simplicity and rapidity of operation, versatility, real-time to near-real-time monitoring capabilities, and miniaturization that has evolved down to a "lab-on-a-chip" format. Biosensors can therefore often find their niche as continuous monitors of environmental contamination or as bioremediation process monitoring and control tools to provide informational data on what contaminants are present, where they are located, and a very sensitive and accurate evaluation of their concentrations in terms of bioavailability. Bioavailability measurements are central to environmental monitoring as well as risk assessment because they indicate the biological effect of the chemical, whether toxic, cytotoxic, genotoxic, mutagenic, carcinogenic, or endocrine disrupting, rather than mere chemical presence as is achieved with analytical instruments. Despite their benefits, biosensors remain relatively unused in the environmental monitoring/bioremediation fields, primarily due to a lack of real-world, real-sample testing and standardization against conventional analytical techniques. Thus, although biosensors show significant promise, it is clear that more field validation studies need to be performed before regulatory agencies and other end users will gain sufficient confidence to adopt their routine use.

ENZYME-BASED BIOSENSORS

The historical foundation of the biosensor rests with the enzyme glucose oxidase and its immobilization on an oxygen electrode by Leland Clark in the 1960s for blood glucose sensing. Although the research emphasis of enzyme-based biosensors continues to be driven by more lucrative medical diagnostics, there has been a predictable application overlap toward environmental monitoring as well. Enzymes act as organic catalysts, mediating the reactions that convert substrate into product. Since enzymes are highly specific for their particular substrate, the simplest and most selective enzyme-based biosensors merely monitor enzyme activity directly in the presence of the substrate. Perhaps, the most relevant examples are the sulfur/sulfate-reducing bacterial cytochrome c3 reductases that reduce heavy metals. They immobilized cytochrome c3 on a glassy carbon electrode and monitored its redox activity amperometrically in the presence of chromate [Cr(VI)] with fair sensitivity (lower detection limit of 0.2 mg/L) and rapid response (several minutes) (Michel et al., 2003). When tested under simulated groundwater conditions, the biosensor did cross-react with several other metal species, albeit at lower sensitivities, and was affected by environmental variables such as pH, temperature, and dissolved oxygen, thus exemplifying certain disadvantages common to enzyme-based biosensors. Similarly, operated biosensors for the groundwater contaminant perchlorate using perchlorate reductase as the recognition enzyme (detection limit of 10 μg/L) (Okeke et al., 2007), organophosphate pesticides using parathion hydrolase or organophosphorus hydrolase as recognition enzymes (detection down to low μM concentrations) (Trojanowicz, 2003), and environmental estrogens using tyrosinase as the recognition enzyme (detection down to 1 μM) (Andreescu and Sadik, 2004) have also been designed.

   Another type of enzyme biosensor relies on enzyme activation upon interaction with the target of interest. For example, heavy metals in the form of cofactors—inorganic ions that bind to and activate the enzyme—can be detected based on this integral association. Metalloenzymes such as alkaline phosphatase, ascorbate oxidase, glutamine synthetase, and carbonic anhydrase require association of a metal ion cofactor with their active sites for catalytic activity, and can thus be used as recognition elements for heavy metals. Strong chelating agents are first used to strip the enzyme of all metal ion cofactors to form the inactive apoenzyme. Upon exposure to the sample, the apoenzyme binds any metal ions present and is reactivated, and this rate of reactivation can be related directly to the stoichiometric amount of metal complexed to the enzyme's active site. Alkaline phosphatase, for example, can be applied in this regard as a biosensor for zinc [Zn(II)] or ascorbate oxidase for biosensing copper(II) with detection limits down to very low part-per-billion levels (Satoh and Iijima, 1995). However, as various other metals as well as other sample cross-contaminants can act as cofactors and/or inhibitors of the metalloenzyme, selectivity becomes somewhat problematic. To enhance selectivity, molecular techniques such as site-directed mutagenesis or directed evolution can be used to genetically engineer or select for enzymes with superior specificity for the target, as has been accomplished with carbonic anhydrase and its selective biosensing of Zn(II) (Fierke and Thompson, 2001).

Alternatively, and more commonly, target analytes such as heavy metals can also inhibit enzyme activity, thereby diminishing the conversion of substrate to product. By monitoring subsequent Michaelis–Menten rate kinetics, a highly sensitive measurement of target can be obtained, often at picomolar detection limits, with little prerequisite sample processing. However, selectivity cannot be ascertained since the specificity of enzymes toward inhibitors is not target specific. Thus, inhibitor-based biosensors detect the global presence of heavy metals rather than identifying a particular heavy metal ion in a sample. The standard suite of enzymes used in these biosensors includes oxidases, urease, alkaline phosphatase, choline esterases, and invertase for the detection of arsenic, bismuth, beryllium, zinc, mercury, cadmium, lead, and copper.

Cholinesterases (acetylcholinesterase, butyrylcholinesterase) are other well-applied biosensor enzymes geared toward the detection of organophosphorus and carbamate pesticides/insecticides, and tyrosinase (polyphenol oxidase) and peroxidase have been extensively applied for phenols. Other environmentally relevant inhibition enzymes include acid phosphatase for the detection of arsenic(V) and protein phosphatase 2A for algal microcystins (both nonselectively). By co-immobilizing several enzymes on the same transducer, biosensors capable of multiplexed sensing can be achieved. Additionally, as with the activated enzymes, the activity of inhibitory enzymes can be enhanced through genetic engineering, as has been accomplished with acetylcholinesterase, where site-directed mutagenesis yielded enzymes with 300-fold more sensitivity to the organophosphate target dichlorvos (Boubik et al., 2002).

Biosensors can also incorporate non-enzymatic proteins or peptides as sensory elements. For environmental monitoring, this typically involves metal ion sensing and is accomplished either through the use of naturally occurring or engineered proteins. Metallothionein, for example, is a mammalian-derived metal-binding protein that has been incorporated into an optical biosensor for nondiscriminatory detection of cadmium, zinc, or nickel (Wu and Lin, 2004). Glutathione and phytochelatin proteins are also commonly used. Specificity is broad and interference or inhibition by other metals or sample constituents is problematic but can be partially addressed through genetic engineering to add, subtract, or replace amino acids to acquire improved metal-binding motifs, as has been done with phytochelatin (Bontidean et al., 2003).

The most common means for signal transduction in enzyme-based biosensors are electrochemical transducers, where the biological recognition event is converted into an electrical signal and measured as either a current (amperometric) or a potential (potentiometric). Impedance can also be measured, usually as a change in conductivity in the medium as the target and receptor interact. The basic enzyme-based biosensor design relies on the redox nature of the enzyme and its requirement for a natural redox partner such as an electron-transfer protein or a small molecule co-substrate. The catalytic redox reaction occurring upon target substrate addition yields a current that is proportional, after calibration, to the amount of substrate added. Employing an electrochemical working electrode to monitor and measure this current forms the biosensor, with electron transfer between the enzyme and the electrode occurring either directly or through a mediator such as NAD, FAD, or ubiquinone. One of the most critical functional aspects of enzyme-based biosensors is the immobilization of the enzyme and/or mediator on the electrode surface, since this affects enzyme/ mediator integrity and ultimately influences biosensor stability, longevity, and sensitivity. Immobilization techniques include adsorption, covalent attachment, entrapment in polymeric matrices such as sol–gels or Langmuir–Blodgett films, or direct cross-linking using polymer networks or antibody/enzyme conjugates (Lojou and Bianco, 2006). The recent integration of redox active carbon-based nanomaterials (nanofibers, nanotubes, nanowires, and nanoparticles) as transducers and their unique ability to interact with biological materials realizes promising advancements in enzyme biosensor design and sensitivity.

Optical transducers (absorption, reflectance, luminescence, chemiluminescence, evanescent wave, surface plasmon resonance) are also commonly employed in enzyme-based biosensors. These can be as simple as optically registering a pH change using a pH reactive dye; for example, bromocresol purple can be immobilized within an acetylcholinesterase-based biosensor to monitor pH changes related to this enzyme's activity upon exposure to pesticides. The hydrolysis of acetylcholinesterase releases protons (H^+), resulting in a decrease in pH, which in turn instigates a decrease in the absorption spectra of bromocresol purple.

The integration of fiber optic cables among optical transducer methodologies is popular due to their ability to transfer signals over long distances, thus permitting sampling in inaccessible areas and remote sampling and monitoring in areas deemed too hazardous or harsh for personnel entry. Luminol is widely used as an electrochemiluminescent indicator, reacting with the acetylcholinesterase/choline oxidase hydrogen peroxide byproduct to yield luminescent light signals that have also been used to quantify pesticide concentrations.

WHOLE CELL-BASED BIOSENSORS

Whole-cell biosensors consist of two components: a bioreporter strain that functions as the detector of toxicity or a specific pollutant, coupled to a signal transducer that converts the response from the bioreporter to a detectable electric signal. Bioreporters are typically composed of a promoter (responsible for sensing or interacting with the target analyte) fused to a reporter gene (responsible for generation of the detectable biological signal) encoded on a plasmid or genetic construct, which is then transformed into a cell. The combination of a bioreporter with appropriate sensing technology in an integrated biosensor has great potential for environmental contaminant detection, with a multitude of possible pollutants monitored by various biosensors, depending on the reporter gene, promoter element, and detection method chosen.

Many types of reporter genes are employed in chemical detection, the most common being those that use luciferase (luc or lux) or green fluorescence protein (gfp and its derivatives) genes. The luc genes were first isolated from *Photinus pyralis* (firefly) and encode for luciferase, which catalyzes the two-step conversion of d-luciferin (which must be added to the reaction) to oxyluciferin, with light emission at 560 nm. Similarly, lux genes have been isolated from several bacterial sources (e.g., *Vibrio* spp. and *Photorhabdus luminescens*), all of which are encoded on a single operon in the order luxCDABE. Within this operon, luxAB encodes for the luciferase enzyme, which converts a long-chain aldehyde substrate (encoded by luxCDE) and a reduced flavin mononucleotide (FMNH2) to FMN and a long-chain carboxylic acid, with light production at 490 nm. Biosensors that utilize the entire luxCDABE operon are self-contained, in that there is no need for substrate addition, although production and recycling of the long-chain aldehyde substrate requires energy expenditure by the cell. However, for biosensors that use luxAB, without luxCDE, the aldehyde substrate must be continually added to the reaction medium (as is also the case for luc-based luminescence), complicating their use in integrated biosensors. Bioreporters that use the green fluorescent protein (gfp) gene, isolated from the jellyfish *Aequorea victoria*, do not require substrates and encode a 238-amino acid protein that becomes activated by cyclization of a tyrosine ([66]Tyr) residue. The protein fluoresces at about 508 nm upon excitation with ultraviolet or blue light. Derivatives of green fluorescent protein that has altered emission maxima are several, including blue-, red-, and yellow-shifted variants, allowing for detection of multiple contaminants simultaneously. Other fluorescent proteins, including red fluorescent protein (termed DsRed) with fluorescence emission at 583 nm, are also utilized; however, all of the fluorescence-based bioreporters require an external light source for excitation of the fluorescent protein product, also complicating their use as biosensors. Each reporter system has different light emission maxima, optimum temperature, and length of signal generation. For example, firefly luciferase (luc) has a temperature optimum at 25°C and is thermally inactivated above 30°C, whereas bacterial luciferase derived from *P. luminescens* (lux) is stable up to 42°C (Hakkila et al., 2002; Keane et al., 2002; Purohit, 2003).

For contaminant detection, fusion of a reporter gene to an appropriate promoter is of critical importance because the promoter (or response element) is the portion of the biosensor that is actually sensing or interacting with the chemical contaminant. A multitude of promoters have been used, creating a range of bioreporters that may detect contaminants as general as anything that causes toxicity or as specific as those that detect a single contaminant. There are numerous examples of bioreporters that detect toxicity, most of which contain a reporter gene fused to a strong constitutive promoter such that a decrease in signal indicates a toxic response.

Bioluminescent *Pseudomonas putida* and *Escherichia coli*, along with a naturally bioluminescent *Vibrio fischeri* (recently reclassified as *Aliivibrio fischeri*) were used to identify areas of a contaminated groundwater site in southern England that contained toxic substances. The three bacterial toxicity bioreporters differed in their assessment of the toxic potential for the samples, attributed to each strain being more or less sensitive to the primary pollutants in each sample, underscoring the benefit of using multiple bioreporters within biosensors.

A classic example of a functional whole-cell biosensor being used on environmental samples is the Microtox system, which utilizes a naturally bioluminescent *V. fischeri* for the detection of toxicity. This strain, which produces light constitutively, exhibits a decrease in bioluminescence proportional to the level of toxicity in the water sample. It has been tested with numerous classes of contaminants, including heavy metals, pesticides, chlorinated solvents, and many others. However, since this assay is based on a marine bacterium, it is less than ideal for toxicity testing of soil or freshwater samples.

In addition to toxicity-sensing bacterial bioreporters, there are also yeast-based toxicity bioreporters, such as those constructed in *Saccharomyces cerevisiae*

(Eldridge et al., 2007). The *S. cerevisiae*-based EDC bioreporter, referred to as strain BLYES, contains two plasmids, pUTK404, from which the *P. luminescens* aldehyde synthesis genes (luxCDE) and the *Vibrio harveyi* flavin oxidoreductase gene (frp) are expressed from the yeast constitutive promoters GPD and ADH1, and pUTK407, from which the *P. luminescens* luciferase genes (luxAB) are expressed under the control of estrogen response elements (EREs). The human estrogen receptor (hER) is cloned into the yeast genome. When this strain encounters an estrogenic agent, the chemical is bound by the human estrogen receptor protein, forming a complex that binds to the EREs on pUTK407, thereby initiating transcription of the luxAB genes. This genetic expression, in concert with the luxCDE genes expressed from pUTK404, yields a quantifiable light signal, allowing users to discern the estrogen concentration in samples. *S. cerevisiae* BLYAS is an analogous assay used to measure androgen class chemicals and has been shown to detect, for example, testosterone within 4 hours at a lower detection limit of 2.5×10^{-10} M. Another *S. cerevisiae* bioreporter, strain BLYR, generates bioluminescence constitutively for measurement of general sample toxicity. When strains BLYES, BLYAS, and BLYR are combined into one assay, this suite of bioreporters can be used to detect substances that are potential endocrine disruptors in environmental samples in near real time, without the addition of exogenous substrates or cell permeabilization steps.

Whole-cell biosensors may even be used to assess the remediative potential of a population at a contaminated site. To date, the only field-deployed whole-cell biosensor for bioremediation has been *P. fluorescens* HK44 genetically modified with luxCDABE fused in a naphthalene degradation pathway, so that breakdown of polycyclic aromatic hydrocarbons (PAHs) could be monitored in field lysimeters that had been contaminated with naphthalene, anthracene, and phenanthrene (Ripp et al., 2000). This biosensor contained an alginate-immobilized *P. fluorescens* bioreporter (HK44) whose bioluminescence was detected via fiber optic cables and a portable photomultiplier tube in response to degradation of pollutants. Despite an initial reduction in cell numbers, bioremediation by this strain was detected successfully in soil samples as bioluminescence in response to the presence of volatile organic compounds in contaminated soils. HK44 cells were present in soils for more than 444 days, with detectable bioluminescence in biosensors after further pollutant exposure, demonstrating the long-term survivability of this biosensor for the bioremediation of PAHs.

ANTIBODY-BASED BIOSENSORS (IMMUNOSENSORS)

Biosensors that use antibodies as recognition elements (immunosensors) are used widely as environmental monitors because antibodies are highly specific, versatile, and bind stably and strongly to target analytes (antigens). This high affinity for target, however, can also be disadvantageous since the target cannot easily be released from the antibody after the measurement has been made, resulting in many antibody-based biosensors being single-use disposable units [although the release (regeneration) can be promoted by the addition of organic solvents or chaotropic reagents, this requirement for a supplementary assay step is not optimal]. Additionally, the synthesis of antibodies and further testing and optimization of their target affinities can be a long, tedious, and expensive process, cross-reactivity with multiple analytes can occur, and antibody/antigen reactivity can be affected by environmental variables such as pH and temperature. Nonetheless, antibodies can be highly effective detectors for environmental contaminants, and advancements in techniques such as phage display for the preparation and selection of recombinant antibodies with novel binding properties assure their continued environmental application. Perhaps the best introduction to antibody-based biosensing is the AWACSS (Automated Water Analyzer Computer Supported System) environmental monitoring system developed for remote, unattended, and continuous detection of organic pollutants for water quality control (Tschmelak et al., 2005).

AWACSS uses an optical evanescent wave transducer and fluorescently labeled polyclonal antibodies for multiplexed detection of targeted groups of contaminants, including endocrine disruptors, pesticides, industrial chemicals, pharmaceuticals, and other priority pollutants, without requisite sample pre-processing. Antibody binding to a target sample analyte occurs in a short 5-minute preincubation step, followed by microfluidic pumping of the sample over the transducer element, which consists of an optical waveguide chip impregnated with 32 separate wells of immobilized antigen derivatives. As the antibody/analyte complexes flow through these wells, only antibodies with free binding sites can attach to the well surface (in what is referred to as a binding inhibition assay). Thus, antibodies with both of their binding sites bound with analyte will not attach to the surface and will pass through the detector. A semiconductor laser then excites the fluorophore label of bound antibodies, allowing for their quantification, with high fluorescence signals indicating low analyte concentrations and low fluorescence signals indicating high analyte concentrations. A fiber optic array tied to each well permits separation and identification of signals by the well, thereby yielding a simultaneous measurement of up to 32 different sample contaminants. The instrument has been used for groundwater, wastewater, surface water, and sediment sample testing with detection limits for most analytes in the ng/L range within assay times of approximately 18 minutes.

Although not as elaborate as the AWACSS, a multitude of other antibody-based biosensors have been applied as environmental monitors, traditionally serving as biosensors for pesticides and herbicides, but their target analytes have broadened considerably over the past several years to include heavy metals, PAHs, polychlorinated biphenyls (PCBs), explosives (TNT and RDX), phenols, toxins such as microcystin, pharmaceutical compounds, and endocrine disruptors (Farre et al., 2007). Detection at part-per-billion or lower concentrations is the norm. For the most part, immunosensors use electrochemical (amperometric, potentiometric, conductimetric) or optical (evanescent wave, surface plasmon resonance, total internal reflection fluorescence, etc.) transducers to detect antibody/analyte binding. Either the antibody, the analyte, or an analyte derivative complementary to the antibody can be immobilized on the transducer surface, and various binding assays can be used to directly or indirectly monitor analyte concentration, typically using one of four assays: direct, competitive, binding inhibition, or sandwich. Direct binding assays employ a pre- or naturally labeled antigen/analyte that is detectable when bound directly to an immobilized antibody and is applied more for clinical than for environmental monitoring. Competitive binding assays rely on the competition for a limited number of antibody binding sites between an immobilized or labeled analyte derivative and the target analyte.

In binding inhibition assays, a labeled antibody and its complementary analyte are first incubated together and then exposed to immobilized analyte derivative, which binds up unbound antibodies, thus yielding an inverse analyte signal. In sandwich assays, the antibody and antigen are incubated together and then a secondary labeled antibody is added which recognizes another binding site on the analyte. Measurement of this label correlates to analyte concentration. Most environmental immunosensors use competitive or binding inhibition assays in conjunction with an optical label for ease of monitoring. Optical labels are usually fluorescent or chemiluminescent, or may even consist of quantum dots in newer biosensor applications.

DNA-BASED BIOSENSORS

The foundation of nucleic acid-based biosensors relies on a transducer capable of monitoring a change in the nucleic acid's structure occurring after exposure to a target chemical. These structural changes are brought on either by the mutagenic nature of the chemical, resulting in mutations, intercalations, and/or strand breaks, or by the chemical's ability to covalently or non-covalently attach to the nucleic acid. Immobilizing the nucleic acid as a recognition layer on the transducer surface forms the biosensor, and detection of the chemically induced nucleic acid conformational change is then typically achieved electrochemically (i.e., a change in the current) or less so through optical or other means (Fojta, 2002).

Nucleic acid biosensors are generally nonselective and provide an overall indication of a potentially harmful (genotoxic, carcinogenic, cytotoxic) chemical or chemical mix in the test environment and, depending on the biosensor format, an estimate of concentration. A DNA biosensor was used to screen soil samples for genotoxic compounds, using benzene, naphthalene, and anthracene derivatives as model targets. Double-stranded DNA was immobilized on a single-use disposable screen-printed electrochemical cell operating off a handheld battery-powered potentiostat (Sassolas et al., 2008). A 10-μL drop of a pre-processed and pre-extracted contaminated soil sample was placed onto the working electrode for 2 minutes, and resulting electrochemical scans, based on the chemical's propensity to oxidize DNA guanine residues, were measured. (Adenine moieties can be similarly redox reactive.) The magnitude of these "guanine peaks" in relation to a reference electrode was linearly related to their concentration in solution (i.e., the higher the concentration of the target chemical, the more damage imposed on the DNA and the lower the electrochemical measurement of the oxidation signal).

Hydrazine and aromatic amine compounds in fresh and groundwater, hydroxyl radicals in uranium mine drainage waters, herbicides such as atrazine, general toxicity events in wastewaters, industrially contaminated soils, and various other environmental sources have all been screened using DNA biosensors. Comparative studies in wastewater samples against more widely used whole-cell genotoxic bioluminescence inhibition biosensors (Toxalert) have indicated adequate correlations, but more substantial data will probably be required before the user community would consider replacing standard bioluminescent toxicity assays with DNA biosensors.

However, the rapidity (only a few minutes to detect but sample processing is often necessary), sensitivity (typically down to low part-per-billion levels), ease of use, and cost-effectiveness of genotoxic DNA biosensors does hold great potential for screening environmental sites for toxic chemical intrusions or monitoring operational endpoints of bioremediation efforts.

Metals are also relevant detection targets due to their various affinities for nucleic acid. Lead, cadmium, nickel, arsenic, copper, iron, chromium, and others have been detected through DNA biosensing, incorporating both single- and double-stranded DNA as the sensing element, but again, nonselectively. Selectivity, though, has been demonstrated by several groups using deoxyribozymes (DNAzymes) or ribozymes (RNAzymes). These engineered catalytic oligonucleotides can mediate nucleic acid cleavages or ligation, phosphorylation, or other reactions. For example, a DNAzyme biosensor for lead uses a single-stranded DNAzyme absorbed to a gold electrode (Xiao et al., 2007). The DNAzyme incorporates a methylene blue tag at its terminus that is held distant from the electrode by pairing of a complementary oligonucleotide to the DNAzyme to maintain rigidity. Upon exposure to lead at concentrations as low as 62 ppb, the DNAzyme strand is cleaved, allowing the methylene blue tag to approach the transducer and transfer electrons, thereby instigating an electrochemical signal. A similar strategy employing a fluorophore-labeled DNAzyme and quencher-labeled substrate strand (thus mimicking the molecular beacon real-time polymerase chain reaction (PCR) amplification strategy where dissociation of the fluorophore away from the quencher results in optical emission) yielded a lead-specific "catalytic beacon" fluorosensor with nanomolar sensitivity that has been tested successfully in artificially contaminated freshwater (Lu et al., 2003).

Nucleic acid can be manipulated similarly to create target-specific aptamers using a process called SELEX (systematic evolution of ligands by exponential enrichment). By iteratively incubating nucleic acid with the desired target, one can select for oligonucleotide sequences (or aptamers) with the greatest affinity for the target. Predominant aptasensor development and application is in the clinical fields, but it is slowly and inevitably encroaching upon environmental sensing. An aptasensor for the cyanobacterial toxin microcystin (lower detection limit of 50 µg/mL) (Nakamura et al., 2001) and another for zinc based on fluorophore beacons (lower detection limit of 5 µM) have been reported (Rajendran and Ellington, 2008).

BIOMEMs, BIOMIMETICS, AND OTHER EMERGING BIOSENSOR TECHNOLOGIES

Biological microelectromechanical systems (BioMEMs) and biomimetics clearly represent the rapidly approaching future of the biosensor. BioMEMs are an assortment of biomicro, bionanotechnological, and microfluidic interfaces that form lab-on-a-chip, biochip, or micro-total analysis system (µTAS) biosensors. Their objectives are toward miniaturization; portability; redundancy; and a reduction in sample size, time of response, and cost. The majority of these biosensors serve biomedical rather than environmental causes, but they are slowly and inevitably being adapted for the environmental monitoring community. BioMEMs most often utilize optical transducers interfaced with enzyme, whole-cell, antibody, or nucleic acid-type receptors.

BioMEMs also include microcantilever-based biosensors that translate a molecular recognition event into nanomechanical motion that is measured by induced bending in a microfabricated cantilever, similar on a macroscale to identifying a person on a diving board based on the deflection of the diving board by their weight. Optical or piezo-resistive transducers usually measure microcantilever deflections at nanometer-to-subnanometer ranges of motion, and due to their small size, several microcantilevers can be accommodated per transducer for multianalyte sensing.

Biomimetics mimic the attributes of naturally occurring biological materials to synthetically recreate or enhance their properties. Molecularly imprinted polymers (MIPs) are one such example, and their application in biosensors (or more appropriately, in sensors, since the "bio" element has been removed) can deliver more robust, stable, and target-specific receptors. MIPs are essentially created by mixing the target analyte (or template) with a monomer. The monomers self-assemble around the template, and subsequent removal of the template leaves behind a highly cross-linked polymeric matrix containing cavities that fit "lock-and-key" with the template. Resulting MIPs then serve as analyte-specific synthetic receptors (or artificial antibodies or enzymes) that can be associated with transducers to form sensors.

Biosensors based on the use of whole animals or their organs represent a very unique mode of sensing. Insect antennas, for example, are covered with highly sensitive and naturally tuned receptors called sensilla that respond to chemical, physical, and mechanical signals via electrical nerve impulses. By immobilizing the antenna or even the entire insect on a transducer and measuring these induced electrical impulses (or electroantennograms), a biosensor materializes. A multianalyte biosensor can be formed by adhering antennas from several different insects. The current targets for such biosensors are odorants such as those related to smoke (guaiacol and 1-octen) for early-warning fire detection or volatiles emanating from diseased plants, with detection limits in the part-per-billion range. Their parallel application for sensing volatiles associated with environmental contaminants and even non-odor-related compounds is a potential future prospect.

FURTHER READING

Andreescu S, Sadik OA. Correlation of analyte structures with biosensor responses using the detection of phenolic estrogens as a model. *Anal. Chem.* 2004;76:552–560.

Biosensors for the environmental monitoring of aquatic systems. In: Damia B, Peter-Dietrich H (eds.) *Bioanalytical and Chemical Methods for Endocrine Disruptors.* Springer-Verlag Berlin Heidelberg; 2007.

Bontidean I, Ahlqvist J, Mulchandani A, et al. Novel synthetic phytochelatin-based capacitive biosensor for heavy metal ion detection. *Biosens. Bioelectron.* 2003;18:547–553.

Boublik Y, Saint-Aguet P, Lougarre A, et al. Acetylcholinesterase engineering for detection of insecticide residues. *Protein Eng.* 2002;15:43–50.

Deep A, Kumar S, eds. *Advances in Nanosensors for Biological and Environmental Analysis: Book Review.* Elsevier; 2019.

Eldridge ML, Sanseverino J, Layton AC, Easter JP, Schultz TW, Sayler GS. *Saccharomyces cerevisiae* BLYAS, a new bioluminescent bioreporter for detection of androgenic compounds. *Appl. Environ. Microbiol.* 2007;73:6012–6018.

Fierke CA, Thompson RB. Fluorescence-based biosensing of zinc using carbonic anhydrase. *Biometals.* 2001;14:205–222.

Hakkila K, Maksimow M, Karp M, Virta M. Reporter genes *lucFF, luxCDABE, gfp*, and *dsred* have different characteristics in whole-cell bacterial sensors. *Anal. Biochem.* 2002;301:235–242.

Keane A, Phoenix P, Ghoshal S, Lau PCK. Exposing culprit organic pollutants: A review. *J. Microbiol. Methods* 2002;49:103–119.

Lojou E, Bianco P. Application of the electrochemical concepts and techniques to ampero-metric biosensor devices. *J. Electroceram.* 2006;16:79–91.

Lu Y, Liu JW, Li J, Bruesehoff PJ, Pavot CMB, Brown AK. New highly sensitive and selec-tive catalytic DNA biosensors for metal ions. *Biosens. Bioelectron.* 2003;18:529–540.

Mascini M, Palchetti I, eds. *Nucleic Acid Biosensors for Environmental Pollution Monitoring.* RSC Publishing; 2011.

Michel C, Battaglia-Brunet F, Minh CT, Bruschi M, Ignatiadis I. Amperometric cyto-chrome *c*(3)-based biosensor for chromate determination. *Biosens. Bioelectron.* 2003;19:345–352.

Nakamura C, Kobayashi T, Miyake M, Shirai M, Miyake J. Usage of a DNA aptamer as a ligand targeting microcystin. *Mol. Cryst. Liq. Cryst.* 2001;371:369–374.

Nikolelis DP, Krull U, Wang J, Mascini M, eds. *Biosensors for Direct Monitoring of Environmental Pollutants in Field.* Springer Netherlands; 1997.

Okeke BC, Ma GY, Cheng Q, Losi ME, Frankenberger WT. Development of a perchlorate reductase-based biosensor for real time analysis of perchlorate in water. *J. Microbiol. Methods.* 2007;68:69–75.

Preedy VR, Patel V. *Biosensors and Environmental Health.* CRC Press; 2012.

Purohit HJ. Biosensors as molecular tools for use in bioremediation. *J. Clean. Prod.* 2003;11:293–301.

Rajendran M, Ellington AD. Selection of fluorescent aptamer beacons that light up in the presence of zinc. *Anal. Bioanal. Chem.* 2008;390:1067–1075.

Ripp S, Nivens DE, Ahn Y, et al. Controlled field release of a bioluminescent genetically engi-neered microorganism for bioremediation process monitoring and control. *Environ. Sci. Technol.* 2000;34:846–853.

Sassolas A, Leca-Bouvier BD, Blum LJ. DNA biosensors and microarrays. *Chem. Rev.* 2008;108:109–139.

Satoh I, Iijima Y. Multi-ion biosensor with use of a hybrid-enzyme membrane. *Sens. Actuat. B.* 1995;24:103–106.

Tiwari A, Turner APF, eds. *Biosensors Nanotechnology.* Beverly, MA: Scrivener Publishing LLC; 2014.

Trojanowicz M. Determination of pesticides using electrochemical enzymatic biosensors. *Electroanalysis.* 2002;14:1311–1328.

Tschmelak J, Proll G, Riedt J, et al. Automated water analyser computer supported system (AWACSS): I. Project objectives, basic technology, immunoassay development, soft-ware design and networking. *Biosens. Bioelectron.* 2005;20:1499–1508.

Wu CM, Lin LY. Immobilization of metallothionein as a sensitive biosensor chip for the detection of metal ions by surface plasmon resonance. *Biosens. Bioelectron.* 2004;20:864–871.

Xiao Y, Rowe AA, Plaxco KW. Electrochemical detection of parts-per-billion lead via an electrode-bound DNAzyme assembly. *J. Am. Chem. Soc.* 2007;129:262–263.

Part III

Wastes and Waste Management

Managing solid wastes in society has been a challenge for as long as people have gathered together in sufficient numbers to impose a stress on local resources. Household waste and other waste streams needed to be removed from the human environment to avoid nuisance and public health problems, and the wider environment provided an ample sink for these negative effects of human life. Growth in population and in individual prosperity have since combined to put greater pressure on the environment, at the same time as permitting a growth in people's appreciation of that environment.

Municipal solid wastes include a proportion of commercial and non-hazardous industrial waste. Depending on the country, the definition can include some or all of:

- household wastes (collected waste, waste collected for recycling and composting, and waste deposited by householders at household waste disposal sites)
- household hazardous wastes
- bulky wastes derived from households
- street sweepings and litter
- parks and garden wastes
- wastes from institutions, commercial establishments, and offices

Waste management stresses the need for:

- reduced waste movements and improved waste transport regulation
- new and better waste management tools (e.g., regulatory and economic instruments)

DOI: 10.1201/9781003272618-17

- reliable and comparable statistics on waste
- waste management plans
- proper enforcement of legislation

Waste management options after generation and before final disposal comprise:

- waste minimization
- collection and sorting
- reuse
- recycling
- composting
- anaerobic digestion
- energy recovery (incineration or other more advanced thermal treatment techniques)
- incineration (without energy recovery)

WASTE MINIMIZATION

Waste minimization, prevention, or avoidance is the most important management technique to be applied to solid wastes, because waste which is avoided needs no management and has no environmental impact. In order to successfully reduce waste volumes, it is first necessary to establish the composition of that waste and the reasons which prompted its creation. In a domestic situation, those reasons may include or be dictated by life style, for example, if both parents in a household are working full time, necessitating the purchase of more convenience foods, or if there is a young baby in the family using disposable nappies.

It is easy to think narrowly of waste as a solid product left over at the end of a process or action, but it would be wrong to concentrate on reducing the amount of solid waste produced, to the exclusion of considerations about, among other things, wastage of energy or water. It is as wasteful to use several gallons of water unnecessarily, or to drive cars when one could walk or cycle, or to consume energy thoughtlessly, as it is to discard newspapers, cans, or empty wrappings.

Waste minimization is hard to achieve for individuals and households, but there are some contributions which can be made. For example, care should be taken when purchasing goods that appropriate amounts and sizes are chosen. Buying large tins of paint to do a small decorating job and buying larger amounts of food than can be consumed while fresh are two examples of unnecessary waste creation. Since waste is not simply the refuse put out for collection each week, considerable waste reduction in terms of resource use can be made by using electricity sparingly, by cutting down the number of car journeys made, and so on. Individuals can reduce the amount of waste they create by buying less; by buying longer life products; and by reusing items: empty tins and jars make good storage, yoghurt pots are ideal for seedlings, and magazines once read can be passed to neighbors and friends. Mending broken or worn items of clothing or equipment has a further important contribution to make.

WASTE COLLECTION AND SORTING

The way that wastes are collected and sorted influences which waste management options can most effectively be used. The collection method significantly shapes the recovery of materials, compost, or energy; this, in turn, determines whether markets can be found. Collection is also the point of contact between generators (e.g., households and commercial establishments) and the waste management system. Collection is rarely independent of subsequent sorting, since the type of collection affects the sorting needs, and some collection methods themselves involve a level of sorting.

COMPOSTING

The word "compost" conjures up images of an untidy heap of assorted kitchen and vegetable remains at the far end of the garden. A more scientific definition of compost would be the product of natural degradation of botanical and putrescible waste by the action of bacteria, fungi, insects, and animals in the presence of an adequate air supply. The biological decomposition processes break down complex organic substances into carbon dioxide, water, and a residue: compost. The final product is relatively stable and can be used as a soil improver, a mulch, or a component in a growing medium. Composting occurs when there is a plentiful supply of oxygen, moisture, and warmth. Compost must be regularly aerated, by turning the heap or by injecting air into the composting material, for the process to be successful. Under properly controlled conditions, the temperature of composting refuse can rise to levels which are sufficient to kill pests, weed seeds, and pathogenic bacteria.

Anaerobic fermentation occurs when the oxygen supply is restricted. Complex hydrocarbons are broken down into reduced intermediate byproducts; both methane—which has a potential value as a fuel—and carbon dioxide are released. Anaerobic decomposition—called fermentation or digestion—occurs all the time in landfill sites containing household refuse.

Three main classes of microorganisms are involved in the decomposition of refuse: bacteria, fungi, and actinomycetes. Bacteria and fungi predominate as the organic material begins to decompose. If sufficient air is available, the rate of metabolic activity of these microorganisms is so great that temperatures can rise to 70°C or more. At this stage, only heat-tolerant (thermophilic) bacteria and actinomycetes can continue to decompose the waste. Gradually, as the substrate is consumed, the rate of decomposition slows, the compost cools, and the fungi and non-thermophilic bacteria become active again. All microorganisms need water to function; if the moisture content of composting waste falls below 40%, microbial activity slows. If the moisture content is too high, however, air spaces within the composting material fill with water, creating anaerobic conditions which can cause unpleasant odors.

THERMAL TREATMENT

Household waste has up to half the energy potential of coal. Recovering energy from the non-recyclable portion of household waste makes economic and environmental sense. The main virtues of energy recovery are the following:

- waste volume reduction,
- rendering waste inert,
- recovering value from waste,
- biodegradable waste diversion, and
- a practical method to manage increased waste arisings.

ENERGY RECOVERY OPTIONS

LANDFILL GAS RECOVERY

Landfill gas is produced by the decomposition of organic wastes in the airless conditions of a landfill site. Landfill gas typically contains around 55% methane and 40% carbon dioxide together with small amounts of nitrogen, hydrogen, and water. These gases can be collected using a network of horizontal pipes and wells, laid prior to and during filling of the site with waste. Beneficial use of landfill gas for energy evolved as a solution to the problem of potentially explosive gas leaking from landfill sites. Since methane is a greenhouse gas with a higher global warming potential than its combustion product carbon dioxide, using methane for energy recovery has the further benefit of reducing the net potential for global warming.

ANAEROBIC DIGESTION

Organic waste can be broken down using anaerobic digestion (AD), and the methane gas generated can be recovered. AD has been used extensively for sewage sludge and for agricultural wastes. Its use to treat MSW, often with sewage sludge, provides a fuel which can be used, like landfill gas, to directly fire burners and to generate electricity or which can be cleaned and added to gas supplies. A major advantage of AD is that all the gas produced can be collected and used, unlike the gas in a landfill where collection efficiencies are relatively low (50% or less). AD also produces a solid residue or digestate which can be cured and used as a fertilizer.

COMBUSTION

The conventional waste combustion technique is mass burn incineration, which involves the waste being burned as delivered, after removal of bulky items. To aid combustion, there is usually some mixing of the waste. In the past, incineration plants were mainly designed purely to process waste, but today's plants are usually designed to recover energy (as steam, hot water, or electricity) from the waste.

FLUIDIZED BED COMBUSTION

Fluidized bed (FB) combustion technology is based on a system where, instead of the waste being burned on a grate (as in mass burn processes), the fire bed is composed of inert particles such as sand or ash. When air is blown through the bed, the bed material behaves as a fluid. There are several different designs of FB combustors, for example, circulating and bubbling beds. All need waste of uniform size. Compared to mass burn, FB combustion systems have reduced emissions, partly because of the process itself and partly because it is possible to add lime to the bed. Since as much as a third of the cost of mass burn plants is spent on the air pollution control (APC) system, there are savings to be made as FB systems have smaller APC needs. On the other hand, mass burn plants have no need for front-end processing of the waste.

GASIFICATION AND PYROLYSIS

Gasification is the process of reacting carbon with steam to produce hydrogen and carbon monoxide. Gasification converts a solid or liquid feedstock into gas by partial oxidation under the application of heat. Pyrolysis is a complex series of reactions initiated when material is heated (to around 400–800°C), in the absence of oxygen, to produce condensable and non-condensable vapor streams and solid residues. Heat breaks down the molecular structure of waste, yielding gas, liquid, and a solid char, all of which can be used as fuels.

LANDFILL

Landfill is often regarded as the last resort waste management option, an "out of sight, out of mind" solution. While this may be partly true, modern landfilling is an active treatment process applied to most solid wastes. An engineered landfill is designed to contain waste and its decomposition products until they are sufficiently stable and inert to present no significant risks to health or the environment. Other benefits, such as material and energy recovery or land reclamation, may also be derived from properly designed facilities.

15 Domestic Wastewater Treatment

WASTEWATER

Water that has been used by people and is disposed into a receiving water body with altered physical and/or chemical parameters is defined as wastewater. If only the physical parameters of the water were changed, e.g., resulting in an elevated temperature after use as a coolant, treatment before final disposal into a surface water may require only cooling close to its initial temperature. If the water, however, has been contaminated with soluble or insoluble organic or inorganic material, a combination of mechanical, chemical, and/or biological purification procedures may be required to protect the environment from periodic or permanent pollution or damage. For this reason, legislation in industrialized and in many developing countries has reinforced environmental laws that regulate the maximum allowed residual concentrations of carbon, nitrogen, and phosphorus compounds in purified wastewater, before it is disposed into a river or into any other receiving water body.

Wastewater has to be purified by application of standardized, widely experienced state-of-the-art treatment technologies to meet the quality standards of environmental legislation. For this purpose, the wastewater must be collected; transported in public, industrial, or private sewer systems; and treated in domestic sewage or industrial wastewater treatment plants to remove organic and inorganic pollutants, as required by environmental laws and enforced by state control agencies of the respective countries. The concentration limits in the purified wastewater for residual carbon (measured as biological oxygen demand, BOD, or chemical oxygen demand, COD), nitrogen (total nitrogen or ammonia nitrogen), and phosphorus (in particular, soluble orthophosphate) that had to be met for disposal into surface waters became more stringent with time. Improvements and the development of new processes for wastewater purification were stimulated by pressure exerted by legislation.

Due to the complexity of the pollutants in different wastewater types or even in a certain wastewater of defined composition, combined multistage processes for physical, chemical, and biological removal of organic pollutants, nitrogen, and phosphorus were required. Generally, the process development for wastewater purification was always slightly ahead of an exact knowledge of the biological reactions or reaction sequences behind it. A lack of detailed biochemical knowledge on even major metabolic pathways or, in particular, on single reactions within the complex ecosystem called wastewater was always the bottleneck for specific improvements in wastewater purification techniques and treatment efficiencies.

Once the wastewaters have been collected and their characteristics have been determined, the wastewater treatment processes can be evaluated and the best

DOI: 10.1201/9781003272618-18

process can be proposed. Wastewater treatment processes can be physical, chemical, biological, or a combination of all three. Ultimately, wastewater must be treated and returned back into the environment with a minimum of damage to the environment. The treatment process should also be economical to construct and operate. Experience has shown that simplicity of design with a minimum of mechanical equipment and the least number of units provides the best opportunity for success. Ultimately, all biodegradable organic materials must undergo microbial degradation before being returned to the environment. Microbial stabilization of organic contaminants can be either aerobic or anaerobic or a combination of both. The key to success with biological treatment of all types of wastewaters lies in understanding the fundamental concepts of microbiology and the biochemistry of different groups of microorganisms together with sound treatment plant engineering. The basic objective is to remove the contaminants from the wastewater with the least possible effort at the lowest possible cost and to return the water and the residual contaminants back into the environment with the least possible damage to the environment. Considerable progress has been made in the advanced nations of the world; but much effort remains in the developing nations. This will be the real test of environmental pollution control engineers.

STORM WATER

The characteristics of storm water indicate that suspended solids are the primary problem. The simplest method for storm water treatment is to prevent the suspended solids from initially accumulating in the environment. This is easier to say than to do. The wind brings much of the dust and dirt and some of the trash. Cars and trucks stir up the dust and dirt in the streets. Passersby often throw trash into the streets or onto adjacent property that does not belong to them. Domestic animals, together with non-domestic animals and birds, deposit their waste products on streets, sidewalks, and yards. Homeowners use fertilizers, herbicides, and pesticides to produce nice lawns and shrubs to beautify their property. Most municipal storm sewers are located in commercial areas or densely populated residential areas. Suburban areas tend to use drainage ditches to collect the storm waters. Where storm sewers are readily available, people tend to wash the dirt off the sidewalks early in the morning when few people are around. The dirt and trash are washed into the street where they flow to the nearest storm water inlet. The amount of daily wash water is usually not sufficient to produce a discharge from storm sewer unless there is a nearby drainage way. For this reason, the dust, dirt, and trash accumulate in the storm sewer until there is a significant rainfall, generating enough runoff to produce a discharge from the storm sewer. In effect, the rainfall event produces about the same amount of contaminants with or without the daily washing of the sidewalks. The best solution for storm sewer discharges appears to be some type of an end-of-pipe treatment system. The simple drainage ditches in the suburban areas will not require treatment in most instances since the grass lawns retain most of the contaminants except for very large rainfall events. Treatment of storm waters will be more physical treatment than biological treatment.

URBAN SYSTEMS

The simplest form of treatment for storm water is settling in a natural pond with skimming near the inlet to remove floating trash. A natural pond with sufficient holding time to allow for retention of several days flow could be developed as a park area for residents, as well as, for storm water treatment. The natural bacteria in the pond water should stabilize the small amount of BOD_5 in the storm water during its retention time in the pond. One of the problems with retention ponds is excessive algae growth if the storm water contains significant concentrations of nutrient elements from fertilized lawns in the drainage area. Artificial wetlands, ahead of the storm water retention pond, can remove the nutrients from the storm water and limit the growth of algae in the retention pond. Periodic harvesting will be required of the excessive plant growth in the wetlands to remove the nutrients permanently. Another form of treatment is the Swirl concentrator, employing the hydraulic flow of the storm water to create centrifugal forces that assist in separating the suspended solids. Floating solids would best be removed by screening prior to the Swirl concentrator. All of the suspended solids removed from the storm water must be collected at periodic intervals and returned back to the environment, either buried in sanitary landfills or applied to the land surface. The effluent from the Swirl concentrator is returned to the nearby receiving stream.

INDUSTRIAL SYSTEMS

Industrial plants can create serious problems with storm water runoff. Chemical leaks and spills leave residues on the ground that can be removed by storm water runoff from time to time. Industries may be required to capture the surface runoff from critical plant areas and discharge the runoff to process sewers rather than to storm sewers. Treatment of storm water containing soluble organic compounds requires the use of biotreatment systems. Currently, few industrial plants have facilities for capturing storm water containing soluble pollutants. Treating domestic wastewater and industrial process wastewater has been the primary focus for industries to date. The federal EPA has developed regulations affecting the measurement of storm water runoff and its chemical analyses from industrial plants. As information is developed, storm water regulations from industrial plants can be expected to change if serious problems are indicated.

AGRICULTURAL SYSTEMS

Agricultural runoff is difficult to treat since it occurs over such a wide area. It is not surprising that agricultural runoff has been classified as non-point source pollution. It is normally not possible to obtain individual flow measurements or pollutant analyses from each farm. Stream flow measurements and chemical analyses along short stretches of streams and rivers adjacent to farms can provide some measurement of agricultural runoff characteristics.

Soluble fertilizer in the runoff provides the grass with nutrients for growth. It is important to periodically cut the grass and remove it or the nutrients will be released back into the runoff. The construction of retention ponds at the end of the grass waterways can help retain the runoff and permit its slow release into adjacent streams. Since most of the complex agricultural chemicals being used are retained on soil particles, minimizing the loss of soil particles from agricultural areas helps to significantly reduce agricultural contamination discharge.

The biggest problems facing agriculture today are wastes from confined animal operations and from animal slaughtering and meat packing operations. Confined animal operations allow more animals to be grown in less space, creating a waste disposal problem. This is the same waste disposal problem that was created when people began leaving their individual farms for life in the city. More waste is generated in a smaller space and has to be properly processed for return to the environment. Since animal manure can only be applied to crops before the growing season, the animal wastes must be collected and retained until needed.

WASTEWATER TREATMENT

Treatment of domestic wastewater utilizes physical and biological systems. Physical treatment is used first to remove the trash by screening. Screening is considered as preliminary treatment. The captured screenings are normally buried in sanitary landfills. If the domestic sewage is contaminated with large amounts of sand, grit chambers are added as part of the preliminary treatment system to remove the sand. Grit chamber design provides sufficient velocity to maintain the organic solids in suspension while allowing the sand to be separated by gravity or centrifugal force. The sand and grit are washed after collection to insure that the organic matter is returned to the wastewater stream for treatment. The washed sand can be returned to the land environment and used where needed.

PRIMARY TREATMENT

Primary treatment follows preliminary treatment and consists of physical treatment by continuous sedimentation to remove the settleable solids and the floating solids from the wastewater together with either physical or biological treatment to process the sludge solids for return to the environment. Physical treatment requires the addition of chemicals to assist in dewatering the solids and some form of mechanical dewatering to produce a relatively dry cake that can be burned in an incinerator or buried in the land environment. Biological treatment of primary sludge utilizes anaerobic digestion to stabilize the biodegradable organic solids prior to physical dewatering and return to the land environment. Primary treatment reduces the BOD_5 of municipal wastewaters about 30%–35% and the suspended solids about 60%–65%. Anaerobic digestion reduces the primary sludge solids about 50%.

PRIMARY SEDIMENTATION

Primary sedimentation tanks can be circular, square, or rectangular. Circular sedimentation tanks are widely used in small to medium size WWTP where land is readily available. Square and rectangular sedimentation tanks are used in large plants where land is limited and common wall construction helps reduce the overall plant cost. Normally, the wastewater is retained for 2 hours at average design flow in the primary sedimentation tanks. The settleable solids will drop to the bottom of the tank and will be collected with slow moving, mechanical sludge scrapers. The settled sludge will thicken to between 4% and 6% total solids (TS), 40,000–60,000 mg/L. Since biological metabolism began prior to the wastewaters entering the collection system, metabolism will continue to occur in the primary sedimentation tanks. Metabolism will be limited by the number of active bacteria in the incoming wastewater and by the growth of bacteria during the time the primary sludge is allowed to accumulate in the sedimentation tanks. Hydrolytic reactions continue to predominate with proteins hydrolyzed to amino acids that are further hydrolyzed to NH_3-N and short-chain organic acids. The starches are hydrolyzed to simple sugars. The dispersed bacteria and the bacteria associated with small particles are carried out in the primary effluent in a relatively short time. Since the incoming wastewaters are maintained in a relatively quiescent condition without aeration, bacteria metabolism will be anaerobic with the formation of soluble, short-chain organic acids, aldehydes, and alcohols. Most of the bacteria growth will be by facultative bacteria. The settled sludge contains additional bacteria and provides a better opportunity for anaerobic growth with concentrated nutrients. The extent of bacteria metabolism can be estimated from the pH in the settled sludge. The excess organic acids will depress the pH to between 6 and 6.5. The settled sludge is moved by a rotating sludge scraper at the bottom of circular and square tanks and by transverse scrapers moving along the bottom of the rectangular tanks. The sludge is slowly moved to a sludge hopper located either near the center of circular and square tanks or at the influent end of rectangular tanks. Periodically, the settled sludge is pumped from the sludge hopper for further treatment, removing it from the primary sedimentation tanks. If the settled sludge is allowed to remain at the bottom of the primary sedimentation tanks for too long a period of time, anaerobic metabolism can become quite extensive. Excess organic acids will drop the pH below 6.0. Sulfate-reducing bacteria can generate sufficient hydrogen sulfide to react with iron and other metallic ions in the wastewater to produce a black color and give an obnoxious odor to the sludge. Carbon dioxide can collect in the sludge as a gas and lift small clumps of solids to the liquid surface. Significant gasification in primary sedimentation tanks processing only raw wastewaters indicates that the settled sludge is not being removed quickly enough. If waste activated sludge is added to the primary sedimentation tanks, gasification can be caused by denitrification. Facultative bacteria will continue anaerobic metabolism with a conversion of solid organic compounds to soluble organic compounds. Excessive retention of primary sludge can reduce the quantity of primary sludge and increase the load on the next treatment units. Good operation requires rapid removal of the settled sludge without pumping excess water with the sludge. Positive displacement sludge pumps are more effective at removing settled, primary sludge than centrifugal sludge pumps since they remove more sludge solids and less water.

Municipal wastewaters also contain grease and other insoluble hydrocarbons that tend to float in the primary sedimentation tanks, rather than settle to the bottom. A surface skimmer is standard equipment for primary sedimentation tanks to remove the floating scum from the sedimentation tank surface. The scum is collected in a scum hopper and a scum pit before being collected and buried in a sanitary landfill. In some plants, the scum is also pumped to the anaerobic digester for further treatment.

Municipal wastewater treatment plants processing less than 44 L/s (1.0 mg d) flow often eliminate primary sedimentation tanks and anaerobic sludge treatment. The economics of using primary sedimentation and anaerobic digestion allow the raw wastewaters to be directly discharged to aerobic biological treatment units. These small treatment plants utilize either extended aeration activated sludge or aerobic digestion of the excess activated sludge in a separate tank prior to dewatering of the sludge for return to the land environment.

Typical composition of untreated domestic wastewater

Contaminants	Concentration (mg/L)		
	Low	Moderate	High
Solids, total	350	720	1,200
Dissolved, total	250	500	850
Suspended solids	100	220	350
Settleable solids	5	10	20
Biochemical oxygen demand	110	220	400
Total organic carbon	80	160	290
Chemical oxygen demand	250	500	1,000
Nitrogen (total as N)	20	40	85
Organic	8	15	35
Free ammonia	12	25	50
Nitrites	0	0	0
Nitrates	0	0	0
Phosphorus (total as P)	4	8	15
Organic	1	3	5
Inorganic	3	5	10

SECONDARY TREATMENT OF SEWAGE

Several methods exist for the treatment of sewage and they may be grouped into aerobic as shown below:

Aerobic Methods

1. The activated sludge system
2. The Trickling filter
3. The oxidation pond

Anaerobic Methods

1. The septic tank
2. The Imhoff tank
3. The cesspool

AEROBIC BREAKDOWN OF RAW WASTEWATERS

The basic microbiological phenomenon in the aerobic treatment of wastes in aqueous environments is as follows:

1. The degradable organic compounds in the wastewater (carbohydrates, proteins, fats, etc.) are broken down by aerobic microorganisms mainly bacteria, and to some extent, fungi. The result is an effluent with a drastically reduced organic matter content.
2. The materials difficult to digest form a sludge which must be removed from time to time and also treated separately.

THE ACTIVATED SLUDGE SYSTEM

The activated sludge method is the most widely used method for treating wastewaters. Its main features are as follows (Nielsen, 2002; Lindera, 2002):

a. It uses a complex population of microorganisms of bacteria and protozoa.
b. This community of microorganisms has to cope with an uncontrollably diverse range of organic and inorganic compounds some of which may be toxic to the organisms.
c. The microorganisms occur in discrete aggregates known as flocs which are maintained in suspension in the aeration tank by mechanical agitation or during aeration or by the mixing action of bubbles from submerged aeration systems. Flocs consist of bacterial cells, extracellular polymeric substances, adsorbed organic matter, and inorganic matter. Flocs are highly variable in morphology, typically 40–400 mm and not easy to break apart.
d. When floc particles first develop in the activated sludge process, the particles are small and spherical. As the sludge age increases and the short filamentous organism within the floc particles began to elongate, the floc-forming bacteria now flocculate along the lengths of the filamentous organisms. The presence of long filamentous organisms results in a change in the size and shape of floc particles. These organisms provide increased resistance to shearing action and permit a significant increase in the number of floc-forming bacteria in the floc particles. The floc particles increase in size to medium and large and change from spherical to irregular.
e. The flocs must have good settling properties so that separation of the biomass of microorganisms and liquid phases can occur efficiently and rapidly in the clarifier. Sometimes, proper separation is not achieved giving rise to problems of bulking and foaming.
f. Some of the settled biomass is recycled as "returned activated sludge" (RAS) to inoculate the incoming raw sewage because it contains a community of organisms adapted to the incoming sewage.
g. The solid undigested sludge may be further treated into economically valuable products.

MICROBIOLOGY OF THE ACTIVATED SLUDGE PROCESS

The activated sludge process is a biological method of wastewater treatment brought about by a variable and mixed community of microorganisms in an aerobic aquatic environment. These microorganisms derive nourishment from organic matter in aerated wastewater for the production of new cells. Some of the bacteria carry out nitrification and convert ammonia nitrogen to nitrate nitrogen. The consortium of microorganisms, the biological component of the process, is known collectively as activated sludge.

There are three essential requirements for the activated sludge process:

a. A mixed population of microorganisms able to degrade the components of the sewage must be present.
b. The population must be able to grow in the environment of the aeration tank.
c. The organisms must grow in such a way that they will form flocs and settle out in the sedimentation tank.

The activated sludge microbial community is specialized and is less diverse than of the biological or trickling filter. Bacteria make up about 95% of the activated sludge and biomass, and aerobic bacteria are the dominant bacteria. Bacteria are followed by protozoa, fungi, and rotifers; nematodes are few and algae are usually absent.

The organisms involved are bacteria and ciliates (protozoa). It was once thought that the formation of flocs which are essential for sludge formation was brought about by the slime-forming organism, *Zooglea ramigera*. It is now known that a wide range of bacteria are involved, including *Pseudomonas, Achromobacter, Flavobacterium* to name a few.

Various bacteria have been reported in the activated sludge setup depending on the chemical nature of the sludge. Perhaps, the most important one is the rod-shaped bacterium, *Z. ramigera*, which produces a large quantity of extracellular slime matrix. This organism is believed to be the main agent for flocculation, although other organisms can also form flocs. Other bacteria which have been reported are *Pseudomonas, Achromobacter, Flavobacterium, Microbacterium, Micrococci, Bacillus, Comamonas, Azotobacter, Staphylococcus, Bdellovibrio, Nitrobacter, Nitrosomonas, Alcaligenes* spp., *Brevibacterium* spp., *Beggiatoa* spp., *Corynebacterium* spp., and *Sphaerotilus* spp.

Fungi encountered are *Zoophagus* spp., *Arthrobotrys, Geotrichum candidum, Pullularia* spp., *Alternaria* spp., *Penicillium* spp., and *Cephalosporium* spp. Among the protozoa, ciliates dominate, and among ciliates, the following are most common: *Aspidisca costata, Carchesium polypinum, Chilodonella uncinata, Opercularia coarctata* and *Opercularia microdiscum, Trachelophyllum pusillum, Vorticella convallaria* and *Vorticella microstoma*. Most of these are sessile and attached to the sludge flocs. In addition, amoebae and flagellates are also seen. Yeasts and algae are of rare occurrence.

A succession of protozoa has been observed. Flagellates (e.g., *Heteronema, Bodo*) occur only in the immature sewage, being replaced in about 3 weeks first by free-swimming ciliates (e.g., *Paramecium caudatum*), then by crawling ciliates (e.g., *Aspidisca costata*), and finally by attached ciliates (e.g., *Vorticella* and *Epistylis*). Flagellates are found in poor quality systems or in "young" as opposed to mature activated sludge systems.

Rotifers are rarely found in large numbers in wastewater treatment processes. The principal role of rotifers is the removal of bacteria and the development of floc. Rotifers contribute to the removal of effluent turbidity by removing non-flocculated bacteria. Mucous secreted by rotifers at either the mouth opening or the foot aids in floc formation. Rotifers require a longer time to become established in the treatment process. Rotifers indicate increasing stabilization of organic wastes.

BULKING IN ACTIVATED SLUDGE SYSTEMS

"Bulking" is a growth condition in which the sludge has poor setting properties, because of loose cotton wool-like growth of filamentous organisms. Bulking may also create problems of aeration by the trapping of oxygen; the net effect may be inadequate stabilization. Some authors have distinguished bulking into two types. The first type, in which *Sphaerotilus* is always present, is according to these authors, caused by overload. The second type of bulking, in which *Sphaerotilus* is absent, is due to underloading in the aeration chamber. *Streptothrix* and non-sheath-forming bacteria are implicated in this type. Organisms which bring about bulking are regarded as nuisance organisms and include, apart from *Sphaerotilus* and *Streptothrix*, *Geotrichum*, *Beggiatoa*, *Bacillus*, and *Thiothrix*.

NUTRITION OF ORGANISMS IN THE ACTIVATED SLUDGE PROCESS

Provided the wastes are nutritionally satisfactory, sludge will be formed whether the sewage is rich in colloids as in domestic sewage or in soluble materials as in some industrial sewage. While domestic sewage is usually rich in organisms and in nutrients, some industrial sewages may be deficient in the key nutrients of nitrogen and phosphorus. These must therefore be added. Sometimes, the industrial activated sewage must also be seeded with soil in order to develop a population capable of stabilizing it. In some systems, the aerobic nitrogen fixing bacteria, *Azotobacter*, are added to provide nitrogen.

MODIFICATIONS OF THE ACTIVATED SLUDGE SYSTEM

Several modifications of the activated sludge procedure exist:

1. The conventional activated sludge setup: The basic components of the conventional system are an aeration tank and a sedimentation tank. Before raw wastewater enters the aeration tank, it is mixed with a portion of the sludge from the sedimentation tank. The raw water is therefore broken down by organisms already adapted to the environment of the aeration tank. The incoming organisms from the sludge exist in small flocs which are maintained in suspension by the vigor of mixing in the aeration tank. It is the

introduction of already adapted flocs of organisms that gave rise to the name activated sludge. Usually 25%–50% of the flow through the plant is drawn off the sedimentation tank. Other modifications of the activated sludge system are given below.

2. Tapered aeration: This system takes cognizance of the heavier concentration of organic matter and hence of oxygen usage at the point where the mixture of raw sewage and the returned sludge enters the aeration tank. For this reason, the aeration is heaviest at the point of entry of wastewaters and diminishes toward the distal end. The diminishing aeration may be made directly into the main aeration tank or a series of tanks with diminishing aeration may set up.

3. Step aeration: In step aeration, the feed is introduced at several equally spaced points along with length of the tank, thus creating a more uniform demand in the tank. As with tapered aeration, the aeration may be done in a series of tanks.

4. Contact stabilization: This is used when the wastewater has a high proportion of colloidal material. The colloid-rich wastewater is allowed to contact with sludge for a short period of 1–1½ hour, in a contact basin which is aerated. After settlement in a sludge separation tank, part of the sludge is removed and part is recycled into an aeration tank from where it is mixed with the incoming wastewater.

5. The Pasveer ditch: This consists of a stadium-shaped shallow (about 3 ft) ditch in which continuous flow and oxygenation are provided by mechanical devices. It is essentially the conventional activated sludge system in which materials are circulated in ditch rather than in pipes.

6. The deep shaft process: The deep shaft system for wastewater treatment was developed by Agricultural Division of Imperial Chemical Industries (ICI) in the UK from their air-lift fermentor used for the production of single cell protein from methanol. It consists of an outer steel-lined concrete shaft measuring 300 ft or more installed into the ground. Wastewater and sludge recycle are injected down an inner steel tube. Compressed air is injected at a position along the center shaft deep enough to ensure that the hydrostatic weight of the water above the point of injection is high enough to force air bubbles downward and prevent them coming upward. The air dissolves lower down the shaft providing oxygen for the aerobic breakdown of the wastes. The water rises in the outer section of the shaft. The system has the advantage of great rapidity in reducing the BOD and about 50% reduction in the sludge. Space is also saved.

7. Enclosed tank systems and other compact systems: Since the breakdown of waste in aerobic biological treatment is brought about by aerobic organisms, efficiency is sometimes increased by the use of oxygen or oxygen-enriched air. Enclosed tanks, in which the wastewater is completely mixed with the help of agitators, are used for aeration of this type. Sludge from a sedimentation tank is returned to the enclosed tank along with raw water as in the case with other systems. The advantage of the system is the absence, (or greatly reduced) obnoxious smell from the exhaust gases, and increased

efficiency of waste stabilization. This system is widely used in industries the world over.

8. Compact activated sludge systems: These do not have a separate sedimentation tank. Instead, sludge separation and aerobic breakdown occur in a single tank. The great advantage of such systems is the economy of space.

EFFICIENCY OF ACTIVATED SLUDGE TREATMENTS

The efficiency of any system is usually determined by a reduction in the BOD of the wastewater before and after treatment. Efficiency depends on the amount of aeration and the contact time between the sludge and the raw wastewater. Thus, in conventional activated sludge plants, the contact time is about 10 hours, after which 90%–95% of the BOD is removed. When the contact time is less (in the high-rate treatment), BOD removal is 60%–70% and the sludge produced is more. With longer contact time, say several days, BOD reduction is over 95% and sludge extremely low. With systems where oxygen is introduced as in the closed tank system or where there is great oxygen solubility as in the deep shaft system, contact time could be as short as 1 hour but with up to 90% BOD reduction along with substantially reduced sludge.

THE TRICKLING FILTER

The trickling filter consists of round rocks 1–4 in. diameter arranged in a bed 6–10 ft deep. Sewage is uniformly spread on the bed by a rotating distributor which is powered by an electric motor or driven by hydraulic impulse where a hydraulic head is available. The sewage percolates by gravity over the rocks and through the spaces between the rocks, and the effluent is collected in an underdrain. From the underdrain, the liquid is allowed to settle in a sedimentation tank which is an integral part of the system.

The sludge consisting of microorganisms and undecomposed matter is removed from time to time and may be used as manure. Various modifications may be made to this basic design. In one modification, the sewage may be pre-sedimented before filtration; in another, two filters may operate; and in yet others, the effluent may be recycled.

Trickling filters may be low rate or high rate. Low-rate filters handle 2–4 million gallons per acre per day (mgad). They are very efficient in BOD reduction and may attain 85%–90% BOD. The effluent is usually still rich in oxygen; hence, nitrates are produced in abundance by nitrifying bacteria. High-rate filters handle 10–40 mgad and BOD reduction is only 65%–75%. The effluent is usually low in nitrates since much of the O_2 is used in breaking down the organic matter load.

MICROBIOLOGY OF THE TRICKLING FILTER

As the sewage percolates over the stones, microorganisms break down the organic matter in it. Although the process is described as aerobic, most of the bacteria are in fact facultative. Aerobic conditions operate mainly when the filter is fresh and in the outer areas of the coating of microorganisms on the rocks in mature filters.

The innermost portion of the coating of microorganisms may in fact be anaerobic. Bacteria are the most important organisms in the trickling filter, but other organisms are also present.

Fungi, for instance, are to be found in the aerobic zone. The fungi encountered include mainly *Fusarium* and *Geotrichum*. Others are *Trichosporon, Sepedonium, Saccharomyces,* and *Ascoidea*. Fungi are more common in low pH sewages and in some industrial sewages.

Protozoa are present: the flagellates, free-swimming ciliates and, to some extent, amoeba at the upper part of the filter rocks, and the stalked ciliates in the bottom portion. It has been suggested that this stratification is a result of the availability of soluble food at various levels, the greater portion of such food being available in the surface region. Some protozoa encountered in the surface region are *Trepomonas agilis* and *Vorticella microstoma*; in the middle region, *Paramecium caudatum* and *Opercularia*; and in the innermost layer, *Arcella vulgaris* and *Aspidisca costata*. Algae, because of their need for sunlight, are found in the upper layers of the filter and may clog the filter. The only groups of algae found in both trickling filters and oxidation ponds are *Chlorophyceae, Euglenophyceae, Cyanophyceae*, and *Chrysophyceae*. In the outer portions of the filter, particularly in older filters, algae are to be found throughout the coating of the rocks especially in the middle layer.

The algae involved include the unicellular *Chlorella, Phormidium, Oscillatoria*, or the sheet-forming multicellular *Stigeoclonium* and *Ulothrix*. Algal photosynthesis provides only some of the oxygen required by the aerobic bacteria which contribute to the breakdown of organic matter. Worms, snails, and larvae are also to be encountered, but these contribute little to the process of filtration. The microorganisms adhere to the rock by weak (van der Waals) forces and grow in one direction only as liquid flows over the film of microbial coating. As the microbial layer increases in thickness, the innermost organisms die and the microbial layer drops off from the rock. A new growth starts thereafter. Since the breakdown is brought about by aerobes, a filter is most efficient when the microbial layer is thinnest or when the filter stones regularly shed their slime coatings.

ROTATING DISKS

Also known as rotating biological contactors, these consist of closely packed disks about 10 ft in diameter and 1 in. apart. Disks made of plastic or metal may number up to 50 or more and are mounted on a horizontal shaft which rotates slowly, at a rate of about 0.5–15 revolutions/minute. During the rotation, 40%–50% of the area of the disks is immersed in liquid at a time. A slime of microorganisms, which decompose the wastes in the water, builds up on the disks. When the slime is too heavy, it sloughs off and is separated from the liquid in a clarifier. It has a short contact time and produces little sludge.

OXIDATION PONDS

Oxidation ponds (also called or stabilization ponds) are shallow lagoons about three feet deep into which sewage is discharged at a single point, usually at the center but also occasionally at the side. After suitable periods of holding, the effluent which usually has a low BOD is discharged at a single point. The effluent is usually low in coliforms and may be discharged into a river or used as raw water source. Oxidation ponds are especially appropriate in warm, sunny climates. Oxidation ponds are a common sewage treatment method for small communities because of their low construction and operating costs. New oxidation ponds can treat sewage fairly efficiently but require maintenance and periodic desludging in order to maintain this standard. They may be made of one or up to four shallow ponds in series. The natural processes of algal and bacteria growth exist in a mutually dependent relationship. Oxygen is supplied from natural surface aeration and by algal photosynthesis. Bacteria present in the wastewater use the oxygen and feed on organic material, breaking it down into nutrients and carbon dioxide. These are in turn used by the algae. Other microbes in the pond such as protozoa remove additional organic material and nutrients to polish the effluent.

There are normally at least two ponds constructed. The first pond reduces the organic material using aerobic digestion, while the second pond polishes the effluent and reduces the pathogens present in sewage. Sewage enters a large pond after passing through a settling and screening chamber. After retention for several days, the flow is often passed into a second pond for further treatment before it is discharged into a drain. Bacteria already present in sewage acts to break down organic matter using oxygen from the surface of the pond. Oxidation ponds need to be desludged periodically in order to work effectively. Oxidation ponds require large amounts of land and the degree of treatment is weather dependent. The only operation necessary, if at all, is to alter by appropriate valves, the point of the discharge of the raw sewage. A number of problems may however be associated with oxidation ponds. First, they permit the growth of mosquitoes in countries where malaria and other mosquito-borne diseases abound. The mosquito larvae can be controlled by spraying the ponds with oil. Second, aquatic weeds may clog them but if they are up to 3 ft deep weeds do not grow except at the edges. Depending upon the design, oxidation ponds must be freed of sludge approximately every 10 years. They are sometimes used for the primary stabilization of wastes from dairies; often, however, they are employed as secondary or tertiary treatment facilities.

Oxidation ponds are usually employed as secondary or tertiary treatment (Gerhardt and Oswald, 1990). Occasionally, they are used as primary treatment plants in which case anaerobic conditions tend to occur because of the heavy load. This anaerobic condition is particularly apt to occur in the dark when, as is seen later, the photosynthetic organisms largely responsible for the provision of oxygen to the aerobic bacteria are inactive.

Oxidation ponds are most often used as secondary treatment for wastewaters or for waters such as shed wastes which contain heavy loads of organic matter. Oxidation pond systems may be a two-pond system or a multi-pond one. Multi-pond systems are usually described as an Advanced Integrated Pond System (AIPS) in which different ponds fulfill different functions. It consists of a series of ponds in the ground, which use heterotrophic bacteria and algae to treat the wastewater. Wastewater first passes into deep pits in the Advanced Facultative Pond. Here, solids are broken down by fermentation anaerobically in a fermentation pit to produce methane and achieve the removal of many pathogens. The water then passes into a High Rate Pond (HRP) for rapid growth of algae and concomitant production of oxygen; in this pond, organic materials are oxidized. In some systems, paddlewheels which revolve slowly help to increase aeration in the HRP. Small ponds immediately downstream remove algae. The water next passes into Maturation Pond where it is held for a period before being discharged. Maturation tanks provide opportunities for pathogen reduction, and when the aim is to reduce pathogens, the water may be passed through a series of maturation ponds. Effluent from AIPS is typically suitable for aquaculture because the fish feed on the algae in it. Under such conditions, the algal settling and maturation ponds may be eliminated or the water may be held for shorter periods in them.

MICROBIOLOGY OF THE OXIDATION POND

The bulk of the stabilization in an oxidation pond is brought about by aerobic bacteria, but zones of growth by anaerobic and facultative bacteria may also exist depending on the depth of the pond. Oxygen is supplied to the aerobic bacteria by two means: (1) Oxygen released by algae during photosynthesis, and (2) diffusion of oxygen into the water assisted by natural winds and sometimes by floating turbine aerators. In large oxidation ponds with retention periods of 3 weeks to 3 months, algal aeration by photosynthesis is not very effective; but in smaller ponds with retention time of less than a week, algal photosynthetic oxygen effectively supplies the oxygen required by the aerobic bacteria. Turbine aerators which float on the ponds are often used in large oxidation ponds to encourage O_2 diffusion into the pond. The CO_2 released by the aerobes is used by the algae.

Bacteria, protozoa, algae, and rotifers are all to be found in the oxidation pond. The predominant bacteria are *Pseudomonas*, *Flavobacterium*, and *Alcaligenes*, but this depends to some extent on the nature of the sewage. Coliforms die off rapidly because they cannot compete for food and because of the predatory activity of ciliates. Some authors suggest that antibiotics released by algae are effective in killing off bacteria. The algae commonly involved include *Chlorella*, *Spirogyra*, *Vaucheria*, and *Ulothrix*. Some algae are confined to the surface, e.g., *Oscillatoria* while others are benthic, e.g., *Scenedesmus*. Some of the free-swimming ciliates include *Paramecium* and *Colpidium*, whereas the stalked ciliates include *Vorticella*. In some oxidation pond designs, only one pond is used, especially when it is used for treating effluents from other treatments such as activated sludge or the Imhoff tank (see below). When however it is used on its own, say in dairy or abattoir wastes where the organic matter content is high, at least two ponds are involved. The first is usually deep, and decomposition in it is usually anaerobic. The subsequent ponds are usually aerobic and involve algae (Okafor, 2007).

ANAEROBIC SEWAGE SYSTEMS

Treatment of the Sludge from Aerobic Sewage Treatment Systems

Anaerobic breakdown of sludge as has been seen above, sludge always accompanies the aerobic breakdown of wastes in water. Its disposal is a major problem of waste treatment. Sludge consists of microorganisms and those materials are not readily degradable particularly cellulose. The solids in sludge form only a small percentage by weight and generally do not exceed 5%. The goals of sludge treatment are to stabilize the sludge and reduce odors, remove some of the water and reduce volume, decompose some of the organic matter and reduce volume, kill disease-causing organisms, and disinfect the sludge. Untreated sludges are about 97% water. Settling the sludge and decanting off the separated liquid removes some of the water and reduces the sludge volume. Settling can result in a sludge with about 96%–92% water. More water can be removed from sludge by using sand drying beds, vacuum filters, filter presses, and centrifuges resulting in sludges with between 80% and 50% water. This dried sludge is called a sludge cake. Anaerobic digestion is used to decompose organic matter to reduce volume.

Digestion also stabilizes the sludge to reduce odors. Caustic chemicals can be added to sludge or it may be heat treated to kill disease-causing organisms. Following treatment, liquid and cake sludges are usually spread on fields, returning organic matter and nutrients to the soil. The most common method of treating sludge, however, is by anaerobic digestion and this will be discussed below.

Anaerobic digestion consists of allowing the sludge to decompose in digesters under controlled conditions for several weeks. Digesters themselves are closed tanks with provision for mild agitation, and the introduction of sludge and release of gases. About 50% of the organic matter is broken down to gas, mostly methane. Amino acids, sugars, and alcohols are also produced. The broken-down sludge may then be dewatered and disposed of by any of the methods described above. Sludge so treated is less offensive and consequently easier to handle. Organisms responsible for sludge breakdown are sensitive to pH values outside 7–8, heavy metals, and detergents and these should not be introduced into digesters. Methane gas is also produced and this may sometimes be collected and used as a source of energy.

THE SEPTIC TANK

Septic tanks are small-scale sewage systems not connected to main sewage systems. A septic tank generally consists of a tank of between 1,000 and 1,500 gal (4,000–5,500 L) which is connected to an inlet wastewater pipe at one end and to a septic drain field or soak away or soakage pit at the other. These pipe connections are generally made via T pipes which allow liquid entry and outflow without disturbing any crust on the surface, i.e., direct current between the inlet and outlet is prevented using baffles and pipe tees.

Wastewater enters through the inlet pipe and the baffle directs it to the bottom. The heavier materials remain at the bottom where they are broken down by anaerobic bacteria, thereby reducing the volume of solids. The oils and fats are not usually broken down as easily as the other materials and they float to the top of scum. As new sewage is added, the accompanying added liquid displaces the top portion of the liquid in the tank into the pipe leading to the soakage pit, taking with it dissolved broken-down materials.

Anaerobic decomposition is not as efficient as aerobic and the unbroken-down materials form sludge which must be removed from time to time along with the fat scum. Septic tanks have some problems which are the result of the added sewage and its breakdown products remaining in the tank (as opposed to sewers which move the added sewage to the processing location). Thus, excessive addition of fats and oils or flushing down non-biodegradable materials such as sanitary towels may lead to early filling up of the tank. Similarly, the addition of chemicals, such as acid or sodium hydroxide, which can kill off the microorganisms carrying out the decomposition, may also lead to an early filling up of the tank. Roots from trees growing nearby may rupture the tank and shrubbery growing above the tank, or the drain field may clog and/or rupture them.

THE IMHOFF TANK

This is named after its inventor, the German Karl Imhoff. It has an upper flow or sedimentation chamber through which the sewage passes at low velocity, and a lower or digestion chamber into which the heavier sewage particles sediment and are broken down. Methane, which is produced by anaerobic breakdown, is discharged through a pipe and may be collected for use as fuel, while the scum and the undigested sediment are transferred from the sedimentation tank by mechanical means to drying beds. The dried sludge is then used as manure. The effluent from the tanks may be further treated either with a trickling filter or in an oxidation pond. Unpleasant odors which accompany black foams sometimes appear at the gas vents but the odors may be reduced somewhat by introducing lime into the vent. Animal manure (i.e., dung and trash) which may also help in odor reduction is added to the gas vents. The Imhoff tank has no mechanical parts and is relatively easy and economical to operate. It provides sedimentation and sludge digestion in one unit, and when operating properly, produces a satisfactory primary effluent with a suspended solids removal of 40%–60% and a BOD reduction of 15%–35%. The two-storey design requires a deep overall tank. The Imhoff tanks is best suited to small municipalities and large institutions where the population is 5,000 or less, and a greater degree of treatment is not needed.

CESSPOOLS

Cesspools are shallow disposal systems that are generally constructed as a concrete cylinder with an open bottom and/or perforated sides (drywell). They are used by multi-family residential units, churches, schools, public meeting facilities, office buildings, industrial and commercial buildings, shopping malls, hotels and restaurants, highway rest stops, state parks, and camp grounds.

This untreated sanitary waste can enter shallow groundwater and contaminate drinking water resources because they are designed to isolate but not to treat sanitary waste. They may introduce into groundwater the following undesirable items: Nitrates, total suspended solids, and coliform bacteria exceeding the quantities recommended for drinking water, as well as other constituents of concern such as phosphates, chlorides, grease, viruses, and chemicals used to clean cesspools (e.g., trichloroethane and methylene chloride). Breakdown of organic matter is anaerobic as no aeration is introduced. The cesspool resembles a septic tank in which the broken-down materials seep into the ground without going through a T-pipe.

TERTIARY TREATMENT

Tertiary treatment of effluent involves a series of additional steps after secondary treatment to further reduce organics, turbidity, nitrogen, phosphorus, metals, and pathogens. Most processes involve some type of physicochemical treatment such as coagulation, filtration, activated carbon adsorption of organics, reverse osmosis, and additional disinfection. Tertiary treatment of wastewater is practiced for additional protection of wildlife after discharge into rivers or lakes. Even more commonly, it is performed when the wastewater is to be reused for irrigation, for recreational purposes or for drinking water.

REMOVAL OF PATHOGENS BY SEWAGE TREATMENT PROCESSES

Disinfection and/or advanced tertiary treatment is necessary for many reuse applications to ensure pathogen reduction. Current issues related to pathogen reduction are treatment plant reliability, removal of new and emerging enteric pathogens of concern, and the ability of new technologies to effect pathogen reduction. Wide variation in pathogen removal can result in significant numbers of pathogens passing through a process for various time periods. The issue of reliability is of major importance if the reclaimed water is intended for recreational or potable reuse, where short-term exposures to high levels of pathogens could result in significant risk to the exposed population. Compared with other biological treatment methods (i.e., trickling filters), activated sludge is relatively efficient in reducing the numbers of pathogens in raw wastewater. Both sedimentation and aeration play a role in pathogen reduction. Primary sedimentation is more effective for the removal of the larger pathogens such as helminth eggs, but solid-associated bacteria and even viruses are also removed. During aeration, pathogens are inactivated by antagonistic microorganisms and by environmental factors such as temperature. The greatest removal probably occurs by adsorption or entrapment of the organisms within the biological floc that forms. The ability of activated sludge to remove viruses is related to the ability to remove solids. This is because viruses tend to be solid associated, and are removed along with the floc. Activated sludge typically removes 90% of the enteric bacteria and 80% to 90%–99% of the enteroviruses and rotaviruses.

Tertiary treatment processes involving physical–chemical processes can be effective in further reducing the concentration of pathogens and enhancing the effectiveness of disinfection processes by the removal of soluble and particulate organic matter. Filtration is probably the most common tertiary treatment process. Mixed-media filtration is most effective in the reduction of protozoan parasites. Usually, greater removal of Giardia cysts occurs than that of Cryptosporidium oocysts, because of the larger size of the cysts. Removal of enteroviruses and indicator bacteria is usually 90% or less. Addition of coagulant can increase the removal of poliovirus to 99%. Coagulation, particularly with lime, can result in significant reductions of pathogens. The high-pH conditions (pH 11–12) that can be achieved with lime can result in significant inactivation of enteric viruses. To achieve removals of 90% or greater, the pH should be maintained above 11 for at least an hour. Inactivation of the viruses occurs by denaturation of the viral protein coat. The use of iron and aluminum salts for coagulation can also result in 90% or greater reductions in enteric viruses. The degree of effectiveness of these processes, as in other solids separating processes, is highly dependent on the hydraulic design and, in particular, on coagulation and flocculation. The degree of removal observed in bench-scale tests may not approach those seen in full-scale plants, where the process is more dynamic. Reverse osmosis and ultrafiltration are also believed to result in significant reductions in enteric pathogens.

REMOVAL OF ORGANICS AND INORGANICS BY SEWAGE TREATMENT PROCESSES

In addition to microbial pathogens and nutrients such as nitrogen and phosphorus, there are other constituents within sewage that need to be kept at low concentrations. These include inorganics, exemplified by metals, and organic priority pollutants. Metals and organics are normally associated with the solid fraction of sewage and neither are significantly removed by sewage treatment.

SLUDGE PROCESSING

Primary, secondary, and even tertiary sludges generated during wastewater treatment are usually subjected to a variety of processes. Raw sludge is sometimes subjected to screening to remove coarse materials including grit that cannot be broken down biologically. Thickening is usually done to increase the solid content of the sludge. This can be achieved via centrifugation, which increases the solid content to approximately 12%. Dewatering can further concentrate the solid content to 20%–40%. This is normally achieved via filtration or by the use of drying beds. Conditioning enhances the separation of solids from the liquid phase. This is usually accomplished by the addition of inorganic salts such as alum, lime, or ferrous or ferric salts, or synthetic organic polymers known as polyelectrolytes. All of these processes reduce the water content of the sludge, which ultimately reduces transportation costs to the final disposal and/or utilization site. Finally, stabilization technologies are available, reducing both the solids content of the sludge and inactivating pathogenic microbes present in the sludge.

PRIMARY SLUDGE TREATMENT

Primary sludge is normally pumped from the primary sedimentation tanks at periodic intervals and treated further by anaerobic digestion. Municipal wastewater anaerobic digesters consist of circular tanks having an overall retention time of 30–60 days, based on the concentrated sludge volume. While small WWTP may have only a single anaerobic digestion tank, medium size and large WWTP have multiple tanks. These digesters have a brick cover over a layer of insulation around the reinforced concrete digesters to minimize the loss of heat to the atmosphere.

Primary anaerobic digestion tanks are normally heated to 35°C by external heat exchangers and mixed by either gas mixers or mechanical mixers. The increased temperature and the sludge mixing permit bacteria in the anaerobic digesters to metabolize the biodegradable organics in the shortest possible time. The bacteria in anaerobic digesters are the same types of bacteria found in anaerobic lagoons. The anaerobic bacteria begin metabolism by hydrolyzing the biodegradable fraction of primary sludge to simple soluble organic compounds that can be further metabolized by other bacteria in the anaerobic environment. Facultative bacteria are responsible for the initial metabolism of the organic solids. Proteins are easily hydrolyzed to amino acids that undergo further hydrolysis to NH_3-N and short-chain fatty acids. Polysaccharides are hydrolyzed to simple sugars that are metabolized to short-chain fatty acids, aldehydes, ketones, and alcohols. Fats are slowly hydrolyzed to glycerol and long-chain fatty acids. The glycerol is quickly metabolized, while the long-chain fatty acids are slowly metabolized. Solubility is a major factor slowing the rate of metabolism of the long-chain fatty acids.

With an excess of organic nutrients, the sulfate-reducing bacteria grow in proportion to the available sulfates in the fluid sludge. Since sulfates are limited in municipal wastewaters, growth of the sulfate-reducing bacteria will also be limited. For practical purposes, sulfate reduction will be complete in municipal anaerobic digesters. The sulfate-reducing bacteria have an important role in anaerobic digesters, creating the highly reduced environment required by the methane bacteria. Successful operation of anaerobic digesters depends upon the combined metabolism of the facultative bacteria, the sulfate-reducing bacteria, and the methane bacteria. The facultative bacteria convert the biodegradable solids into soluble organic compounds. The sulfate-reducing bacteria lower the oxidation-reduction potential (ORP) in the digester.

The methane bacteria convert the short-chain fatty acids into methane gas and carbon dioxide. Metabolism of long-chain fatty acids results in the production of hydrogen gas that can be used either by a second group of methane bacteria or by the acetogenic bacteria. The acetogenic bacteria use hydrogen and carbon dioxide to produce acetic acid and water. Normal methane bacteria will then metabolize the acetic acid produced as an end product by the acetogenic bacteria. The hydrogen-utilizing methane bacteria compete with the acetogenic bacteria for both hydrogen and carbon dioxide and produce methane and water as their end products. It is important to recognize that specific sequences of bacteria are required to convert the complex organic solids to methane, carbon dioxide, and water. About 50% of the TS in the primary sludge will be destroyed in anaerobic digesters, since only 60%–70% of the VS in primary sludge are biodegradable. Half of the primary sludge added to the anaerobic digesters is composed of inert material that remains unchanged through the digestion process. Most of the inert solids settle out in the anaerobic digestion tanks, requiring periodic removal and dewatering before being returned to the land. Examination of the data in anaerobic digesters indicates that the COD of the VS destroyed by the bacteria appears as methane gas COD. Continuous measurement of the methane gas production is a useful tool in evaluating the operational efficiency of anaerobic digesters on a daily basis. A decrease in the daily rate of methane gas production indicates a potential problem with the sludge treatment system that can become quite serious if action is not taken immediately to resolve the problem. An increase in volatile acids and a decrease in pH or an increase in toxic compounds in the feed sludge will result in a decrease in methane gas production. The acetate-utilizing methane bacteria are the most sensitive group of bacteria in anaerobic digesters and will be the first group of bacteria to indicate problems.

Municipal sludge digesters normally produce gas with about 65% CH_4 and 35% CO_2. Digester gas contains very little hydrogen sulfide, even though it can be detected by smell. Examination of the metabolism of the methane bacteria indicates why the anaerobic digestion process is controlled by the growth of the methane bacteria. The acetate-utilizing methane bacteria obtain very little energy from the metabolism of acetate to methane and carbon dioxide. The net effect of the low energy yield for the acetate-utilizing methane bacteria is limited by cell growth. Each cell must process large quantities of acetate to obtain the energy necessary to produce a new cell. The large concentration of inert solids and inadequate mixing in most anaerobic digesters make it difficult for the methane bacteria to come into contact with sufficient acetate to grow enough new cells for rapid metabolism. The addition of large quantities of primary sludge allows the rapid formation of organic acids with a localized pH decrease that adversely affects the methane bacteria. Better operations can be achieved by adding small quantities of primary sludge at frequent intervals with good digester mixing. Methane bacteria require an environmental pH level between 6.5 and 9.0 for their maximum rate of metabolism. The pH in municipal digesters is held between 6.5 and 8.0 by bicarbonate alkalinity. Good mixing is also important for rapid reaction between the short-chain organic acids and the bicarbonate alkalinity. Municipal digesters normally have bicarbonate alkalinity levels between 2,000 and 4,000 mg/L. The bicarbonate alkalinity is formed primarily by the reaction of NH_3, CO_2, and HOH. The NH_3 is released from the neutralized organic acids formed during the metabolism of proteins. The CO_2 is released from the organic acid metabolism. Ammonium bicarbonate is the natural pH buffering system found in all municipal biological wastewater treatment systems. If the anaerobic digester does not have sufficient bicarbonate alkalinity to neutralize the organic acids, it is necessary to add a chemical source of alkalinity for good operations. Fortunately, metabolism of the acetate to methane results in an increase in bicarbonate alkalinity. A well-operating digester produces volatile acids that are quickly neutralized by the bicarbonate alkalinity and then metabolized to methane with the production of the previously used bicarbonate alkalinity. Problems with the methane bacteria can disrupt this operation and allow the volatile acids to accumulate. One of the problems with volatile acids lies in the fact that volatile acid salts titrate the same as alkalinity in the alkalinity test. A buildup of volatile acid salts will create the impression of having considerable alkalinity when there is very little alkalinity to neutralize additional volatile acids. As soon as there is an excess of volatile acids, the pH will plunge downward. For this reason, volatile acids have been used as a major operating parameter for anaerobic digesters together with pH, bicarbonate alkalinity, and methane gas production.

Many digesters depend upon the formation of gas bubbles for natural mixing. In large, flat digesters, the gas bubbles are released over a large surface area, creating limited turbulence for liquid mixing. Recent digester designs have looked at tall, cylindrical tanks with small cross-sectional areas. The same volume of rising gas bubbles in the tall, cylindrical tanks produces better mixing than in the flat, shallow, circular tanks. Although the importance of mixing has long been recognized in biological systems to bring the nutrients together with the bacteria for rapid metabolism, engineers have resisted the addition of good mixers in anaerobic digesters. The original reason for this was the failure of mixing to produce additional digestion of the VS in anaerobic digesters with long retention times.

The ability to accumulate methane bacteria in anaerobic digesters controls the overall operations. Since anaerobic digesters are operated as single pass systems, it is not possible to increase the population of methane bacteria above the numbers supported by the organic loading rate. As the digester retention time is increased, endogenous respiration keeps reducing the active mass of methane bacteria. At 35°C, the maximum methane bacteria mass will be about 1.5 times the daily increase in the entire system. As the digester retention time increases, the increased liquid volume dilutes the active microbial concentrations in a well-mixed system. Limited mixing in the active metabolism zone may allow the concentration of methane bacteria to remain high, permitting complete metabolism with poor mixing and long retention periods. Balancing mixing with the rate of metabolism of primary sludge and retention time in anaerobic digesters could be a productive area for future research. It has also been shown that trace metals affect the growth of anaerobic bacteria. If the bacteria do not have access to all the trace metals they require for metabolism, their metabolic activities will be retarded. Iron is the most important trace metal required for all bacteria. Nickel, cobalt, molybdenum, copper, zinc, and manganese in trace quantities also stimulate the metabolism of the methane bacteria. Selenium has been found to be helpful for a few species of methane bacteria. It is important to recognize that bacteria must have all the nutrients required to produce normal cell protoplasm if the bacteria are to grow at their maximum rate. In trace metal deficient environments, the bacteria are limited by their ability to recycle the trace metals in the desired quantities. These heavy metal ions form metallic sulfide precipitates in anaerobic digesters, reducing the toxic effect that excess concentrations of the heavy metals produce while allowing the bacteria to obtain sufficient amounts of trace metals for metabolism.

Digested sludge is fluid and relatively inert. It can be applied onto agricultural land if the water and the inorganic salts in the water phase can be absorbed without damaging the soil, the water table, or the adjacent waterways. Soil fungi can slowly metabolize some of the residual VS in the digested sludge, creating soil humus over time. The nitrifying bacteria in the surface soil will metabolize the soluble NH_3-N to NO_3-N. The soluble phosphate will be adsorbed onto soil particles depending upon their active chemical bonds. Surface plants can use these nutrients during their normal growth periods. The soil is the ultimate receptor for the digested sludge. The sludge components originally came from the soil and must eventually return to the soil to keep the natural cycle in balance. Pathogenic bacteria survival in digested sludge is dependent upon the pathogens ability to form spores. Few vegetative cells of pathogenic bacteria are able to survive the environmental conditions created in the anaerobic digestion of primary sludge with 20–30 days sludge retention. Viruses that are adsorbed onto sludge particles will pass through the anaerobic digesters unless they are metabolized by bacteria needing nutrients for survival.

STABILIZATION TECHNOLOGIES

Aerobic Digestion

This consists of adding air or oxygen to sludge in a 4–8 m/L feet deep open tank. The oxygen concentration within the tank must be maintained above 1 mg/L to avoid the production of foul odors. The mean residence time in the tank is 12–60 days, depending on the tank temperature. During this process, microbes aerobically degrade organic substrate, reducing the volatilize solids content of the sludge by 40%–50%. Digestion temperatures are frequently moderate or mesophilic (30°C–40°C). By increasing the oxygen content, thermophilic digestion can be induced (>60°C). By increasing the temperature and the retention time, the degree of pathogen inactivation can be enhanced. Pathogen concentrations ultimately determine the treatment level of the product.

Class B biosolids can contain many human pathogens. Class A biosolids, which result from more stringent and enhanced treatment, contain very low or nondetectable levels of pathogens. The degree of treatment, Class A versus Class B, has important implications on the reuse potential of the material for land application. Aerobic digestion generally results in the production of Class B biosolids.

Anaerobic Digestion

This type of microbial digestion occurs under low redox conditions, with low oxygen concentrations. Carbon dioxide is used as a terminal electron acceptor, and organic substrate is converted to methane and carbon dioxide. This process reduces the volatile solids by 35%–60% and results in the production of Class B biosolids.

Advantages of Anaerobic Digestion

- No oxygen requirement which reduces cost digestion.
- Reduced mass of biosolids due to low energy yields of anaerobic metabolism.
- Methane produced, which can be used to generate electricity.
- Enhanced degradation of xenobiotic compounds.

Disadvantages of Anaerobic Digestion

- Slower than aerobic.
- More sensitive to toxicants.

SLUDGE PROCESSING TO PRODUCE CLASS A BIOSOLIDS

Class B biosolids that arise following digestion can be further treated to Class A levels prior to land application. The three most important technologies here are composting, lime treatment, and heat treatment.

COMPOSTING

This process consists of mixing sludge with a bulking agent that normally has a high C:N ratio. This is necessary because of the low C:N ratio of the sludge. The mixtures are normally kept moist but aerobic. These conditions result in very high microbial activity and the generation of heat that increases the temperature of the composting material. There are three main types of composting systems:

1. The aerated static pile process typically consists of mixing dewatered digested sludge with wood chips. Aeration of the pile is normally provided by blowers during a 21-day composting period. During this active composting period, temperatures increase to the mesophilic range (20°C–40°C), where microbial degradation occurs via bacteria and fungi. Temperatures subsequently increase to 40°C–80°C, with microbial populations dominated by thermophilic (heat-tolerant) and spore forming organisms. These high temperatures inactivate pathogenic microorganisms and frequently result in a Class A biosolid product. Subsequently, the compost is cured for at least 30 days, during which time temperatures within the pile decrease to ambient levels.
2. The windrow process is similar to the static pile process except that instead of a pile, the sludge and bulking agent are laid out in long rows of the dimension 2 m by 3 m by 80 m. Aeration for windrows is provided by turning the windrows several times a week. Once again, if the composting process is efficient, Class A biosolids are produced.
3. In enclosed systems, composting is conducted in steel vessels of size 10–15 m high by 3–4 m diameter. For this type of composting, aeration via blowers and temperature of the composting are carefully controlled. This results in a high-quality Class A compost, with little or no odor problems. However, costs of enclosed systems are higher.

LIME AND HEAT TREATMENT

Lime stabilization involves the addition of lime as $Ca(OH)_2$ or CaO, such that the pH of digested sludge is equal to or greater than 12 for at least 2 hours. Liming is very effective at inactivating bacterial and viral pathogens, but less so for parasites (Bitton, 2011). Lime stabilization also reduces odors, and can result in a Class A biosolid product.

Heat treatment involves heating sludge under pressure to temperatures up to 260°C for 30 minutes. This process kills microbial pathogens and parasites, and it also further dewaters the sludge.

LAGOONS

Lagoons are the simplest method of holding the animal wastes until applied to land. Improperly constructed lagoons tend to leak into the soil and create groundwater pollution, as well as creating obnoxious odors. Lagoons can fill with wastes and overflow before the wastes can be applied to land. Excess animal waste production often results in improper addition of those waste materials onto unprepared land. Sudden storms can produce runoff that carries the manure into adjacent streams and rivers, creating fish kills and serious environmental pollution problems. Failure to understand the basic problems created by animal wastes has been a major problem worldwide. Even the advanced, industrial countries with all their technology have not developed a simple solution for the animal waste problems. Confined animal wastes are directly related to the animal feed since all of the feed that is not used by the animals will appear as wastes. Feed and water are taken in by the animals and used in their metabolism. Part of the feed is oxidized for energy and part of the feed is used to create new animal mass. The remainder of the feed appears in the manure and urine.

Wastewater treatment engineers have found by experience that concentrated organic wastes are easier to treat anaerobically than aerobically. Anaerobic wastewater treatment requires a complex consortium of bacteria working together to bring about maximum stabilization of the organic contaminants in the wastewater. The complex organic suspended solids must be converted by bacteria to soluble organic acids, aldehydes, ketones, and alcohols as the first step in metabolism. The soluble organic acids, aldehydes, ketones, and alcohols are then metabolized to volatile organic acids by other bacteria. The sulfate-reducing bacteria metabolize some of the soluble organic compounds with sulfate to produce reduced sulfides that lower the oxidation-reduction potential (ORP) to the proper level for the methane bacteria to convert the volatile acids to methane gas for discharge to the atmosphere. Metabolism of the urea from the urine results in the production of ammonium carbonate and a rise in pH unless sufficient carbon dioxide is available to convert the ammonium carbonate into ammonium bicarbonate with a lowering of pH. Metabolism of the proteins will result in the release of ammonia ions and the formation of ammonium salts of the volatile acids, keeping the pH from dropping.

Since the acetate-utilizing methane bacteria are essential for the production of the highest quality effluent from anaerobic systems, it is necessary to keep the unionized NH_3 concentration below the toxic level. The pH of the anaerobic environment will be a primary factor in determining the unionized NH_3 concentration. If the pH is kept between 6.5 and 7.5, there is little chance that the unionized NH_3 concentration will reach toxic levels.

Anaerobic lagoons have been designed with liquid retention times ranging from 3 months to well over a year. Too often, anaerobic lagoons have been sized to fit the available land and economics rather than being designed for the wastes to be treated. Current design criteria for anaerobic lagoons are based on field experience collected from numerous anaerobic lagoons over several decades. Anaerobic lagoons can be single-cell lagoons or multi-cell lagoons. Multi-cell lagoons will have two, three, or more cells in series, depending upon site configuration. Single-cell lagoons are more popular than multi-cell lagoons, primarily from an economic point of view. The characteristics of hog manure indicate that most of the pollutants are suspended solids that will settle out near the discharge from the inlet wastewater pipe, creating a sludge mound. The biodegradable organic matter in the suspended solids will stimulate the growth of different groups of bacteria. The relatively dense environment in the settled sludge mound limits the movement of bacteria, retarding rapid metabolism of the biodegradable organic compounds. Micro-environmental pockets form throughout the settled sludge mound. Some of the micro-environmental pockets will contain high concentrations of organic acids, depressing the pH well below the 6.5 level required by the methane bacteria. Some of the micro-environmental pockets will have high NH_3 concentrations with pH values above 8.5. Trace metals are not uniformly dispersed in the sludge mound, slowing the rate of metabolism of different groups of bacteria. As the methane bacteria slowly metabolize the salts of organic acids, bicarbonate alkalinity is produced along with the methane gas. The insolubility of the methane gas allows the methane to form tiny bubbles that rise by the path of least resistance up through the settled sludge mound. As the bubbles move upward around the solids, they create localized turbulence that moves the soluble end products of metabolism and tiny bacteria upward to points where new organic compounds are available for metabolism. Excess carbon dioxide will also be formed by metabolism and discharged as a gas along with the methane gas. While the total gas production is not adequate to completely mix the settled solids, the limited mixing action helps the total bacteria population to increase. It can take anywhere from one month to a year for anaerobic lagoons to develop sufficient numbers of all types of bacteria to permit optimum metabolism of the daily load. Temperature is also an important factor in the overall rate of metabolism. Warm temperatures will speed the rate of bacterial metabolism and assist the bacteria in reaching equilibrium in the shortest time. The rising gas bubbles may be sufficient to cause some of the suspended solids to rise to the liquid air surface and form a layer of solids over the surface of the anaerobic lagoon. The surface solids layer will act as a lid over the anaerobic lagoon and prevent oxygen transfer from the air into the anaerobic lagoon.

Anaerobic lagoons are started by filling the lagoon about half full with freshwater. As wastewaters are added to the anaerobic lagoon, the suspended solids settle to the bottom of the lagoon as the wastewater velocity slows on entering the lagoon. The soluble organic compounds and the soluble inorganic salts tend to diffuse into the freshwater that was in the lagoon. The NH_3-N concentration is quickly dropped below the toxic level by simple dilution. As the bacteria population increases, the methane gas production will soon reach the expected daily production rate for the level of organic matter contained in the influent wastewaters. For a short time period, the gas production will exceed the addition of daily organic compounds, creating the impression that more methane gas is being produced than is theoretically possible. The excess gas production is generated from the accumulated organic solids that had not been metabolized during the acclimation period. The daily gas production will reach a peak and then decrease down to the normal equilibrium value. The slow rate of buildup of bacteria in anaerobic lagoons tends to lengthen the time required for the treatment system to reach equilibrium. Engineers and wastewater treatment plant operators who are used to rapid changes created in microbial populations in trickling sludge wastewater treatment plants are easily frustrated by the long time period required for equilibrium conditions in anaerobic lagoon wastewater treatment systems.

FURTHER READING

APHA. *Standard Methods for Water and Wastewater.* Washington, DC: American Public Health Association; 1998.

Bitton G. *Wastewater Microbiology*, 4th ed. New York: Wiley-Liss; 2011.

Bitton G. *Wastewater Microbiology.* New York: Wiley-Liss; 1999.

Eckenfelder WW. *Industrial Water Pollution Control*, 3rd ed. New York: McGraw-Hill; 2000.

Gerhardt MB, Oswald WJ. Advanced integrated ponding system in sewage reuse. In: Edwards P, Pullin RSV (eds.) *Wastewater-Fed Aquaculture. Proceedings of the International Seminar on Wastewater Reclamation and Re-use for Aquaculture, Calcutta, India, 6–9 December, 1988.* Bangkok: Environmental Sanitation Information Centre, Asian Institute of Technology; 1990.

Grant WD, Long PE. *Environmental Microbiology.* Springer US; 1981.

Gray NF. *Biology of Wastewater Treatment.* 2nd ed. London: Imperial College Press; 2004.

Lindera KC. Activated sludge – the process. In: *Encyclopedia of Environmental Microbiology* Vol. 1: 74–81. New York: Wiley-Interscience; 2002.

McKinney RE. *Environmental Pollution Control Microbiology.* New York: M. Dekker; 2004.

Nielsen PH. Activated sludge – the floc. In: *Encyclopedia of Environmental Microbiology* Vol. 1: 54–61. New York: Wiley-Interscience; 2002.

Okafor N. *Environmental Microbiology of Aquatic and Waste Systems.* Springer Netherlands; 2011.

Okafor N. *Modern Industrial Microbiology and Biotechnology.* Enfield: Science Publishers; 2007.

Pepper IL, CP Gerba, TJ Gentry. *Environmental Microbiology.* Academic Press; 2014.

U.S. Environmental Protection Agency. *Handbook for Sampling and Sample Preservation of Water and Wastewater; EPA-600/4-82-029.* Cincinnati, OH: Environmental Monitoring and Support Laboratory; 1982.

Wainwright M. *An Introduction to Environmental Biotechnology.* Springer US; 1999.

Water Environment Federation. *Industrial Wastewater Management, Treatment, and Disposal,* 3rd ed. Manual of Practice-3-McGraw-Hill Professional; 2008.

16 Wastewater from Food Industries

WASTEWATER COMPOSITION AND TREATMENT STRATEGIES IN THE FOOD PROCESSING INDUSTRY

In view of the great variety of industrial branches, this section is by no means comprehensive in examining all areas of industry, but concentrates on the most important branches of the food processing industry.

SUGAR FACTORIES

The most important wastewater components are the high organic loads derived from the sugar. In contrast, the amounts of nitrogen and phosphorus are comparatively low; because growing ecological awareness has led to a more controlled fertilization of sugar beet fields, the amounts of these pollutants are ever decreasing.

For the treatment of sugar factory wastewater, the following processes are common:

- soil treatment
- long-term batch processing
- small-scale technical processes, usually consisting of anaerobic pretreatment and aerobic secondary treatment

Agricultural soil treatment is primarily geared toward utilization of the contained nutrients and water. The simultaneous degradation of the organic substances is of less importance; it does not form the basis of dimensioning. If large quantities of ion-exchange waters occur, the salt tolerance of the irrigated plants has to be considered. In recent years, mainly anaerobic wastewater treatment processes have become established in the sugar industry. In comparison to the processes to the aerobic biological treatment, anaerobic methods have a number of distinct advantages: less area demand, lower sludge production, low energy consumption, energy generation from biogas, and no noxious odors.

STARCH FACTORIES

Starch is produced from potatoes, corn, and wheat. Depending on the original raw material, the wastewater fractions, amounts, and loads emerging during starch production vary considerably. Wastewater from starch factories has a high organic load and usually consists of easily degradable matter. The undissolved organic contents are mainly carbohydrates and proteins. The fat content is normally ~10%. Usually,

DOI: 10.1201/9781003272618-19

this kind of wastewater does not contain any toxic substances. The substrate ratio COD:N:P is satisfactory; actually, there may even occur an excess of N and P, so that it is not necessary to add nutrient salts.

The following processes are used for the treatment of starch factory wastewater:

- soil treatment
- pond processing
- small-scale technical processes (anaerobic, aerobic, evaporation)

Pond processing in batch operation is possible only for production of potato starch, since this is a campaign operation. Continuously operating ponds are rare, since they require additional aeration, due to the high load concentration. They can only be recommended as a secondary treatment method, because the demands for area and aeration are too high.

In the potato starch industry, a further possibility for fruit water treatment is reduction of the wastewater by evaporation to a solid content of 70%. The dried solids can then be utilized, e.g., as fertilizer. To save evaporation energy, a membrane process can be used to increase the concentration prior to evaporation, which would be done as a cascade operation under vacuum conditions. Aerobic secondary treatment is always required if the company intends to discharge its wastewater directly into waterways.

VEGETABLE OIL AND SHORTENING PRODUCTION

In the vegetable oil industry, one differentiates between the following branches: extraction plants for extraction of raw fat and oil, refining plants for refining of raw fat and oil, plants for further processing of the nutrient fats, e.g., into margarine, which consists of approx. 80% fat, 20% water, plus some other ingredients. The process technology for nutrient fat and oil production largely depends on the raw materials, depending on whether oil is extracted from seeds (soy, sunflower, coconut, rapeseed, etc.), animal fats from tallow and fat liquefiers (tallow or lard), and fruit flesh fats (palm oil and olive oil, which due to its poor storage stability is usually extracted near the site of production), and fats from marine animals.

Wastewater treatment processes usually consist of a physical–chemical pretreatment by fat separators or flotation systems to decrease the amounts of undissolved solids and lipophilic substances. Whereas installation of fat separators is only a minimal pretreatment concept, because the efficiency of these systems may be completely reduced by a hot water surge, a more expensive flotation system using additives also allows the elimination of emulsified, mainly lipophilic, substances.

Often a mixing and equalizing tank is installed behind the flotation system to equalize pH value and temperature peaks and to allow a more constant loading of the subsequent biological treatment stage. In biological wastewater treatment, aerobic activated sludge treatment with a sludge load of approx. 0.1 kg BOD kg/MLSS day has proved to be effective. To prevent flotation of sludge and development of scum layers, the fat contents of the wastewater should be <200 mg/L.

POTATO PROCESSING INDUSTRY

Potatoes are used as fodder or feed potatoes, as industry potatoes (starch production, distilleries), or as market potatoes. Potato processing can be divided into three main production methods: fresh products (peeled potatoes, precooked potatoes, potato salads, sterilized potatoes, potato products), dried products (dehydrated potatoes, instant mashed potatoes, raw dehydrated potatoes), and fried products (French fries, chips, sticks).

For each of different production methods, wastewater occurs during washing, peeling, cutting, sorting, blanching, and steam drying. Wastewater from the potato processing industry contains comparatively much nitrogen and phosphorus. Nutrient salt limitations occur only in exceptional cases.

Wastewater treatment systems often consist of the following sections:

- grit chamber, screening system, settling tanks for purification of the flume, and washing water recirculation
- production-integrated screening systems and separators to recover organic solids that have been separated from the production wastewater and dewatered, to recycle valuable substances (e.g., cattle forage)
- fat separators for wastewater containing fat, when deep-fat fryers are used in the production process
- biological treatment for pretreatment and full treatment.

SLAUGHTERHOUSES

Slaughterhouses and meat processing plants can be divided into four categories according to their different production processes:

- slaughterhouses for cattle
- slaughterhouses for poultry
- meat-cutting plants
- meat-processing industry

The primary wastewater and waste product sources are divided into three production areas:

- truck washing and animal sheds
- slaughter and cutting
- stomach, intestine, and entrails cleaning—the latter, however, are often not done on the slaughterhouse grounds.

In slaughterhouses, wastewater pretreatment is mainly done with mechanical procedures. Up to now, the number of plants for which physicochemical or biological operational steps have been added is comparatively small. The main cleaning effect of mechanical and physical procedures consists of retention and separation of solids with the help of stationary strainers, rotating screening drums, separators, fine rakes, screening catchers, or fat separators with a preceding sludge catcher. In wastewater from slaughterhouses, part of the organic matter consists of oils and fats in emulsified form. Thus, it may be wise to add a physical–chemical stage to the mechanical pretreatment, which would consist of a precipitation/flocculation stage and a flotation unit.

The following biological methods have been used successfully on an industrial scale:

- large space biological methods (oxidation ponds, frequent formerly)
- various activated sludge systems (single-stage, cascade, two-stage)
- anaerobic biological methods

For activated sludge systems, the sludge load should not exceed 0.15 kg BOD_5 kg/ MLSS day. Direct anaerobic treatment in contact sludge reactors or joint treatment in municipal digestion tanks is particularly suitable for fats, floating materials, stomach and gut contents, and the liquid phase from the dewatering of rumen contents.

DAIRY INDUSTRY

Dairy products are classified into drinking milk, cream products, sour milk and milk mix drinks, butter, curd products, hard and soft cheeses, condensed milk, and dried milk. The wastewater produced in milk processing plants consists of cooling water, condensation water, sanitation water, and process water. The process water consists of the wastewater from the pretreatment, water losses during production, those residues that can no longer be used economically, washing water, detergents, rinsing and cleansing water, and water processing. In dairies, the wastewater derives almost exclusively from cleaning of the conveyance and production implements. More than 90% of the organic solids in the wastewater result from milk and production residues. For the milk processing industry, wastewater discharge is almost always identical with loss of products that could otherwise be utilized or sold.

Dairy wastewater has only a small ratio of settleable solids. Thus, conventional mechanical procedures, such as settling tanks, are ineffective. Inevitable losses of fats can be retained in fat separators, which, however, have only a limited efficiency if the wastewater temperatures are comparatively high. One major problem with dairies is the considerable variation in wastewater volume and concentration. Thus, as a first step after straining and a sand trap, it is recommended to install a mixing and equalizing tank (M+E tank). In the M+E tank, wastewaters of different concentrations and pH are mixed, the wastewater flow is equalized, and partial biological degradation occurs, which can also result in a biological neutralization.

As a further pretreatment stage, a flotation implement is recommended, which can either replace the M+E tank or be downstream from it; this facility allows for the removal of fats and proteins, i.e., the major part of the organic pollutants. Since dairy wastewater has a tendency to develop bulking sludge, it is recommended to design the plant so that the microorganisms in the activated sludge are intermittently subjected to high loads, which can be achieved by installing an activated sludge system in plug flow design, by a preceding contact tank (selector), or by a SBR method (sequencing batch reactor: the process steps of filling, denitrification, aeration, and sedimentation happen one after the other, but in the same tank).

Fruit Juice and Beverage Industry

Natural mineral waters and spring waters are collected and bottled at the location of the source spring. Table water consists of drinking water or natural mineral water to which salts are added. Refreshment beverages are produced from water, flavoring substances, sugar or sweetener, and carbon dioxide. The technology of fruit juice production can, in a simplified manner, be divided into the production stages of washing, grinding, refining, filtering, heating, recooling, and bottling. The wastewater produced in these three industrial branches consists of the following streams (some do not occur in every branch): wastewater from cleaning bottles and containers and from bottling, rinsing and washing water, exhaust vapor condensate, wastewater from the production facilities, wastewater from surface cleaning (floors of the production sheds and the parts of the yards where production takes place), and wastewater from cleaning the conveyor facilities.

Wastewater produced in the mineral water industries contains the following components: adhesive materials and fibrous substances, cleaning alkalis and acids, and soiling from the deposit bottles. For soft drinks, one has to consider the fact that the wastewater additionally contains organic pollutants (with a high ratio of carbohydrates, a large part of it being sugar), which derive from residues and product losses. For the fruit juice industry, product losses—in particular, the loss of fruit concentrates—and the sugar, which is often added, are a considerable part of the wastewater pollution.

Extensive wastewater pretreatment can be successfully done with anaerobic reactors in the fruit juice industry. Two-stage implements (first stage: acidification reactor with mixing and equalization function; second stage: methane reactor) have proved to be advantageous. At volumetric loads in the methane reactor of up to $>10\,kg$ COD/m^3 day, the COD elimination rate amounts to about 80%. For reuse of deposit bottles, denitrification may be necessary because, as part of the label glue, nitrogen is added to the water.

Breweries

Beer of course contains both alcohol and carbonic acid. Different kinds of beer (lager, stout, top-fermented, bottom-fermented) are produced mainly by varying the original wort concentrations and by using different kinds of malt and yeast. Residues and wastewater flow fractions occur in the brewing room, in the fermentation and storage cellars, in the filter and pressure tank cellar, during dealcoholization, and during bottling (bottle, barrel, other containers). Brewery wastewater is prone to heavy variation with regard to volume and concentration in the single flow fractions.

Brewery wastewater has a comparatively high temperature (25°C–35°C). With decreasing wastewater volume, the temperature tends to rise to 40°C. It is likely that pH values vary strongly. In companies that reuse deposit bottles, the wastewater from the bottle washing generally has alkaline pH values. Acidic wastewater may at times result from cleaning processes and from regeneration by ion exchange techniques (for water processing). The nitrogen consists mainly of organic nitrogen (albumen, yeast) and to some extent of nitrate (nitric acid). Furthermore, the wastewater is likely to be contaminated by cleaning and detergent agents, as well as by kieselguhr and by particles arising from abrasion of bottles and shards.

For wastewater pretreatment, the first step should be the removal of settleable solids, such as shards, labels, spent hops, and bottle caps, by suitable screens and strainers. To neutralize the mainly alkaline wastewater, it is common to use carbonic acid from the fermentation or flue gas. It is also possible to biologically neutralize the alkalis with the carbon dioxide that is produced during the BOD degradation. Because of the heavy variations, it is always recommended to use an equalization tank, which can be run as an aerated mixing and equalizing tank with a biological partial purification or which may serve as an unaerated pretreatment tank or acidification reactor for the anaerobic plant.

For the full-scale purification of brewery wastewater to direct discharge quality, aerobic activated sludge systems have proved to be best, because of the need to eliminate nitrogen and phosphorus. The activated sludge system can either be the sole treatment stage or be post-positioned to an anaerobic plant or an aerobic trickling filter unit. Another reasonable solution is the use of SBR methods (sequencing batch reactor) in which all treatment steps are run one after the other, but in the same tank.

DISTILLERIES

Distilleries produce alcohol for human consumption by fermentation and distillation of agricultural products that contain sugar or starch. Some of this alcohol is also used for vinegar production and in the pharmaceutical and cosmetics industries. In companies that produce spirits, the alcohol is diluted to make it potable and is enhanced with flavor additives.

For the production process, it is important that raw materials containing starch be turned into sugar by enzymes, fermented into ethanol, and then distilled, whereas raw materials containing sugar are only fermented and then distilled. Wine is only distilled. Normally, the first steps are mechanical disintegration of the fruit and mashing with water. In some instances, the saccharification must be artificially boosted. After fermentation is finished, the raw spirit (approx. 80 vol.% alcohol) is cleaned of its distillation residues (slops) by a first distillation. In a further refining step, either a so-called fine spirit (approx. 86 vol.% alcohol) is produced by a second discontinuous distillation, or a fine spirit or neutral alcohol (approx. 96 vol.% alcohol) is produced by continuous rectification. The residues of this second refinement step are called singlings.

Depending on the raw product and production methods, distilleries produce washing water, steaming water or fruit water, slops, and cleaning water. Slops contain a high amount of organic acids, proteins, minerals, trace elements, unfermentable carbohydrates, or—especially with fruit slops—high solids ratios (cores, stalks, stones, skins), which result in very high COD and BOD_5 values as well as in low pH.

In wastewater treatment, the slops are of particular importance, since the other fractions of the wastewater can normally be discharged into the wastewater sewage without further treatment. Thus, slops should, if possible, be collected and utilized separately from the wastewater flow. The most common utilization method for slops from grain, potatoes, or fruit is direct feeding in cattle farming (if necessary, after thickening with decanters). If direct feeding is not possible, one should consider spreading on agricultural fields (if necessary, after anaerobic treatment of the slops in a factory-owned biogas plant or after injection as co-substrate into a municipal digestion tank). Only if these two utilization methods are not possible, may the slops wastewater be mixed with the other production wastewater flow. Direct discharge, however, is then possible only with sufficiently powerful municipal wastewater treatment plants. In any case, the wastewater should be neutralized as a major pre-treatment step. Another point one has to consider is the danger of bulking sludge development and hydrogen sulfide emission (corrosion, odors).

Slops from molasses distilleries often retain very high residual COD ratios after biological treatment. Here, evaporation of slops to a dry solid content of approx. 75% with ensuing separation of potassium sulfate (fertilizer) and utilization of the evaporated slops as an additive for cattle feed has proved to be a suitable utilization method. The condensed exhaust vapors from the evaporation unit are often subjected to anaerobic secondary purification.

FURTHER READING

Andreadakis A, ed. *Pretreatment of Industrial Wastewaters II. Water Science and Technology*; 1997:36.

Craveiro AM, Soares HM, Schmidell W. Technical aspects and cost estimation for anaerobic systems treating vinasse and brewery/soft drink wastewaters. *Water Sci. Technol.* 1996;18:123–134.

Jordening HJ, Winter J, eds. *Environmental Biotechnology. Concepts and Applications.* WILEY-VCH Verlag GmbH & Co; 2005.

Ruffer HM, Rosenwinkel KH. The treatment of wastewater from the beverage industry. In: Barnes D, Forster CF, Hrudey SE (eds.) *Food and Allied Industries*, Vol. 1. Boston, MA: Pitman Advanced Publishing; 1984.

Speece RE. *Anaerobic Biotechnology for Industrial Wastewater.* Nashville, TN: Archae Press; 1996.

17 Solid Waste Management

SOLID WASTES

Solid wastes are one of the more interesting environmental contaminants to deal with. The primary characteristic of importance for solid wastes lies in the fact that they are solid. Unlike gaseous wastes that flow into the vast atmosphere around us or liquid wastes that flow downhill until they ultimately reach the ocean, solid wastes tend to stay put until a major effort is made to move them to another spot. Solid wastes tend to remain as such until someone finds a use for the solid wastes and converts them into something of value. Suddenly, the solid wastes disappear until the new product loses its value and becomes solid wastes once again. The lack of value of solid wastes has resulted in limited information on their interaction with various microorganisms.

SOLID WASTE CHARACTERISTICS

One of the frustrating problems confronting environmental scientists and environmental engineers lies in trying to determine the characteristics of solid wastes. Solid wastes are simply solid materials that have lost value for their owner and are discarded. It does not mean that the solid wastes being discarded have no value. It simply means that the solid wastes have no value for the current owner. It may well have value for another owner. If the solid wastes have value for a new owner, these materials are no longer solid wastes, but rather raw materials for further use with renewed value until the new owner decides to discard them as solid wastes. All material goods that society makes and uses will become solid wastes in time.

Solid wastes cannot be characterized by their chemical composition or their size or their weight alone. Solid wastes are characterized by many parameters. Chemical composition is an important parameter, along with size and weight. Biostability is also an important parameter for solid wastes that has largely been ignored. Biostability is the parameter that defines how the solid waste materials react to microorganisms and the rate of that reaction. Most solid wastes currently being produced are largely biostable, showing little to no reaction with microorganisms. Food wastes are the least biostable solid wastes, undergoing rapid reaction with microorganisms. Grass clippings and leaves are seasonal solid wastes that are not very biostable. From a practical point of view, environmental microbiologists are only interested in the solid wastes having the least biostability.

There are many ways to characterize solid wastes. Classifications tend to start with the major sources of solid wastes:

1. Residential SW—all solid wastes produced by people living in residences within the classification area. Residences include single-family houses, duplexes, and apartments with multiple family units.

DOI: 10.1201/9781003272618-20

2. Commercial SW—all solid wastes produced by commercial establishments within the classification area.
3. Industrial SW—all non-hazardous solid wastes produced by industrial manufacturing plants within the classification area.
4. Construction and Demolition SW—all solid wastes generated during the normal construction of houses, apartments, commercial establishments, and industrial factories or from the destruction of houses, apartments, commercial establishments, or industrial factories.
5. Street Sweepings SW— all solid wastes collected by street sweepers operating in urban communities.
6. Water & Wastewater Treatment Plant Sludge SW—all sludge solid wastes produced by water and wastewater treatment plants within the classification area.
7. Automotive SW—all solid wastes generated when automobiles and trucks are junked.
8. Bulky SW—all large solid waste items from residences and commercial establishments that require special collection and handling. Bulky solid wastes include washing machines, refrigerators, stoves, and sofas.
9. Trees—large trees that die or trees that are cleared for construction projects require special handling.
10. Agricultural SW— all solid wastes produced from farming operations.
11. Mining SW—all residues remaining from mining and mineral processing.

These broad classifications of solid wastes have value for looking at the total solid waste problem at the national level. Regional classifications normally look at municipal SW, rural SW, agricultural SW, and mining SW. Few people were concerned enough about solid wastes to determine their characteristics. The major problem was simply that it was too difficult to take representative samples of solid wastes and make complete analyses. Although every person was concerned with the generation and disposal of solid wastes, no one cared enough to characterize those solid wastes.

MUNICIPAL SOLID WASTES

Municipal solid wastes include the residential SW, the commercial SW, and limited industrial SW generated within the jurisdiction of the municipality. Examination of different groups of municipal solid wastes indicates that glass and plastics are the most biostable ones with metals being relatively biostable. Paper, wood, rubber, leather, and textiles are slowly biodegradable under specific conditions. Food wastes and yard wastes are biodegradable. One of the key parameters for biodegradability is moisture content. Food wastes contain up to 80% moisture, averaging about 70% moisture. Yard wastes average about 60% moisture with grass clippings having the most moisture and dead branches having the least moisture. Paper and cardboard normally contain less than 10% moisture. Wood can contain up to 40% moisture. Rubber, leather, and textiles usually contain less than 10% moisture. These materials are not biodegradable without the addition of water to increase their moisture level. Dry metal objects are biostable except in the presence of water. Microorganisms can slowly react with wet metal products, if the end products of the microbial–metal reactions are not toxic to the microorganisms. Glass and plastics have very low moisture content and are biostable even when immersed in water. For the most part, municipal solid wastes reflect the lifestyle of the people within the municipal jurisdiction and the changes in the local economy.

INDUSTRIAL SOLID WASTES

Industrial solid wastes consist of hazardous solid wastes and non-hazardous solid wastes. The hazardous solid wastes are dangerous for the environment and are controlled by separate federal legislation than the non-hazardous solid wastes. The characteristics of industrial solid wastes are highly variable, depending upon the specific industrial processes and the workers within specific plants. Industrial solid wastes range from inorganic materials to organic materials.

CONSTRUCTION AND DEMOLITION SOLID WASTES

The construction of new houses and new commercial buildings results in the production of construction solid wastes in direct proportion to the materials employed in the construction project. The characteristics of the construction solid wastes will vary with each project. Normally, construction wastes will include the residual materials that could not be incorporated into the finished building. The construction of houses will include pieces of wood, wallboard fragments, partial shingles, and excess roofing paper. If the house has a cinder block foundation and/or a brick front, broken cinder blocks and broken bricks will also be included. The amounts of construction wastes produced are directly proportional to the number of houses and buildings constructed. As a net result, the construction waste quantities and characteristics are highly variable and not easily estimated.

Eventually, every house and every building will become solid wastes when their value is too low and the maintenance costs are too high. Under good economic conditions, old buildings are bulldozed to allow new replacement structures on the same site. Demolition of old structures results in complete destruction of the old structures. Demolition debris will consist of all the materials remaining in the structure. These materials are simply bulldozed and placed in a truck for removal from the site. The entire weight of all the materials in the old structure comprises the demolition wastes. Major urban renewal projects create large amounts of demolition wastes over a specific time period. The amount and characteristics of demolition wastes require separate evaluation of every project and cannot be generalized.

STREET SWEEPINGS

Many cities use street sweepers to collect dirt and trash that accumulates on streets in the major commercial and industrial areas. Occasionally, the residential streets will also be swept. The shape of road surfaces allows the dirt and trash to accumulate next to the curb and the road. For this reason, street sweepers normally clean the area of the street next to the curb. The quantity of street sweepings collected is a function of the size of the city and the rainfall characteristics. Large cities produce more street sweepings than small cities. Large cities tend to attract dirty industries that can operate with a minimum of complaints. Cities tend to keep the dirty industries together, making it easier to keep the streets reasonably clean. Industrial areas tend to have large impervious surface areas, allowing maximum runoff during and after every rain event. The rapid runoff of rainwater carries the contamination on the streets into storm sewers for collection and removal. Areas with heavy rains will find that the streets are cleaned of all but the heaviest solid materials. Areas with light rains, occurring frequently, will find only the lightweight materials removed by the runoff water.

Examination of street sweepings indicates they are composed of dirt, sand, grit, paper products of various types, plastic materials, broken glass, rubber particles, and miscellaneous materials too difficult to identify and measure. Most of the street sweepings are relatively inert material that can be used as fill or placed into a sanitary landfill. One of the problems with street sweepers is their tendency to suspend fine particles around the street sweeper rather than capturing the fine particles.

WATER AND WASTEWATER SLUDGES

Every community that has a water treatment plant and/or a wastewater treatment plant will have water treatment sludge and/or wastewater treatment sludge. Emphasis on water treatment plants and wastewater treatment plants is on the removal of pollutants from the water being treated. The treatment processes produce a clean water and waste sludge. The water treatment plants generate alum sludge and calcium carbonate sludge, if the treatment plant softens the water. The clean water, produced in the water treatment plant, is distributed to every user in the community through a complex pipe network. Once the clean water has been used, it is returned as wastewater. The wastewater collection system is also a network of complex pipes that services every house and building in the community. The wastewaters are collected and discharged into a wastewater treatment plant. The pollutants in wastewater are classified as suspended pollutants and soluble pollutants. The majority of the suspended pollutants are removed by gravity sedimentation. The remaining suspended pollutants and the soluble pollutants are treated biologically with the pollutants converted into suspended solids that form settleable floe particles that are removed by gravity sedimentation. The two types of wastewater sludge are normally mixed together and treated in an anaerobic digester. The anaerobic digester destroys the readily biodegradable suspended solids, leaving a relatively inert residue for disposal on land.

The characteristics of water treatment plant sludge are quite similar from plant to plant as far as chemical characteristics are concerned. The quantity of sludge produced varies with the magnitude of treatment required to remove the contaminants. Alum treatment of surface water produces an aluminum hydroxide floe that removes the suspended solids in the surface water. River water will contain both organic and inorganic particles. During periods of heavy rainfall and runoff in rural areas, the river water will contain soil particles and any materials washed from the soil surface. The net result is for wide variations in alum use and in sludge production. Lake water contains colloidal suspended solids and various microorganisms. Understanding the source of the water being treated is essential to knowing the type and magnitudes of contaminants in the water plant sludge. Water plants that soften the water will produce calcium carbonate sludge. A few plants may produce magnesium hydroxide in addition to the calcium carbonate. Softening sludge is an inorganic sludge. Each water treatment plant must determine its own sludge production rates as they are variable from plant to plant.

Wastewater treatment plants produce screenings and grit in addition to the sludge generated every day. Screenings are large solid wastes that are collected on bar screens at the head of the treatment plant. Most plants collect the screenings and dispose them with the other municipal solid wastes in sanitary landfills. A few plants grind the screenings and return them to the liquid stream. Logic indicates that it is simpler to handle the screenings as solid wastes once they have been collected. Grinding the screenings and returning them to the liquid wastes means that the screenings have to be removed twice from the wastewater. Grit is largely sand and dense organic particles that could damage mechanical equipment in the units that follow the grit chamber. Grit is usually washed as it is collected and used as fill material. The primary sludge and the biological sludge are combined and anaerobically digested before being dewatered and placed on the land. Wastewater treatment plants handling large quantities of industrial wastes will produce even more excess sludge.

AUTOMOTIVE SOLID WASTES

Every automobile solid will eventually become solid waste. Old automobiles are processed by automobile scrap dealers. With an average life of 7–8 years, the number of automobiles scraped each year can be estimated by looking at the production data 7–8 years ago. With increased emphasis on waste recycling, the automobile manufacturers have tried to use materials that could be processed and reused with a minimum of effort. The automobile scrap dealers remove all the parts of value before shipping the residue to a metal grinder and separator to recover the metal for reuse. Automobile tires pose one of the major processing problems for recycling.

BULKY SOLID WASTES

Every community produces bulky solid wastes from time to time. Bulky solid wastes are all large household items that are discarded. Stoves, refrigerators, washing machines, sofas, large chairs, beds, mattresses, and other large items require special collection. Usually, the person discarding the item is responsible for seeing that the bulky waste is collected and delivered to the proper disposal site. Often, when a large item is replaced, the dealer selling the new item collects the old item and delivers it to the disposal site. Reclamation dealers remove any useful parts before sending the residue for recycling or for burial in a sanitary landfill.

TREES

As communities age, there will be a steady production of dead trees and trees removed for residential or commercial expansion. The dead trees are usually ground up and used for mulch in the community. Living trees removed from private or public property will be used commercially, where possible. Large trees can be cut for lumber, while small trees are mulched for use in paper production or in parks and gardens. Every effort is made to reuse the wood rather than burying it in sanitary landfills.

AGRICULTURAL SOLID WASTES

Farms produce large quantities of solid wastes. The quantities of solid wastes produced are a function of the specific crops grown on the farm. Farmers have long recognized that manure from farm animals and crop residues after harvest must be returned to the land to help maintain the soil quality for future crop production. Farmers are among the oldest of the recyclers in our society. The major problems in recent years have come from the large confined animal farms. The manure from the confined animal buildings is handled as a liquid waste rather than as a solid waste. Operators of the large confined animal farms have had to learn new methods of handling the liquid manure to minimize environmental pollution. The basic problem is processing the liquid manure for return to the land environment. It is not surprising that there are no simple solutions for processing the liquid manure as previously indicated.

MINING SOLID WASTES

The mining industry produces large quantities of solid wastes that must also be returned to the land environment. Mining results in the removal of non-valuable materials along with the valuable materials. The failure of the mining industry to properly handle the non-valuable solid materials at the same time they handled the valuable materials has produced localized environmental damages and created a negative image for the mining industry. Too often, mining companies simply closed down and walked away from the mined area when the mining operation stopped being profitable. Some of the mining residues contain hazardous chemicals that slowly wash into nearby streams, creating even more environmental damage. The net effect has been for the mining industry to have created a very negative impression on the public at large.

PROCESSING SOLID WASTES

Normal processing of solid wastes has always followed the path of least resistance. Solid wastes have always been handled with the least effort and the lowest cost. Biological treatment of solid wastes falls into three general categories: sanitary landfills, composting, and soil stabilization. Although sanitary landfills have been extensively used, the lack of understanding of how microorganisms react in sanitary landfills has created problems in both design and operation of sanitary landfills. Composting is another biological process that has not been fully evaluated from a microbiological point of view. Soil stabilization of solid wastes has long been used for treating agricultural solid wastes and has potential for treating any readily biodegradable solid waste.

SANITARY LANDFILLS

By definition, sanitary landfills are engineered burial of solid wastes. Unfortunately, sanitary landfills grew up at a time when municipal government paid little attention to solid waste processing. Although municipal government is the primary beneficiary of well-designed and well-operated sanitary landfills, municipal government must accept full responsibility for all the problems related to sanitary landfills. Sanitary landfills should be carefully designed for maximum operating efficiency with a minimum of discomfort and objections from the citizens living around the landfill. Fundamentally, all sanitary landfills should be located within the boundaries of the community producing the solid wastes. This location allows the citizens of the community to continuously observe how their elected officials manage one of the most important municipal functions. It is completely possible to design and operate a sanitary landfill within the municipality without creating problems for the citizens or the municipal officials.

BASIC CONCEPTS

A sanitary landfill should be constructed in dry soil on flat terrain capable of holding at least 20–30 years production of solid waste from the contributing population. The natural water drainage should be away from the landfill site or should be diverted around the landfill site to prevent excess runoff from crossing the landfill and eroding the cover soil. Because of concerns about leachate formation and its effect on groundwater below the landfill, the bottom of the landfill should be sealed with clay or an impermeable membrane. A series of perforated pipes should be placed at intervals across the bottom of the landfill to collect the leachate and to convey it to a concrete sump where the leachate can collect and be pumped to a leachate storage facility prior to treatment. A layer of coarse sand is placed over and between the leachate pipes to prevent possible damage to the collection pipes by the tractor or tractors used for compacting the solid wastes in the landfill. The solid wastes should be placed first at the far end of the landfill. Bulldozers push the solid wastes into a corner and compact the solid wastes to their maximum concentration. By moving back and forth over the solid wastes, the bulldozer is able to compact the solid wastes to a concentration close to 593 kg/m³ (1,000 Ibs/cy). At the end of each day, the compacted solid waste is covered with about 0.15 m (6 in.) of topsoil to act as a temporary cover. As more solid wastes are added to the landfill, the landfill slowly fills to the design level. As the landfill fills with solid waste, perforated gas collection pipes are placed in the upper part of the landfill to collect the gas produced in the landfill. The landfill is covered with a layer of clay or an impermeable membrane to minimize infiltration of water. The top of the impermeable layer is then covered with at least 0.61 m (2 ft) of topsoil. The topsoil is sloped to allow the natural storm water runoff to flow away from the landfill into catch basins and a storm water collection sewerage system to minimize the infiltration of rainwater into the landfill. Sanitary landfills can be constructed as a series of trenches or as areas, depending on the specific terrain. Trench landfills are built into the soil surface, while area landfills are built against a hill. The compaction and burial process for the solid wastes continues until the landfill has been filled and the landfill site has reached its design capacity for solid wastes. A new site must be designed and developed before the old landfill site is filled. The production and processing of solid wastes is a never-ending process that must be continuously handled.

BIOLOGICAL ACTIVITY

The municipal solid wastes placed into the sanitary landfill contain materials that are readily biodegradable. In fact, biological activity is well underway by the time the solid wastes are collected and transported to the sanitary landfill. The food wastes that contain over 60% moisture will allow bacteria to metabolize the organic matter in the food waste materials. Fungi will be able to metabolize organic matter with only 40% moisture content, provided oxygen and other required nutrient elements are available for aerobic metabolism. Even though solid wastes have been compacted to a density of $593\,kg/m^3$ (1,000 Ibs/cy), there are still plenty of void spaces in the compacted landfill. Both bacteria and fungi are able to grow on the moist food wastes initially. The microbes grow directly on the surface of the food materials being metabolized.

The metabolic reactions are aerobic with carbon dioxide, water, new cell mass, and heat as the end products. As the initial microbial layer spreads across the organic surface, some microbes are forced to grow on top of the existing microbial layer. Soon a second layer of microbes becomes a third layer and then a fourth layer. Metabolism of the food material shifts from being aerobic to being oxygen limiting. The microbes at the surface of the food material are unable to obtain sufficient oxygen for aerobic metabolism. The bacteria and the fungi continue to metabolize to the best of their ability. As oxygen becomes limiting, the microbes slow their rate of metabolism and organic end products increase in solution. Soon oxygen becomes exhausted and only the facultative bacteria are able to continue metabolism at the food surface. Fungi and strict aerobic bacteria on the food surfaces cease growing. The bacterial layer next to the food surfaces slow their growth as metabolism shifts to the production of organic acids and other partially oxidized end products rather than carbon dioxide and water. The pH in the liquid around the bacteria decreases, as the organic acids accumulate. Further metabolism causes the pH to drop sufficiently low that metabolism stops and the microbes die.

If the food waste is in contact with a metal can, the organic acids will cause the metal to dissolve and neutralize some of the organic acids. The presence of alkaline materials in the vicinity of the food wastes will also neutralize the organic acids, raise the pH, and allow metabolism to continue. As the neutralized organic acids accumulate around the organic matter being metabolized, methane bacteria will begin to metabolize the neutralized organic acids, producing methane gas and carbon dioxide as primary end products. Metabolism of the neutralized organic acids releases the cations, allowing the cations to form bicarbonate alkalinity in the water that can neutralize more organic acids. Slowly, the methane bacteria metabolize the neutralized organic acids, producing methane, carbon dioxide, water, new cell mass, and heat. Sulfate-reducing bacteria will also grow under anaerobic conditions where there are sulfates in the solid wastes. The sulfate-reducing bacteria can compete with the methane bacteria for organic nutrients in the right environment and may reduce the methane gas production. The metabolic process is a slow process since there is no mixing of the organic matter and the microorganisms in the sanitary landfill. The localized environment determines what microbes can grow and to what extent they grow. Temperature, moisture, and alkaline materials are the three major factors that determine the rate of biodegradability of the organic matter in the landfill. A rising temperature will increase the rate of metabolism.

Moisture within the landfill will determine how far the microorganism will be able to move to find new food for metabolism. The presence of alkaline materials will determine the pH of the water around the bacteria and the food wastes being metabolized. The production of water forms a thin liquid film for the microorganisms to move to a new location for metabolism. For the most part, microbial movement inside the sanitary landfill is restricted to the immediate environment around the rapidly biodegradable organic matter. The gas movement in the sanitary landfill depends upon pressure differentials and available void spaces. The heat released by metabolism is quickly absorbed by the materials in the sanitary landfill. Overall, microbial metabolism within a sanitary landfill is limited, and the environment within the sanitary landfill is simply not conducive to rapid microbial metabolism.

There are few organic compounds in municipal solid wastes that bacteria can metabolize under the best of circumstances. Most paper in solid wastes contains lignin as well as cellulose. Yard wastes also have a considerable amount of lignin with the cellulose. The lignin prevents bacteria from metabolizing the cellulose, allowing the waste paper and much of the yard wastes to remain untouched once the oxygen in the void spaces has been removed. Fungi have the ability to metabolize lignin as well as cellulose, but the fungi must have dissolved oxygen available for metabolism. Bacteria are able to metabolize the cellulose materials that do not have the cellulose combined with the lignin. Completely sealing the solid wastes in the sanitary landfill, as currently recommended, creates an environment that limits the overall metabolism to 10%–15% of the municipal solid wastes no matter how long the solid wastes are contained in the sealed sanitary landfill. It also means that 85%–90% of the municipal solid wastes placed into a completely sealed sanitary landfill will be available for reuse whenever conditions favor recycling municipal solid wastes.

STIMULATING BACTERIAL ACTIVITY

It is important for everyone to recognize that completely sealing a sanitary landfill, as currently recommended to minimize groundwater contamination, limits biodegradation within the landfill. The simplest method to stimulate biodegradation inside the sanitary landfill is to pass water through the landfill at periodic intervals to allow the bacteria to migrate throughout the landfill and find all the readily biodegradable materials. Metabolism in the landfill will continue to be anaerobic with the production of methane gas. Recirculation of liquid around the sanitary landfill on a continuous basis has been recommended to speed the stabilization of the biodegradable organic matter. The problem with recycling the leachate from the bottom of the landfill around the landfill is the accumulation of soluble, non-biodegradable contaminants in the leachate. The soluble non-biodegradable materials accumulate in the recycled leachate on each pass through the solid wastes in the landfill. Eventually, the accumulated non-biodegradable materials may adversely affect metabolism of the biodegradable organic compounds. Under the anaerobic environment inside the sanitary landfill, iron and other metallic ions, released by reaction with organic acids, tend to remain partly soluble. Some of the iron and metallic ions may form insoluble sulfide salts that move with the leachate as colloidal precipitates. When the leachate is exposed to air, oxygen in the air reacts with the ferrous iron to form insoluble ferric oxide. Some of the other metallic elements also form insoluble metallic oxides. The organic compounds are unaffected by the changes in the metallic elements and remain in solution. Leachate tends to be on the acidic side and may need additional alkalinity to provide a suitable pH, pH 7–8, when the leachate is recycled back to the landfill surface. It is important to maintain a suitable environment within the landfill for maximum stabilization of the organic matter.

Recycling of leachate will also remove any salts that are soluble in water, increasing the mineral content of the leachate with time. Maintaining a good environment within the sanitary landfill is not an easy task. The solid waste materials are fixed within the landfill and cannot move to produce a better opportunity for biodegradation. The recycled leachate always takes the path of least resistance and flows through the void spaces between the solid particles by gravity to the bottom of the landfill for collection.

Continuous recirculation of leachate should result in moving the bacteria throughout the entire volume of the sanitary landfill and the maintenance of anaerobic conditions within the sanitary landfill. Care must be taken that the recycled leachate is not applied by spraying in the air. Spraying the leachate will create small droplets and allow oxygen in the air to enter the leachate before it re-enters the sanitary landfill. A high rate of continuous recirculation of leachate will be a waste of energy and will produce a limited increase in treatment efficiency. The basic problems in the biological degradation of solid wastes lie in the large mass/surface area ratio and the complexity of the chemical composition of different solid waste components.

Periodic application of leachate to the landfill surface, followed by sufficient time to allow the free water to drain through the sanitary landfill should produce adequate dispersion of the microbes throughout the sanitary landfill without slowing the metabolic reactions with the solid waste. As the upper layers of organic material are stabilized, the upper environment will shift from anaerobic to aerobic, allowing the fungi and the actinomycetes to metabolize the complex paper products in proportion to the availability of oxygen to the microorganisms on the surface of the solid wastes. Slowly, aerobic conditions will be established throughout the entire volume of the sanitary landfill.

The lack of nitrogen, phosphorus, and trace elements can also limit both the rate of biodegradation and the total amount of biodegradation. Remember that the metabolism of organic matter by microorganisms results in the production of new cell mass. Without the synthesis of new cell mass, the microorganisms will produce very limited metabolism, only endogenous respiration, and the synthesis of new cells from the nutrients released. Adding treated domestic wastewater to the sanitary landfill on a periodic basis is one method that can be used to add water and to increase the nitrogen, phosphorus, and trace metals that the microorganisms can use in their metabolism of the biodegradable materials in the sanitary landfill. Storm water runoff can be collected and stored for use as needed during the dry season in water-limited regions. There is no single method for stimulating microbial growth in a sanitary landfill. It is possible to design the top of the landfill to allow natural rainfall to enter the landfill as the primary water source while still capturing the gas and the leachate. In every sanitary landfill, there is a fixed amount of biodegradable materials that can be metabolized. Normally, there will be a slow increase in biological activity to a maximum, followed by a slowing rate of increase to a plateau and a slowing rate of metabolism that eventually ceases to be significant. It normally takes several years for the biodegradable materials in a sanitary landfill to be metabolized.

GAS PRODUCTION

In recent years, efforts have been made to collect the gas from sanitary landfills for power generation. The concept of gas production and energy generation has attracted considerable attention from landfill owners and landfill operators as it is a method for recovering some of the operating costs of the sanitary landfill. While sanitary landfills handling biodegradable solid wastes will produce methane gas that can be collected and burned for energy, there is a fixed amount of potential energy available that can be produced. The actual energy recovery will be something less than the potential energy available from the solid waste analyses. The value of the landfill gas must be sufficient to defray the cost of the extraction equipment and the energy conversion equipment along with the operating costs for the energy system. The landfill gas will contain methane, carbon dioxide, and low concentrations of hydrogen sulfide, the same as contained in gas from an anaerobic wastewater sludge digester. The methane fraction is important in determining the extent of biodegradation and in energy recovery. The amount of methane gas that can potentially be produced is related to the biodegradable COD (BCOD) of the organic matter metabolized. If the BCOD of solid wastes in the sanitary landfill averages 1.2 times the weight of the biodegradable organic fraction of solid wastes, the maximum amount of methane that could be expected from a municipal solid waste sanitary landfill would be about $40\,m^3\ CH_4/m^3$ (1,080 cf CH_4/cy) solid waste. The theoretical energy yield from the biodegradable organic compounds in the solid wastes contained in the sanitary landfill should be about $1.5 \times 10^6\ kJ/m^3$ (1.1×10^6 Btu/cy). The actual energy capture will depend on the efficiency of the gas collection system. As gas pressure builds in the encapsulated landfill, punctures or cracks in the liner system will allow loss of the landfill gas to the outer environment. The value of the gas produced from biodegradation of municipal solid wastes is not sufficient to warrant complex gas detection equipment to detect all gas leaks from the sanitary landfill. Most of the gas production will occur within the first 3–5 years after the solid waste has been placed in the sanitary landfill. Gas production will slow and continue to be produced for a number of years.

Metabolism of the solid wastes in the sanitary landfill will result in loss of solid materials and the creation of increased void spaces. The weight of soil and solid wastes above the metabolized organic matter will often cause the surface of the sanitary landfill to settle over time. Since the surface settling is not uniform, it will be necessary to add soil to the landfill surface to maintain the desired surface for proper rainfall runoff. Since sanitary landfills may contain useful materials for future generations, it is important to keep the surface of finished sanitary landfills free of complex structures that would have to be removed before the solid wastes could be mined for reuse. Completed sanitary landfills should be designed as neighborhood parks when the communities expand. Park areas provide suitable open spaces for the people to enjoy in their residential areas. Too often, the finished sanitary landfill areas are used for industrial buildings, commercial buildings, and dense residential structures. These structures will find the slowly settling sanitary landfill unsuitable as a solid foundation and will retard future recovery of the solid wastes. Proper urban planning can insure these potential problems do not occur.

NUTRIENT DEFICIENT

One of the problems with biological treatment of municipal solid wastes lies with the deficiency of nutrient elements for proper metabolism. Not only are municipal solid wastes deficient in nitrogen and phosphorus, they are also deficient in trace metals needed for proper enzyme development. The daily soil cover added to the compacted solid wastes in the sanitary landfill is the primary source of nutrient elements for many of the bacteria. If leachate is not recycled through the surface soil, the nutrient elements in the soil will not be available for the bacteria. Even the nutrient elements in food wastes will not be readily available unless leachate is recycled or infiltration of surface water is allowed to enter the landfill. The food wastes with excess nutrient elements will undergo metabolism first and will be degraded to the greatest extent. The lack of sufficient nitrogen and phosphorus as well as trace metals will seriously limit the degradation of cellulose materials in the sanitary landfill. Metabolism of organic matter in the solid wastes requires sufficient bacteria, acclimated to the organic matter being metabolized, together with water, nutrient elements, and trace metals.

FURTHER READING

Chandrappa R, Brown J. *Solid Waste Management: Principles and Practice.* Springer-Verlag Berlin Heidelberg; 2012.

Christensen TH (eds.) *Solid Waste Technology & Management.* Blackwell Publishing Ltd; 2010.

Kumar S. *Solid Waste Management.* Northern Book Centre; 2009.

Ludwig C, Hellweg S, Stucki S (eds.) *Municipal Solid Waste Management: Strategies and Technologies for Sustainable Solutions.* Springer-Verlag Berlin Heidelberg; 2003.

McKinney RE. *Environmental Pollution Control Microbiology.* M. Dekker; 2004.

Wong JWC, Surampalli RY, Zhang TC, Tyagi RD, Selvam A (eds.) *Sustainable Solid Waste Management.* American Society of Civil Engineers; 2016.

18 E-Waste Management

E-WASTE

In recent years, there has been growing concern about the negative impacts that industry and its products are having on both society and the environment in which we live. The concept of sustainability and the need to behave in a more sustainable manner have therefore received increasing attention. With the world's population growing rapidly and generally improving wealth, the consumption of materials, energy, and other resources has been accelerating in a way that cannot be sustained.

The electronics industry is one of the most important industries in the world. It has grown steadily in recent decades, generates a great number of jobs, promotes technological development, and at the same time, fuels a high demand for raw materials that are considered scarce or rare (e.g., precious metals and rare earth elements). This development affects the environment in two ways: first through the large and growing amount of equipment that is discarded annually and second through the extraction of natural raw materials to supply the demand of the new equipment industry. Both can be measured by the amount of equipment that is produced and discarded annually by many countries.

One area in which there has been much concern about the lack of sustainable behavior is in the manufacture, use and disposal of electrical and electronic products. The electronics industry provides us with the devices that have become so essential to our modern way of life and yet it also represents an area where the opportunities to operate in a sustainable way have not yet been properly realized. In fact, much electrical and electronic equipment (EEE) is typically characterized by a number of factors, including improved performance and reduced cost in each new generation of product, that actually encourage unsustainable behavior.

MATERIALS USED IN ELECTRONICS

Lead

The proscription of lead has caused significant concerns for electronics manufacturers and the need to replace it before July 2006 gave the industry a huge task in terms of evaluating, testing, and qualifying potential substitutes, particularly in soldering applications. Although there are numerous "lead-free" solders available, there are many issues that need to be addressed before they can be successfully implemented as viable alternatives. These issues include, among others, the compatibility of new solders with printed circuit board and component finishes, the ability of existing materials and equipment to handle higher soldering temperatures and the selection and supply of components that can survive higher soldering temperatures and give long-term reliability. Lead can also be found in other components, products, and parts of the electronics assembly process. A good example is in PCB manufacturing

DOI: 10.1201/9781003272618-21

where lead was once widely used as a solderable final finish. The key will be to know what finish is used, since it is important that the combination of solderable finish, solder and component finish are compatible in order to avoid solderability and subsequent reliability issues. Lead is also used in the solderable finishes of component connectors. A challenge for electronics assemblers is the identification of the finish on component leads and terminations since there is, as yet, no agreed standard convention for identifying whether or not a component finish is lead-free. Given that components with and without lead-free finishes may appear visually identical, it is clear that careful attention needs to be paid to component sourcing, selection and storage, etc.

BROMINATED FLAME RETARDANTS

Brominated flame retardants are widely used in electronics products; for example, they are found in PCBs, semiconductor encapsulants, cables and connectors, as well as in the polymers used in equipment housings and enclosures as discussed below. Reactive brominated flame retardants, such as the tetrabromobisphenol A-based compounds used in most PCB laminates and semiconductor encapsulants are not currently addressed by this legislation. However, suppliers of some polymers have already commercialized new bromine-free materials and many companies have already begun to avoid the use of brominated flame retardants in their products. It is also worth noting that pentabromodiphenyl ether is reportedly still used in a small percentage of Far East produced FR2 printed circuit board laminates.

CADMIUM, MERCURY, AND HEXAVALENT CHROMIUM

These three metals can occur in a wide range of applications in electronics, and in this respect, they are perhaps more troublesome from a compliance perspective since there are so many potential, and sometimes unexpected, applications. Cadmium has been widely used in specialist electroplated parts but it is most likely to be found in electronics applications in nickel–cadmium batteries. Cadmium is also found in cadmium sulfide-based photodetectors and as a component of the green and blue phosphors used in older color television cathode ray tubes. Importantly, it has sometimes been found in pigments used to color plastics and can thus be found in various electronic products.

It has been estimated that 22% of all mercury is employed in electrical and electronic equipment. It is used in thermostats, sensors, relays, switches, medical equipment, fluorescent lamps, mobile phones, and batteries. Mercury is also used in the backlights of flat-panel displays and this usage has increased as liquid crystal displays increasingly replace conventional cathode ray tubes. Laptop computers, flat-panel displays, and digital cameras may all, therefore, contain small amounts of mercury. Use in batteries, however, has been decreasing for some years and it is now found only in button cells that power relatively small electronic goods such as watches, toys, and cameras.

Hexavalent chromium compounds are widely used in electroplating and metal treatment processing. They will passivate the surface of zinc and zinc alloy electrodeposits with a thin film that provides end-user benefits such as color, abrasion resistance, and increased corrosion protection. Hexavalent chromium, although not widely used directly in the electronic components, is thus often found as a coating on various brackets, fittings, and other metal parts used in many actual products. It may be used to provide corrosion protection to steel parts, as well as screws and nuts, and thus, may be found in both consumer goods and those destined for operation in harsher environments. It also finds use as an anti-corrosion material for the protection of carbon steel cooling systems in absorption refrigerators. The difficulty for producers responsible for end-of-life electronics is that, for some large products with a significant non-electronic content, there may well be individual component parts that have been plated or treated in some way that means they contain hexavalent chromium.

Waste Electrical and Electronic Equipment (WEEE) Components

One of the key challenges in the recycling of end-of-life electronics is the selection of the appropriate recycling technology for the type of product and waste stream being treated. For example, there are clearly very different material compositions in a mobile phone and a washing machine, and thus, the optimization of the recycling processes will need to take these differences into account. This section considers three very different specific products (mobile phones, televisions, and washing machines) from a materials perspective and highlights their differences and the varying approaches that will be necessary in order to optimize their recycling.

Mobile Phones

Mobile phones represent what are, in terms of mass and volume, the most valuable electronic products currently found in large numbers in Waste Electrical and Electronic Equipment (WEEE) streams. The rapid introduction of new and improved technology, coupled with increasing functionality such as global positioning systems, wireless, cameras, and music players, means that mobile phones have relatively short life cycles and are often obsolete within a year.

When it is not possible to refurbish phones or to reuse individual components, the next level of recycling will involve the recovery of materials. Traditionally, this has tended to mean recovery of valuable metals via a smelting process with little regard being paid to the other materials such as plastics, which could be recovered and recycled for use in "new" polymers. The polymers used in mobile phone casings are typically materials such as ABS/PC and they are a potentially useful source of polymer that could be reused. Although it is known that the physical and mechanical properties of polymers degrade both during their service lives and when recycled, the condition of the materials can be determined (see the chapter "Rapid Assessment of Electronics Enclosure Plastics" by Baird, Herman and Stevens) and the problem can be overcome by blending with virgin material and/or by adding suitable plasticizers and additives.

Mobile phone batteries also contain valuable materials. For example, the cobalt in lithium-ion batteries can be recycled for use in magnetic alloys and the nickel and iron from nickel–metal hydride and nickel–cadmium batteries can be used in stainless steel. Nickel–cadmium (containing 16%–20% nickel) and nickel–metal hydride batteries (28%–35% nickel) used to be the main power sources for phones but lighter lithium-ion batteries (1%–1.5% nickel) are now favored.

Televisions

Televisions have a unique combination of materials, as well as specific challenges in terms of materials recovery and recycling. They also have long lifetimes and unique attributes in terms of the materials used in their manufacture. Television technology has also undergone a paradigm shift as manufacturers have switched from cathode ray tube (CRT)-based units to newer liquid crystal (LC) and plasma flat-panel displays. Thus, from a materials perspective, there are many interesting factors that need to be considered in terms of materials selection at the design stage in order to realize the benefits of recycling.

In terms of materials composition, plasma and LCD TVs tend to have less glass and more metals and plastics than their CRT equivalents. Modern TVs also feature multiple input and output connectors that are often gold coated, making recovery and recycling more financially attractive. Relative to CRT-based TVs, those with LC displays also have a simplicity and low number of internal components, making them easier to disassemble. The major waste materials issue for this technology is the presence of mercury in the backlights and the liquid crystal materials themselves. A typical LCD unit will contain around 45 mg of mercury.

Printed Circuit Boards (PCBs) represent a significant proportion of the overall value obtainable from end-of-life flat-screen TVs. Large concentrations of copper, tin, and lead are found in PCBs. Nickel and zinc are also present in significant quantities, although levels of hazardous materials such as arsenic, cadmium, and mercury are generally very low or below detection limits.

Another materials-related challenge with the recycling of conventional CRT-based televisions relates to the glass that is used in the picture tube. This glass is essentially of two compositions, with the funnel glass containing lead while the front-panel glass is lead-free. There are large quantities of this glass appearing as increasing numbers of CRT televisions reach end of life and there are questions about the best routes for recycling the glass and subsequent reuse opportunities, especially as closed-loop opportunities are diminishing with the demise of CRT manufacturing. It is also interesting to note that only two material sources in a CRT-based television have a positive net value after dismantling at end of life and these are the copper-bearing yoke and the electronics.

Overall, it seems that the only truly cost-effective approach to disposing of CRT-based televisions is via landfill, but this will no longer be possible in the traditional sense, because of the introduction of legislation such as the WEEE Directive.

Washing Machines

Washing machines represent the opposite end of the WEEE spectrum in terms of size, mass, material composition, and value relative to mobile phones and TVs. They are regarded as typical white goods and traditional recycling activities have been focused on metals recovery. Washing machines, even today, contain only a relatively small amount of electronics and processing power, and unlike the high value circuit boards found in mobile phones, these are typically made from low-cost laminates containing fewer valuable materials. Most of the environmental impact of a washing machine occurs during the use phase. Solid waste is generated at a number of stages, such as when the original packaging is removed and disposed of and also at the end of life when disposal takes place. However, while the solid waste levels at these stages are significant, they were found to total less than 15% of the total solid waste produced by the washing machine. This is because of the large number of washing powder packets and other aids that are consumed and disposed of during the machine's life. It is also interesting to note how small the contribution is at end of life to the overall impact of the washing machine. Clearly, significant opportunities exist at the use phase, such as the use of lower washing temperatures and the development of new detergents that require less material to achieve a given function. During the use phase, the washing machine consumes large quantities of water and produces polluted water. Consequently, there are also opportunities for water consumption reductions during use that would have a significant impact on the overall environmental impact.

WEEE MANAGEMENT

The WEEE Directive directly controls the disposal of end-of-life equipment and the percentage going to landfill, as well as setting targets for the percentages of a product that have to be recovered and recycled. Several environmental protection agencies around the world consider WEEE to be hazardous waste because they have chemical compounds in their composition that are toxic and harmful to human health and to the environment (Jang et al., 2010).

The chemical composition of WEEE varies according to each product. For example, LED TVs have a higher amount of polymer, while stoves and microwaves have a larger number of metals. In fact, the chemical composition depends on several factors, such as the type of WEEE, year of manufacture, manufacturer's brand, and country of origin. In general, a mixture of metals can be found in WEEE, such as copper, iron, aluminum, brass, and even precious metals, such as gold, silver, and palladium, in addition to a mixture of polymers, such as polyethylene, polypropylene, and polyurethane. WEEE may also include ceramic materials, such as glass, and other inorganic, organic, and even radioactive materials.

WEEE is considered toxic to human health and to the environment because it often has inorganic compounds in its composition, such as mercury, lead, cadmium, nickel, antimony, arsenic, and chromium, in addition to organic compounds such as polychlorinated biphenyls, chlorofluorocarbons, and polycyclic and polyhalogenated aromatic hydrocarbons.

The incineration of WEEE is an alternative to landfill, but can also cause environmental problems. Many components of WEEE have organic compounds in the composition that, when incinerated, can generate dioxins and furans. The presence of halogens in WEEE can be explained by the addition of flame retardants containing bromine. Approximately 12.5% of all types of WEEE contain halogenated compounds. If the burning of WEEE is performed without proper environmental precautions, the release of polybrominated polyphenyl compounds, and others like it, will occur. The primary means of preventing the release of dioxins and furans during the incineration of equipment is a gas treatment system. However, these are very expensive pieces of equipment for the companies that perform these services. As a result, many companies are burning without an adequate treatment of gases.

In order to evaluate the environmental compatibility of waste and scrap that will be used as raw material or that is to be disposed in landfills, different characteristics must be evaluated. In general, a characterization of waste and scrap should include, among other characteristics, the evaluation of the items listed below:

- Heterogeneity.
- Chemical composition, considering total elemental composition, concentration of salts, oil and grease content, pH, redox potential, etc.
- Physical characteristics: density, particle size, permeability, porosity, humidity, etc.
- Hazardousness characteristics: corrosivity, reactivity, radioactivity, toxicity, etc. Regarding toxicity, leaching and solubilization tests can assist in the waste characterization.

The recycling conditions can only be established after a complete characterization of the waste and scrap, i.e., only after thoroughly understanding the waste, should the recycling processes and routes be chosen and applied.

Extensive research is currently under way into e-waste management in order to mitigate problems at both the national and international levels. Several tools have been developed and applied to e-waste management, including LCA (Life Cycle Assessment), MFA (Material Flow Analysis), MCA (Multi Criteria Analysis), and EPR (Extended Producer Responsibility).

These management tools, combined with the existing laws in different countries, can help improve the disposal of electronic waste in the world, increasing the reuse of materials and reducing environmental impacts.

PROCESSING TECHNIQUES

Electronic waste processing is very complex due to the great heterogeneity of its composition and its poor compatibility with the environment. The first step is usually manual disassembly, where certain components (casings, external cables, CRTs, PCBs, batteries, etc.) are separated. Following disassembly, the technologies used for the treatment and recycling of electronic waste include mechanical, chemical, and thermal processes. For metals recovery, there are four main routes: Mechanical Processing, Hydrometallurgy, Electrometallurgy, and Pyrometallurgy.

MECHANICAL PROCESSING

The mechanical processing of WEEE is used to select and separate materials, and the separation is based on steps of mineral processing techniques. As far as scraps are concerned, mechanical processing is generally seen as a pretreatment for the separation of materials and associates with different separation stages of WEEE components (Hayes, 1993; Zhang and Forssberg, 1999; Tenorio et al., 1997; Zhang and Forssberg, 1997). The metal fraction obtained after the mechanical processing step is sent to hydro-, electro-, and/or pyrometallurgical processes.

HYDROMETALLURGY

The initial steps in hydrometallurgical processing consist of a number of acidic or caustic attacks to dissolve the solid material. In the following steps, the solutions are subjected to separation processes such as solvent extraction, precipitation, cementation, ion exchange, filtration, and distillation to isolate and concentrate the metals of interest (Pozzo et al., 1991; Gloe et al., 1990).

The main advantages of hydrometallurgical processing of electronic waste, when compared to pyrometallurgical methods, are as follows:

- Reduced risk of air pollution.
- Higher selectivity to metals.
- Lower process costs (e.g., low power consumption and reuse of chemical reagents).

The disadvantages are as follows:

- Difficulty in processing more complex electronic scraps.
- Need for mechanical pretreatment of the scrap to reduce volume.
- The chemical dissolution is effective only if the metal is exposed.
- Large volume of solutions.
- The wastewater can be corrosive, toxic, or both.
- Generation of solid waste.

BIOTECHNOLOGY

The use of bacteria in metals recycling has been described in the literature specifically in the dissolution of metals and the recovery of gold from electronic waste (Brandl et al., 2001; Nakazawa et al., 2002). This process is still quite restricted in terms of large-scale implementation, but several studies are being conducted that should be able to produce processes with low operating costs and low investment in equipment, in addition to generating little waste, effluents, or toxic gases.

Biohydrometallurgical processes include the leaching of metals by different bacterial cultures. The main limitations of biohydrometallurgical processes are the long periods necessary for the leaching and the need of the metal to be exposed, i.e., the metals content must be mainly located on the surface layer.

ELECTROMETALLURGY

Most electrometallurgical processes associated with the recycling of electronic waste are steps of the electrowinning process that ultimately seeks to recover a pure metal (Hoffmann, 1992). Electrochemical processes are usually performed in aqueous electrolytes or molten salts and can be used to recover metals from various types of waste. Metal concentrates obtained by hydrometallurgical processes (e.g., selective dissolution, ion exchange, or solvent extraction) can be electrodeposited from aqueous solutions on the cathode. The advantages of electrometallurgical processes are:

- Few steps
- Higher selectivity for desired metals
- The electrolyte can be reused
- Pure metals can be obtained.

The main limitation is the need of a pretreatment (usually based on mechanical and hydrometallurgical processes).

PYROMETALLURGY

Pyrometallurgical processes, notably smelting, have become a traditional method to recover metals from e-waste (Cui and Zhang, 2008). The conventional pyrometallurgical processing mechanism consists essentially of concentrating metals in a metallic phase and rejecting most other materials in a slag and/or gas phase. Pyrometallurgical processing has some advantages, such as applicability to any type of electronic waste, no need for pretreatment, and few steps in the process. Some of the methods involving thermal processing of electronic waste, however, can cause the following problems:

- Polymers and other insulating materials become a source of air pollution through the formation of dioxins and furans.
- Some metals can be lost through volatilization of their chlorides.
- Ceramic and glass components present in the scrap increase the amount of slag in the furnace, increasing the losses of precious and base metals.
- Recovery of some metals is low (e.g., Sn and Pb) or almost impossible (e.g., Al and Zn).

RECYCLING OF E-WASTES

Processes that reduce and reuse generated waste in a closed circuit are in everyone's interest. The foundations for an environmental oriented process, therefore, are the following:

a. Primary Measures: Avoid or reduce the generation of waste;
b. Secondary Measures: Reuse of waste;
c. Tertiary Measures: Disposal with environmental compatibility, i.e., waste that cannot be reused must be inert, so that it can be disposed without impacting adjacent areas.

As such, two alternatives remain for the future of the consumer goods industry:

a. Set up processes to reduce and reuse the generated waste in a closed loop;
b. Produce a residue that can be viably inserted into the recycling industry.

The recycling of obsolete or defective materials is important because it represents an economic gain of the material itself, which is quite significant for metals like aluminum, lead, and copper, and especially, for noble metals like gold, silver, platinum, and indium. Another advantage of reuse lies in energy savings. In the primary production of metals, the metal is obtained by ore reduction with high energy consumption.

In the secondary process, the metal is obtained primarily from molten scrap, which is already in the metallic state, and the energy consumption is significantly lower.

It is important to characterize the products to define the recycling options. A characterization of waste and scrap consists of several stages and has as main objective the determination of the physicochemical properties of the materials and the evaluation of hazardousness. In fact, it is important to determine the chemical composition and physical properties of the materials to define the recycling possibilities, enabling their use as secondary raw material for different industries.

FURTHER READING

Brandl H, Bosshard R, Wegmann M. Computer-munching microbes: Metal leaching from electronic scrap by bacteria and fungi. *Hydrometallurgy.* 2001;59:319–326.

Cui J, Zhang L. Metallurgical recovery of metals from electronic waste: A review. *J. Hazard. Mater.* 2008;158:228–256.

Gloe K, Mühl P, Knothe M. Recovery of precious metals from electronic scrap, in particular from waste products of the thick-layer technique. *Hydrometallurgy.* 1990;25:99–110.

Hayes PC. *Process Principles in Minerals and Materials Production.* Brisbane: Hayes Publishing Co.; 2003:292.

Hoffmann JE. Recovering precious metals from electronic scrap. *JOM.* 1992;44(7):43–48.

Jang YC. Waste electrical and electronic equipment (WEEE) management in Korea: Generation, collection, and recycling systems. *J. Mater. Cycles Waste Manag.* 2010;12:283–294.

Nakazawa H, Shouming W, Kudo Y. Bioleaching of waste printed wiring board using *Thiobacillus ferrooxidans.* In: *Recycling and Waste Treatment in Mineral and Metal Processing: Technical and Economic Aspects.* Lulea:16–20.

Pozzo RL, Malicsi AS, Iwasaki I. Removal of lead from printed circuit board scrap by an electrodissolution-delamination method. *Resour. Conserv. Recycl.* 1991;5:21–34.

Tenório JAS, Menetti RP, Chaves AP. Production of non-ferrous metallic concentrates from electronic scrap. In: *TMS 2022 Annual Meeting & Exhibition.* Orlando: EUA, 1997:505–509.

Zhang S, Forssberg E. Mechanical separation-oriented characterization of electronic scrap. *Resour. Conserv. Recycl.* 1997;21:247–269.

Zhang S, Forssberg E. Intelligent liberation and classification of electronic scrap. *Powder Technol.* 1999;105:295–301.

19 Industrial Waste Management

Industrial solid wastes are created by modern societies as unavoidable byproducts of mining, industrial production, and the requirements of today's modern consumers. During the nineteenth and twentieth centuries, and now the twenty-first century, industries concentrated on the quality of the final product, while discarding the residues and waste products generated during increasingly complicated manufacturing processes.

At the beginning of the Industrial Revolution, mining of coal and metal ores became both the fuel and the building blocks of Western society's industrialization. With the discovery of petroleum and natural gas deposits, chemical processing industries quickly followed. From the early 1900s to the late 1960s, hundreds of thousands of petroleum-derived and synthetic chemicals were used in the production of goods.

Although carbon-based plastic materials, together with such organic chemicals as pesticides and solvents, have dominated industrial production since the early 1950s, metal-based goods remain fundamental to modern industry. Numerous modern goods—from cars to paints—require the use of common metals such as iron, aluminum, and copper. In addition, less abundant but more toxic metals like lead, cadmium, nickel, mercury, arsenic, and selenium are also used by industry. Metallic elements are therefore commonly found in industrial wastes, where they have complex and incompletely understood effects on the environment. Consumers also discard massive amounts of hazardous wastes. The advent of the development of countries such as China and India have exacerbated the production of industrial wastes. We buy, use, and dispose of increasing quantities of goods that have hazardous characteristics including paints, solvents, pesticides, batteries, lightbulbs, and electronic goods.

The disposal of wastes can result in adverse effects not only on the environment but also on human health. Early in the twentieth century, industries, cities, and towns began to use abandoned or natural depressions to dispose of the ever-growing amounts of municipal waste residues. Today, there are an increasing number of waste minimization programs recycle and reuse programs, and numerous solid waste disposal programs, which not only regulate hazardous industrial wastes but also provide guidelines for the safe recycling and disposal of hazardous household wastes.

MAJOR FORMS OF INDUSTRIAL WASTES

Regardless of how wastes are technically categorized by regulatory agencies, their pollutant-releasing capacities depend on their physical characteristics, which determine their mode of transport into the environment. The four major waste types based on physical characteristics—combustible wastes, solid wastes, sludge and slurry

DOI: 10.1201/9781003272618-22

wastes, and wastewaters, which show how each of these types can eventually release pollutants into the atmospheric, terrestrial, and aquatic environments. Combustible wastes yield byproducts that can be released directly into the atmosphere as gases or particulate pollutants when they are not properly filtered. Solid wastes can release pollutants into the atmosphere via dust or particular transport, or when these wastes come into contact with water, their soluble constituents can be leached out into the soil surface or below. Sludge and slurry wastes can release pollutants into the soil and groundwater from both their solid and liquid phases. Wastewaters, owing to their liquid state, are always potential sources of pollution if discharged directly into the aquatic or terrestrial environment without proper treatment. In general, we can see that the pollutants whose impact on the environment is most severe are most usually found in liquid or gas phases. By their fluid nature, these pollutants can be transported large distances and thus affect large segments of the environment.

Except for a few isotopic forms, none of the 40 elements that are economically important to modern industrial society can be made synthetically. Thus, these elements have to be mined, extracted from the natural state, and subsequently purified. Because of such processes, as well as the processes used in manufacturing, these elements—including many precious and strategic metals such as gold (Au), platinum (Pt), cobalt (Co), antimony (Sb), and tungsten (W)—can be found in significant amounts in some industrial wastes. A few elements, such as mercury (Hg) and arsenic (As), can be found in gaseous, liquid, or solid wastes. Yet others, such as lead (Pb) and chromium (Cr), can be discharged into the atmosphere as particulate matter associated with other elements such as sulfur (S) and carbon (C).

However, most of the elements, including metals, metalloids, and salts, can be dissolved in the liquids (usually the aqueous phase) found in wastewaters, sludges, and solid industrial wastes. That is, with few exceptions, most industrial liquid, slurry, and sludge wastes have a water phase that can range from 99% to less than 10% by weight. Therefore, the process of dewatering can be used to reduce much of the mass of these wastes. Moreover, many forms of organic pollutants are also water soluble to various degrees and are therefore found in the water phase of these wastes. These aqueous phases, once separated from solids, must be processed as wastewaters prior to discharge into the open environment.

TREATMENT AND DISPOSAL OF INDUSTRIAL WASTES

Technologies and practices used for the treatment and disposal of metal- and salt-containing wastes vary widely, but ultimately, the byproducts of these technologies end up being released into the air, land, or water environments. Methods that separate metals from the other waste constituents are driven by the need to reduce waste disposal costs and ultimately by the potential costs associated with liability. Organic wastes that contain low residual concentrations of metals and salts can often be degraded by using thermal or biological destruction processes that completely transform the waste into carbon dioxide and water. Conversely, wastes that contain significant amounts of metals and salts always leave indestructible residues that may or may not be recycled economically. Therefore, these wastes that cannot be eliminated and must be disposed of in a manner that minimizes their impact on the environment.

GAS AND PARTICULATE EMISSIONS

Gases and dust particulates that are generated in the thermal destruction of wastes and smelting can be prevented from escaping into the air using one or more of the following processes: electrostatic precipitators, baghouse and cyclone separators, and wet scrubbers. These technologies are expensive and difficult to operate efficiently. However, without this equipment, smelters and incinerators would discharge large quantities of toxic metals into the atmosphere, thereby contaminating large tracts of land.

A very controversial issue is the reduction or trapping of greenhouse gases like carbon dioxide (CO_2) that are released by power-generating plants that use fossil fuels (coal, oil, and natural gas). These gases are known to be linked to global warming. Technologies being evaluated include trapping and liquefying CO_2 gas from smokestacks, followed by storage by injection into depleted deep oil or geologic gas reservoirs.

CHEMICAL PRECIPITATION

Wastewater and slurried sludges can be treated with chemical agents to precipitate and remove metals from the rest of the waste components. For example, solutions containing alkaline materials such as calcium carbonate, sodium hydroxide, aluminum oxide, or sodium sulfide are commonly used to precipitate metals from waste streams. These chemicals help form insoluble metal hydroxides, carbonates, and sulfides. Similarly, aluminum and iron oxides sorb metals such as cadmium and metalloids such as arsenic, thus removing them from solution. General precipitation reactions can be represented by the following:

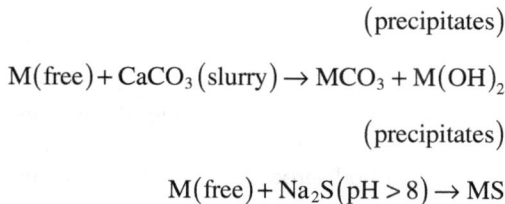

$$\text{(precipitates)}$$

$$M(\text{free}) + CaCO_3(\text{slurry}) \rightarrow MCO_3 + M(OH)_2$$

$$\text{(precipitates)}$$

$$M(\text{free}) + Na_2S(\text{pH} > 8) \rightarrow MS$$

where $M = Cd^{2+}$ or Zn^{2+} or Cu^{2+} or Pb^{2+}.

FLOCCULATION, COAGULATION, DEWATERING–FILTRATION–DECANTING–DRYING

Particulate pollutants in wastewaters can be made to settle out quickly by using chemicals known as flocculants and coagulants. Flocculants, which react with dissolved chemicals, facilitate the formation of aggregates or clumps, which can be decanted or filtered out of solution. Conversely, coagulants destabilize colloids, thereby permitting suspended particles to form aggregates that can settle out of solution. These chemicals can be useful in removing all kinds of pollutants from wastewaters, including metals and organic constituents. Flocculants and coagulants include iron and copper sulfates and chlorides, as well as complex synthetic organic polymers.

Dewatering and drying is an acceptable option for waste reduction where the liquid components of a liquid waste, such as a slurry or wastewater, have very low levels of pollutants, and these can be treated using conventional wastewater treatment methods or even discharged directly into the environment. In this case, liquids are separated from solids via several dewatering options that include air drying, particle filtration, sedimentation, and decanting. The separated solids are then treated and disposed of as solid wastes.

STABILIZATION (NEUTRALIZATION) AND SOLIDIFICATION

Sludges and slurries containing metals must be chemically and physically stabilized prior to final disposal. Since metals are more soluble in water at low pH, acidic wastes must be neutralized with such basic compounds as calcite ($CaCO_3$) and hydrated lime ($Ca[OH]_2$) to form low water solubility metal carbonate and metal hydroxide complexes.

The process of waste solidification usually involves trapping or encapsulating the waste into a physically stable matrix. For example, when wet cement is mixed in with sludges, it forms a stable block after a few days of curing (drying). This solidification method encapsulates wastes in a matrix that is relatively low in porosity and cannot easily be deformed or cracked under typical landfill overburden pressures. Consequently, percolating water does not readily infiltrate into the matrix, and metals or salts are less likely to leach out. In some cases, waste products like fly ash, collected from electrostatic precipitators during coal burning, can also be used to neutralize and encapsulate other wastes like metal waste streams, due to their strong cementing (pozzolanic) properties.

OXIDATION

Carbon-based waste streams can be oxidized to detoxify and destroy organic pollutants. There are two major types of treatment processes: thermal and chemical. The overall goals are the same in both processes. Thermal oxidation reactions can be described as follows:

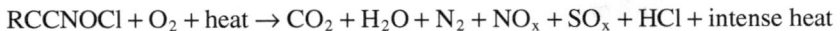

$$RCCNOCl + O_2 + heat \rightarrow CO_2 + H_2O + N_2 + NO_x + SO_x + HCl + intense\ heat$$

Thermal oxidation processes include incineration, which use conventional fuel-driven burners. Solid or liquid organic wastes having low water content but high heat values are good candidates for thermal oxidation because they burn hotly enough to sustain the energy these processes require. The major disadvantages of this process include high costs, limited reliability, and a negative public perception of their safety. In addition, large emissions of pollutants can result when the systems are not operated and maintained properly. Incinerators operate most efficiently when designs are tailored to specific waste stream characteristics. Thus, incinerators are not suitable for the efficient oxidations of mixed waste streams.

Chemical oxidation generally proceeds via the following reaction pathway:

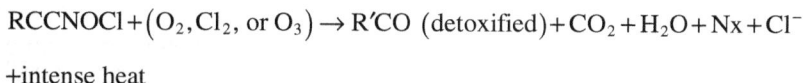

$$RCCNOCl + (O_2, Cl_2, \text{ or } O_3) \rightarrow R'CO \text{ (detoxified)} + CO_2 + H_2O + Nx + Cl^-$$

+intense heat

where R or R' is the remaining part of the organic molecule.

For example, chemical oxidation is used in the destruction of cyanide-containing wastewaters. This process involves chlorination, which can be summarized as follows:

$$2CN^- (\text{liquid}) + 3Cl_2 (\text{gas}) + 6NaOH \rightarrow N_2 \text{ gas} (pH > 8 - 9) + 6Cl^- + 2HCO_3^- + 2H_2O$$

Many forms and combinations of chemical oxidation processes utilize chlorine (Cl_2), chlorine dioxide (ClO_2), ozone (O_3), or ultraviolet (UV) radiation to eliminate low concentrations of organics from industrial wastewaters. The chemicals oxidized using these processes include acids, alcohols, and aldehydes (such as oxalic acids and phenols), as well as more stable chlorinated pesticides and some petroleum-derived solvents such as DDT or xylenes.

LANDFILLING

Industrial wastes that contain significant concentrations of metals that cannot be recycled or recovered economically are usually good candidates for landfill disposal, which is the disposal of waste materials within soil or the vadose zone. However, sludges high in metals must be neutralized and solidified prior to landfill disposal, and only landfills approved for the disposal of solidified hazardous wastes may be used for the disposal of these wastes. Similarly, wastes high in salts may also be buried in special landfills or used as fill materials for road and dam construction. Although such wastes are not considered toxic, they are very soluble in water. Therefore, it is important to minimize water contact to prevent the leaching of salts into the environment.

LANDFARMING OF REFINERY SLUDGES AND OILFIELD WASTES

Landfarming techniques have been used to treat hazardous wastes, particularly oily sludges, whose major constituents are oil, sediments, and water. Land disposal restrictions now prohibit land treatment of hazardous oily wastes. These restrictions have compelled the petroleum industry to look at alternative disposal methods, including landfilling and incineration. On the positive side, petroleum refineries have also been spurred to develop waste minimization strategies. Oilfield wastes generated during well drilling activities are treated in treatment facilities that have to be permitted. These facilities dewater and lower the salinity and oil content of these wastes. In addition to residual amounts of oil (~1%–5%), these wastes often contain significant amounts of metals such as zinc, and minerals like barite, and sodium chloride.

Deep-Well Injection of Liquid Wastes

Deep-well injection of liquid wastes into the subsurface is another waste disposal method. This method greatly reduces the potential hazard posed by wastes through disposing of them in the deep subsurface. Examples of deep-well injection of liquid wastes include oilfield wastes (brines) metal-containing wastewaters and slurries and wastewaters with high concentrations of toxic organic chemicals such as chlorinated hydrocarbons, pesticides, and radioactive wastes.

Deep-well injection is most suitable for handling large volumes of liquid or slurry wastes that have a water-like consistency (low viscosity). Watery liquids are usually pumped down into confined, aquifer-like zones composed of highly water-permeable material, such as sandstone or limestone. Similarly, oily wastes can be deep-well injected into high permeability subsurface zones. The depth of the injection zones, usually hydrologically confined, ranges from about 200 to 4,500 m (600–13,000 ft), and most zones are located between 700 and 2500 m (2,000 and 7,000 ft).

Drilling and constructing wells for these depths can be very expensive; therefore, deep-well injection is primarily used by the petroleum industry, which already possesses in service oilfield wells. Once constructed, these systems are comparatively cheap to operate and maintain.

The practice of deep-well injection of wastes still poses some potential problems. Possible clogging of the injection zone due to solid particles or bacterial growth may occur, and the injected waste may contaminate resident groundwater. This possibility is of major concern if the groundwater is a current or potential potable water source. Waste injection can cause groundwater contamination in several ways: (1) direct injection into the aquifer used for drinking water; (2) leaking wells (waste leaks from the well bore into an aquifer used for drinking water); and (3) movement of waste to a zone that supplies drinking water.

Incineration

The process of incineration that destroys highly toxic and hazardous organic wastes differs from MSW incineration, where energy is often produced. In general, low-temperature (up to 850°C) and high-temperature (~1,200°C) incinerations use energy to oxidize carbon- and water-containing wastes to CO_2, H_2O vapor, and HCl and NO_x gases. However, some incinerators can serve as heat-energy sources, especially when oily wastes, such as spent oils, are used as fuel. But these incinerators must meet stringent emissions levels. Incineration efficiencies are also closely regulated, and destruction of all organic compounds must exceed 99.99%.

Incineration cannot be used with wastes that have high concentrations of water and non-combustible solids, nor can it be used for radioactive materials. Moreover, incinerators are very expensive to build and maintain. Thus, this technology is mostly limited to low-volume wastes such as medical wastes. Finally, besides the gases previously listed, incinerators emit small but significant amounts of numerous toxic chemicals including dioxins, furans, polynuclear aromatic hydrocarbons, and metals such as mercury, beryllium, and cadmium. Often incinerators produce ash residues (bottom and fly ash) that have hazardous characteristics (particularly fly ash) and must be buried in landfills approved to handle such wastes. Thus, incinerators do not totally eliminate the problems of environmental contamination, and this has resulted in the technology being condemned by the public. Grassroots organizations are currently fighting for the elimination of this technology as a waste treatment option, even as a renewable energy recovery source.

STOCKPILING, TAILINGS, AND MUDS

Mining activities produce vast quantities of tailings, which are usually stockpiled in the form of terraces on or near the ore-processing mills. The environmental impacts of mining activities, which vary widely from site to site, are usually associated with runoff or percolation of waters contaminated with sediments. They are also related to pH, metal solubility, salt concentrations, and the quantities of wind-blown particulates that are contaminated with metals. Oil and gas fields produce large quantities of well cuttings (muds). These well cuttings, which contain barite, salts, and crude oil, are usually stockpiled or treated until they can be used as fill materials. The potential for offsite releases of pollutants from these sites is normally associated with water and wastewater releases that contain varying concentrations of these chemicals.

REUSE OF INDUSTRIAL WASTES

Because metal- and salt-containing wastes cannot be destroyed, when improperly disposed, they can be a hazard to humans and the environment. At the same time, these wastes have a potential economic value that is becoming more evident as the quality of their natural sources diminishes. Precious and strategic metals, such as gold, platinum, cobalt, antimony, and tungsten, are routinely mined out of wastes that contain significant concentrations of these elements. However, wastes that contain fewer valuable metals, such as aluminum and iron, are far less likely to be treated for the removal of these elements. Even more improbable is the extraction of highly reactive metals, non-metallic elements, and their soluble salts, such as magnesium and calcium sulfates and carbonates, which are found in most industrial wastes.

METALS RECOVERY

Economically valuable elements such as precious metals can be recovered from waste streams by using complex chemical reactions, including chemical separation or precipitation. Silver, for example, can be recovered from photographic wastes by acidifying the liquid wastes and separating the Ag sludge that precipitates. Subsequently, the supernatant can be neutralized and disposed of safely, while the Ag sludge is sent to a smelter for purification. Metals can also be made to react selectively with synthetic organic chemicals known as chelates and synthetic zeolites, which are porous aluminosilicate minerals. In such reactions, metals are grabbed or trapped by the chelates or sequestered in the internal structure of the zeolites, while other materials pass by. Once reacted, the metal of interest is either precipitated or extracted out of the waste solution or is removed with a sorbent. The recovered metal sludge can be smeltered and refined into pure solid metal. Metal-containing residues resulting from the smelting and refining of ores can be further refined by using a combination of chemical and physical separation techniques and high-temperature furnaces.

ENERGY RECOVERY

The majority of wastes containing organic carbon forms have large quantities of stored energy. Thus, wastes containing high concentrations of reduced carbon (usually organic carbon), such as organic liquids, woody materials, oils, resins, or asphalts, can be used in incinerators and electrical generators as energy sources. For example, kerosene and natural gas have about 44×10^6–49×10^6 Joules of energy/kg. Many solvents such as hexane, xylenes, and paraffins, which are found in oils and alcohols, have similar stored energies (42×10^6–58×10^6 J/kg). Therefore, many waste mixtures of organic chemicals have energy values approaching those of commercial fuels.

INDUSTRIAL WASTE SOLVENTS

Industrial wastes high in solvents are being successfully recycled using recovery systems that include distillation techniques and chemical and physical fractionation processes. For example, spent solvents used to clean and paint metal parts can be redistilled as a means to separate the heavy impurities from the volatile solvents. Some examples of solvents that can then be reused via solvent recovery stills include chloroform, acetone, benzene, xylene, hexane, and methylene chloride.

INDUSTRIAL WASTE REUSE

Some mine tailings can be used as fills for earthworks. However, economic and potential liability issues related to transportation costs and potential releases of contaminants usually limit their reuse. Consequently, mine tailings are most often stockpiled in place. Oilfield wastes that are high in salts may be leached with water to remove the soluble salts. The salt-free muds are then dried and stockpiled for use as fill material. These treated solid oilfield wastes have been shown to be safe in earthworks and release no residual salts into the surrounding environment. Coal burning wastes such as fly ash and flue gas desulfurization sludges have been used successfully as soil amendments. Coal-burning fly ash has also been used as a fill material for earthworks. These materials can be good sources of gypsum and calcite, which can neutralize acidity in soils and replenish macronutrients such as Ca, Mg, and S, and even trace elements like Zn, Cu, Fe, and Mn. Fly ash additions to acidic agricultural soils have also been shown to improve the soil pit, porosity, and overall structure.

FURTHER READING

Brown KW, Carlile BL, Miller RH, Rutledge EM, Runge ECA. *Utilization, Treatment, and Disposal of Waste on Land*. Madison, WI: Soil Science Society of America; 1986.
Cope CB, Fuller WH, Willets SL. *The Scientific Management of Hazardous Wastes*. Cambridge: Cambridge University Press; 1983.

LaGrega MD, Buckingham PL, Evans JC. *Hazardous Waste Management.* New York: McGraw-Hill; 1994.

National Research Council. *Waste Incineration and Public Health.* Washington, DC: National Academy Press. National Research Council; 1999.

Pepper IL, Gerba CP, Brusseau ML. *Environmental & Pollution Science.* USA: Elsevier Academic Press; 2006.

Wang LK, Wang MHS. *Handbook of Industrial Waste Treatment.* New York: Marcel Dekker, Inc.; 1992.

20 Composting of Organic Wastes

Composting is the biological decomposition of the organic compounds of wastes under controlled aerobic conditions. In contrast to uncontrolled natural decomposition of organic compounds, the temperature in waste heaps can increase by self-heating to the ranges of mesophilic (25°C–40°C) and thermophilic microorganisms (50°C–70°C). The end product of composting is a biologically stable humus-like product for use as a soil conditioner, fertilizer, biofilter material, or fuel. The objectives of composting can be stabilization, volume and mass reduction, drying, elimination of phytotoxic substances and undesired seeds and plant parts, and sanitation. Composting is also a method for decontamination of polluted soils. Almost any organic waste can be treated by this method. The pretreatment of organic waste by composting before landfilling can reduce the emissions of greenhouse gases.

In any event, composting of wastes is conducted with the objective of high economic effectiveness and has the goal of compost production with the lowest input of work and expenditure. The consequence of this approach is the effort to optimize the biological, technical, and organizational factors and elements that influence the composting process. The factors that influence the composting process are well known and have been published in numerous reviews and monographs. The period since 1970 has been characterized by the development of new strategies, composting processes, and technologies and the optimization of existing processes against the background of an expanding market for composting technology. Among others, reasons for these developments are rising costs for sanitary landfills, improved environmental protection requirements, as well as new laws, ordinances, and regulations. The realization that resources are limited and the idea of recycling refuse back to soil have also provided important impetuses for developments in this field.

WASTE MATERIALS FOR COMPOSTING

The origins of organic waste for composting are households, industry, wastewater treatment plants, agriculture, horticulture, landscapes, and forestry. The amount, composition, and physical characteristics of plant wastes are influenced by numerous factors such as the origin, production process, preparation, season, collecting system, social structure, and culture. The wide range in the amount and composition of waste requires analyses for planning a composting plant and for estimating the compost quality in each individual situation.

The content of heavy metals and organic compounds in the waste is of great importance, particularly with regard to the use of compost as soil conditioner and fertilizer. Ways to reduce the heavy metal content in compost from biowaste are to collect the waste separately and to obtain detailed information of the producers of

DOI: 10.1201/9781003272618-23

the waste. The fact that the concentrations of heavy metals increase as the organic compounds are degraded also must be taken into consideration.

FUNDAMENTALS OF COMPOSTING PROCESS

Degradation of the organic compounds in waste during composting is initiated predominately by a very diverse community of microorganisms: bacteria, actinomycetes, and fungi (Chamuris et al., 2000; El-Din et al., 2000; Hart et al., 2002; Barrington et al., 2003). Just as in biological wastewater treatment, an additional inoculum for the composting process is not generally necessary, because of the high number of microorganisms in the waste itself and their short generation time. Invertebrate animals play no role in the rotting process during the first phase at a high temperature level. Nevertheless, earthworms are sometimes used in waste management and to produce a high-value compost (Edwards and Neuhauser, 1988; Singh and Sharma, 2002).

Rotting waste material, even during well-aerated composting, is characterized by aerobic and anaerobic microbial processes at the same time (Schuchardt, 2000). The relation between aerobic and anaerobic metabolism depends on the physical properties of the waste/compost, including the structure of the heap, its porosity, its water content and capacity, its free airspace, and the availability of nutrients. The aerobic microorganisms in the rotting material need free water and oxygen for their activity. End products of their metabolism are water, carbon dioxide, NH_4 (or, at higher temperature and pH > 7, NH_3), nitrate, nitrite (nitrous oxide as a product of nitrification), heat, and humus or humus-like products. The waste air from the aerobic metabolism in compost heaps contains evaporated water, carbon dioxide, ammonia, and nitrous oxide. The end products of the anaerobic microorganisms are methane, carbon dioxide, hydrogen, hydrogen sulfide, ammonia, nitrous oxide, nitrogen gas (both from denitrification), and water as liquid (Fukumoto et al., 2000).

Mature compost consists of components that are difficult to digest or undegradable components (lignin, lignocellulosics, minerals), humus, microorganisms, water, and mineral nitrogen compounds. The organisms that take part in the composting process are microorganisms (bacteria, actinomyces, mildews) in the first phase of composting. They each have optimal growing conditions at different temperatures: psychrophilic between 15°C and 20°C, mesophilic between 25°C and 35°C, and thermophilic between 55°C and 65°C. In mature compost at temperatures below 30°C–35°C, other organisms such as protozoa, collembolans, mites, and earthworms join in the biodegradation.

A pile of organic wastes consists of solid, liquid, and gaseous phases, and the microorganisms depend on free water for their metabolism. Dissolved oxygen, from the gas phase in the heap, must be available for the activity of aerobic microorganisms. To make sure that oxygen transfer from the gas phase to the liquid phase and carbon dioxide transfer from the liquid phase to the gas phase occur, a permanent partial-pressure gradient must be maintained, which is possible only by a permanent exchange of the gas phase by forced or natural aeration. Specifications about the optimal water content for composting are meaningful only in combination with the knowledge of the specific type of waste to be composted, its structure and volume of air pores. In general, the water content can be higher when the waste structure, air

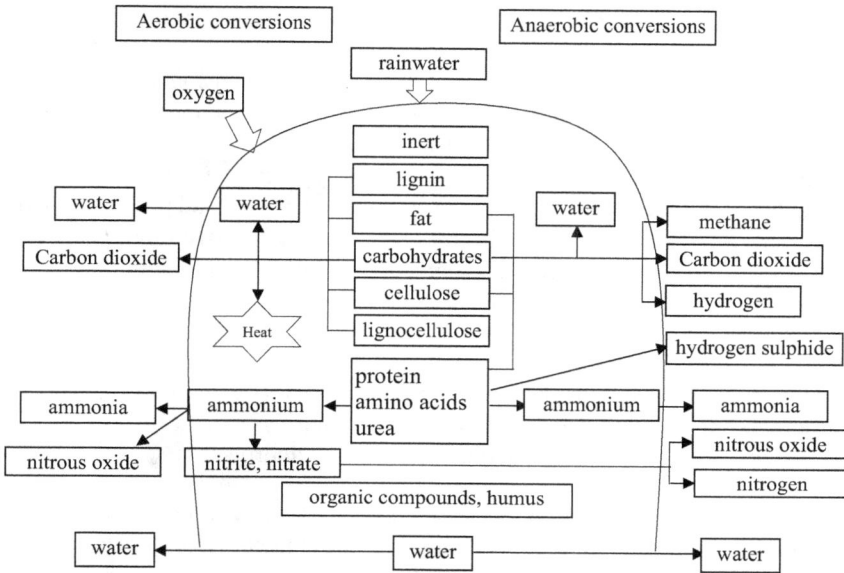

FIGURE 20.1 Substrates and products of microbial activity in a compost heap (Schuchardt, 2000).

pore volume, and water capacity are higher and more stable (also during the rotting process). Theoretically, the water content for composting can be 100%, provided the oxygen supply is sufficient for microbial activity (Figure 20.1).

In addition to a sufficient content of free water, the microorganisms need a C/N ratio in the substrate of 25–30 for optimal development and fast enough rotting process, and the carbon should be readily available. At C/N ratios below the optimum, there is a danger of nitrogen loss as ammonia gas increases (especially when the temperature rises and the pH is >7). If the C/N ratio is higher than optimum, the composting process needs a longer time to stabilize the waste material.

The structure of the waste, i.e., its consistency and the configuration and geometry of the solids, determines the pore volume (whether filled with water or air) and the air flow resistance of a compost heap. These in turn influence the gas exchange and the oxygen and carbon dioxide concentrations in the air pores and liquid phase. When these factors are optimum, exothermic microbial activity is rapid, leading to increasing temperature by buildup of heat within the heap. Microbial activity is affected by the water content, nutrients (C/N ratio, availability), and pH. The mass and volume of the heap influence the temperature according to the heat capacity and heat losses by irradiation. Heat convection within the heap, which is conditioned by the temperature difference between the material and the atmosphere, affects the gas exchange. The gas exchange and the temperature influence the evaporation of water and thus also the proportion of water-filled pores. One effect of the activity of different microbial groups is a characteristic temperature curve during composting. After a short lag, the temperature increases exponentially to 70°C–75°C. At 40°C, there is often a lag during the changeover from mesophilic to thermophilic microorganisms. After reaching a maximum, the temperature declines slowly to the level of the atmosphere. The progression of the temperature curve depends on numerous factors such as the kind and preparation of the waste, the surface/volume ratio of the heap, air temperature, wind velocity, aeration rate, C/N ratio, processing technique, and mixing frequency. The first phase of the composting process, up to a temperature up to about 60°C, is called the pre- and main composting; the second phase is called the post-composting or mature phase. Both phases are characterized by different processes. Frequently, the designers of a composting facility must consider both phases by dividing the entire composting process into different technical stages, especially when the wastes have a risk of strong odor emissions:

- Pre- and main composting occurs in closed reactors or in roofed facilities, and in frequently mixed or forced-aerated windrows.
- The post-composting/mature phase is done in windrows.

The consequences for the composting process are basically to optimize the factors that influence the rotting process. The most important factor is, for a given composition of waste, to ensure gas exchange in the heap. This can be done by taking the following measures:

- Adapting the height of the heap to the structure, water content, and oxygen demand (high during pre- and main composting, low during the mature phase)
- turning (mixing, loosening) the windrows
- constructing windrows in thin, ventilatable layers
- mixing and loosening the rotting material in reactors (in rotating drums, with tools)
- using forced aeration
- decreasing the streaming resistance by adding bulking material having a rough structure or in the form of pellets

TABLE 20.1
Phases and Characteristics of the Composting Process (Jordening and Winter, 2005)

Pre- and Main Composting	Post-Composting, Mature Phase
Degradation of easily degradable compounds: sugar, starch, pectin, protein	Degradation of difficult-to-decay degradable compounds: hemicellulose, wax, fat, oil, cellulose, lignin
Inactivation of pathogenic microorganisms and weed seeds	Composition of high-molecular weight compounds (humus)
High oxygen demand	Low oxygen demand
Emissions of odor and drainage water	Low emissions
Time: 1–6 weeks	Time: 3 weeks–1 year

COMPOSTING TECHNOLOGIES

The production of compost consists of preparing and conditioning the raw material, followed by the actual composting. To produce a marketable product, it is necessary to convert the compost to an end product. The aim of raw material preparation and conditioning is to optimize conditions for the following composting process, to remove impurities so as to protect the technical equipment, to reduce the input of heavy metals and hazardous organic components (if the impurities contain these components), and to meet quality requirements for the finished compost. The basic steps of raw material preparation and conditioning are:

- disintegration of rough wastes (e.g., wood scraps, trees, brush, long grass) by chopping, crushing, or grinding to increase the surface area available for microbial activity
- dehydration or (partial) drying of water-rich, structureless wastes (e.g., sludge, fruit remains) if they are too wet for the composting process
- addition of water (freshwater, wastewater, sludge) if the wastes are too dry for the composting process
- mixing of components (e.g., wet and dry wastes, N-rich and C-rich wastes, wastes with rough and fine structure)
- manual or automatic separation of impurities (glass, metals, plastics)

The products of preparation and conditioning of the wastes are waste air (depending on the composition and the conditions of storage, it may include bad smells and dust) and possibly drainage water beneath the raw material. The basic steps of the subsequent composting process may be:

- aeration to exchange the respiration gases oxygen and carbon dioxide and to remove water (the only essential step during composting)
- mixing to compensate for irregularities in the compost heap (e.g., dry zones at the surface, wet zones at the bottom, cool zones, hot zones) and to renew the structure for better aeration
- moistening of dry material to improve microbial activity
- drying of wet material by aeration or/and mixing to increase the free air pore space for microbial activity or to improve the structure of the compost for packaging
- manual removal of impurities

The products of the composting process are a biologically stabilized compost, waste air, and drainage water (when the material is very wet). It may be necessary to prepare the compost for transport, storage, sale, and its application. When post-preparation is needed, the basic steps can be:

- sieving the compost to obtain different fractions for marketing or to remove impurities
- manually or automatically removing impurities

- drying wet compost to prevent formation of a clumpy, muddy product and drainage of water during storage
- disintegrating clumps in the compost by crushing or grinding to prevent problems that may occur when the fertilizer is packaged
- mixing the compost with additives (soil, mineral fertilizer) to produce potting mixes or gardening soils

Disintegration (crushing, chopping, grinding), especially of bulky wastes containing wood pieces, is necessary to increase the surface area available for the microorganisms and to ensure the functioning of the machines and equipment used in subsequent stages of the process (e.g., turning machine or tools, screens, belt conveyor). The intensity of disintegration depends on the velocity of the biodegradation of the waste, the composting process, the dimensions of the heap, the composting time, and the intended application of the final product. For disintegration of organic wastes, chopping machines or various kinds of mills (cutting, cracking, hammer, screw) are mainly used.

FACTORS AFFECTING EFFICIENT COMPOSTING

Temperature	Adequate aeration and moisture must be maintained to ensure temperatures reach 60°C, to inactivate microbial pathogens
Aeration	Must be provided via blowers or by turning
Moisture	Must be neither too moist, which promotes anaerobic activity, nor too dry, which limits microbial activity
C:N ratio	Should be maintained around 25:1, to ensure adequate but not excessive amounts of nitrogen for the microbes
Surface area of bulking agent	Shredded material should be used to increase substrate surface area for microbial metabolism

The raw waste or compost is screened to separate particles with a required granule size. These particles can be the organic raw material for composting, the compost itself, or impurities. In practice, drum- and plain-screens (with hole plates, wire grates, stars, or profile iron) are usually used. The size of the sieve holes depends on the subsequent use of the compost or on whether impurities are being removed (>80 mm: removal of impurities; 80–40 mm: production of mulching material; 10–25 mm: production of compost for landscaping, agriculture, and gardening).

COMPOSTING SYSTEMS

NON-REACTOR COMPOSTING

Field composting: During field composting, which is the simplest way of composting organic wastes, all microbial activity takes place in a thin layer at the soil surface or within a few centimeters of the soil surface (arable land or grassland). This system is useful for treating both sludge and green wastes (grass, straw, brushwood). To ensure rapid and uniform decomposition, green wastes need to be chopped. Mulching machines can be used if the wastes are growing in the same area (e.g., vineyard pruning); otherwise, collected wastes are spread out with a manure spreader after chopping. Because the waste material surface exposed to the atmosphere is large, self-heating does not occur, and therefore, neither do thermal disinfecting nor killing of weed seeds. Therefore, only wastes without problems of hygiene or weed seeds can be utilized in this kind of composting. In the narrower sense of the definition of composting, field composting is not composting, because there is no self-heating and no real process control (Figure 20.2).

Windrow composting: The main characteristic of non-reactor windrow composting is direct contact between the waste material and the atmosphere, and therefore, interdependence between the two. The composting process influences the atmosphere by emitting odors, greenhouse gases, spores, germs, and dust. The atmosphere, which carries the respiration gas oxygen, can influence the composting process by

Supplying rain water
  Advantage: adds water, which is needed if the material for composting is or
    has become too dry, thus resulting in more rapid biodegradation
  Disadvantages: blocks free airspace, favors anaerobic conditions and associ-
    ated odor emissions, decreases compost quality, increases drainage water

Changes in air temperature
  Advantages: high air temperatures can increase the evaporation rate of very
    wet wastes, increasing the amount of free airspace; high temperatures
    can shorten the lag phase at the start of the process
  Disadvantages: high air temperatures can increase the evaporation rate,
    leading to insufficient moisture; low air temperatures can delay or
    inhibit self-heating

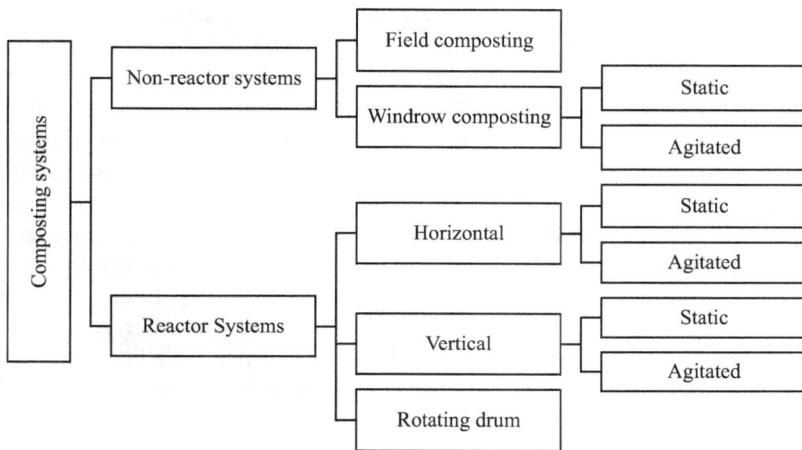

FIGURE 20.2 Composting systems can be classified into non-reactor systems and reactor or vessel systems (Haug, 1993; Thome-Kozmiensky, 1995; Krogmann and Korner, 2000).

Changes in air humidity
    Advantages: low air humidity can increase the evaporation rate of very wet
        wastes; high air humidity reduces the evaporation rate
    Disadvantages: low air humidity can increase the evaporation rate, leading
        to insufficient moisture; high air humidity can decrease the evaporation
        rate, leading to too much moisture

Supplying wind
    Advantages and disadvantages: wind can intensify the effects of air tem-
        perature and humidity

The extent of contact between waste material and atmosphere can be influenced by covering the piles with mature compost material, straw, or special textile or fleece materials that allow gas exchange but reduce the infiltration of rain water. The cross-section shape of a windrow compost pile can be triangular or trapezoidal. The height, width, and shape of a windrow depend on the waste material, climatic conditions, and the turning equipment.

Natural aeration in windrows can be supported by (1) addition of bulking material to the waste, (2) using bulking material as an aeration layer at the bottom of the windrow (20–30 cm), (3) aeration pipes from the bottom of the windrow, and (4) perforated floor.

To ensure a high quality of the compost, windrows are disturbed from time to time by turning. The effects of turning are (1) mixing of the material for homogenization (dry or wet zones at the surface, wet zones at the bottom) and for killing pathogenic microorganisms and weed seeds, (2) renewing the structure and free airspace, and (3) increasing evaporation to dry the waste material or the mature compost. The turning frequency depends on the kind and structure of the waste and the quality requirements of the finished compost. It can vary from several times a day (at the start of the process when the oxygen demand is high or for drying mature compost) to once every several weeks. Machines and equipment for turning include tractor mounted front-end loaders, wheel loader shovels, manure spreaders, tractor-driven windrow turning machines, and self-driven windrow turning machines. The mixing quality of frontend loaders and wheel loaders is relatively poor and requires an experienced driver. Compression of the (wet) wastes by the weight of the machinery can be a disadvantage.

Tractor-driven turning machine consists of a concrete or asphalt floor area with an open shed for storing the finished compost. All the drainage and rain water are collected in a tank or basin. Large pieces of waste (branches, trees) are chopped periodically by a machine that is driven from one place to another. After separating the impurities from the biowaste, windrows are formed from both components with a wheel loader. The windrows are turned frequently with a tractor-driven turning machine (or a self-driven machine). The finished compost is screened with a mobile screening device. The oversize fractions from the screening are reused in the compost plant, and the finished compost is stored under a roof until it is sold.

REACTOR COMPOSTING

Every method of composting in an enclosed space (e.g., container, box, bin, tunnel, shed) with forced exchange of respiration gases is a type of reactor composting. Composting reactors can be classified according to the manner of material flow as

- horizontal-flow reactors having
  a static solids bed or
  an agitated solid bed
- vertical-flow reactors
- rotating-drum reactors

With few exceptions, all reactors have a means of controlled forced aeration and enable the waste air and the drainage water to be collected and treated. Addition of water or other additives is possible only when the waste material is mixed in the reactor. If possible, composting processes are used only for pre-composting, because of the high costs of reactor composting compared to windrow composting.

Horizontal-flow reactors with static solids bed: This is a batch system in which the waste material is loaded with a wheel loader or transport devices into a horizontal reactor or is covered by a foil or textile material. The forced aeration, in positive or negative mode or alternating, issues from pipes at the bottom of the material or from holes in the floor. The waste air, with its odor components and water, can be treated in a biofilter or biowasher. In some systems, some of the waste air is recycled. The air flow rate can be controlled according to the temperature in the material or the oxygen/carbon dioxide concentration in the air. The retention time in the reactor is between several days (pre-composting) and several weeks (mature compost). The end product can be inhomogeneous (partially too dry) and not biologically stabilized, because there is no turning/mixing and no water addition to the waste material and is forced aeration from only one direction.

Horizontal-flow reactors with agitated solids bed: In this reactor type, the waste material can be turned mechanically and water can be added, in contrast to a horizontal-flow reactor with a static solid bed. The devices for mixing the wastes can be horizontally or vertically operating rotors or screws, scraper conveyors, or shovel wheels. Fully automated functioning of the whole process is possible.

Vertical-flow reactors: In this reactor type, the waste material flows vertically, with or without stages, with mass flow from the top to the bottom as a plug flow system, or if the material from the outlet at the bottom is loaded back in at the top, as a mixed system. Forced aeration occurs from the bottom or from vertical pipes in the material. This process can also be run in a fully automated way.

Rotating-drum reactor: The waste material is loaded into a horizontal slowly rotating drum with forced aeration. The filling capacity is approximately 50%. The material is transported (plug flow) in the helical pathway from one end of the drum to the other and is mixed intensively. Self-heating starts after a short time. Water addition is possible. Rotating-drum reactors can also be used as mixing equipment. This composting plant is characterized by an enclosed building for receiving the biowaste, preparing it for the rotting process (disintegration, separation of impurities), pre-rotting in a drum, and main rotting in turned windrows (composting I). All highly contaminated waste air can be collected and treated with a biofilter. Only the maturing phase of the compost (composting II) occurs under natural climatic conditions in windrows under a roof. In a fully enclosed composting plant, even the maturing phase takes place in a closed shed.

COMPOST QUALITY

To be used as a fertilizer and soil conditioner, compost must meet certain quality requirements, such as (1) optimal maturity, (2) favorable contents of nutrients and organic matter, (3) favorable C/N ratio, (4) neutral or alkaline pH, (5) low contents of heavy metals and organic contaminants, (6) no components that interfere with plant growth, (7) mostly free from impurities, (8) mostly free from germinatable seeds and living plant parts, (9) low content of rocks, (10) typical smell of forest soil, and (11) dark brown to black.

"Maturity" and "stability" are different properties of compost. Stability is defined in terms of the bioavailability of organic matter, which relates to the rate of decomposition. Maturity describes the suitability of the compost for plant growth and has been associated with the degree of humification. Several methods are used in practice to determine maturity and stability, e.g., simple field tests in Dewar flasks, tests on plants, respiration activity, chemical analyses, nuclear magnetic resonance (NMR). Various procedures for determining the maturity and stability of compost are discussed in the literature; general, final tests have not been agreed upon. Concerning the compost quality factors that relate to human health, such as the contents of heavy metals, hazardous organic compounds, parasites, and other disease organisms, several countries have passed laws, ordinances, regulations, and norms.

FURTHER READING

Barrington S, Choiniere D, Trigui M, Knight W. Compost convective airflow under passive aeration. *Bioresour. Technol.* 2003;86:259–266.

Chamuris GP, Koziol-Kotch S, Brouse TM. *Compost Sci. Util.* 2000;8(1):6–11.

Edwards CA, Neuhauser EF (eds.) *Earthworms in Waste and Environmental Management.* The Hague: SPB Academic; 1988.

El-Din SMSB, Attia M, Abo-Sedera SA. Field assessment of composts produced by highly effective cellulolytic microorganisms. *Biol. Fert. Soils.* 2000;32(1);35–40.

Fukumoto Y, Osada T, Hanajima D, Haga K. Patterns and quantities of NH_3, N_2O and CH_4 emissions during swine manure composting without forced aeration: effect of compost pile scale. *Bioresour. Technol.* 2003;89(2):109–114.

Hart TD, De Leij FAAM, Kinsey G, et al. Strategies for the isolation of cellulolytic fungi for composting of wheat straw. *World J. Microb. Biot.* 2002;18:471–480.

Haug RT. *The Practical Handbook of Composting.* Boca Raton, FL: Lewis; 1993.

Insam H, Riddech N, Klammer S (eds.) *Microbiology of Composting*. Berlin/Heidelberg: Springer; 2002.

Jordening H-J, Winter J (eds.) *Environmental Biotechnology: Concepts and Applications*. Wiley-VCH; 2005.

Krogmann U, Körner I. Technology and strategies of composting. In: Rehm HJ, Reed G (eds.) *Biotechnology. Vol. 11c: Environmental Processes III*. Weinheim: Wiley-VCH; 2000, 127–150.

Kutzner HJ. Microbiology of composting. In: Rehm HJ, Reed G (eds.) *Biotechnology. Vol. 11c: Environmental Processes III*. Weinheim: Wiley-VCH; 2000, 35–100.

Schuchardt F. Composting of plant residues and waste plant materials. In: Rehm HJ, Reed G (eds.) *Biotechnology. Vol. 11c: Environmental Processes III*. Weinheim: Wiley-VCH; 2000, 101–125.

Singh A, Sharma S. Composting of a crop residue through treatment with microorganisms and subsequent vermicomposting. *Bioresour. Technol.* 2002;85(2):107–111.

Thome-Kozmiensky KJ (ed.) *Biologische Abfallbehandlung*. Berlin: Erich Freitag Verlag; 1995.

21 Vermicomposting

Earthworms belong to the order Oligochaeta, which includes more than 8,000 species from about 800 genera. Earthworms are common all over the world in natural forests and grasslands as well as agroecosystems. Earthworms are found in most regions of the world, except those with extreme climates, such as deserts and areas that are under constant snow and ice. Earthworms are arguably the most important components of the soil biota in terms of soil formation and maintenance of soil structure and fertility. Although not numerically dominant, their large size makes them one of the major contributors to invertebrate biomass in soils. Their activities are important for maintaining soil fertility in a variety of ways in forests, grasslands, and agroecosystems. Aristotle was one of the first people to draw attention to the role of earthworms in turning over the soil; he aptly called them "the Intestines of the Earth." However, it was not until the late 1800s that Charles Darwin, in his definitive work *The Formation of Vegetable Mould through the Action of Worms*, really brought attention to the extreme importance of earthworms in the breakdown of dead plant and animal matter that reaches soils and in the continued turnover and maintenance of soil structure, aeration, drainage, and fertility.

EARTHWORMS AND SOIL FERTILITY

SOIL FORMATION

Earthworms are extremely important in soil formation, principally through activities in consuming organic matter, fragmenting it, and mixing it intimately with soil mineral particles to form water stable aggregates. During feeding, earthworms promote microbial activity by an order of magnitude, which in turn also accelerates the rates of breakdown and stabilization of humic fractions of organic matter. Different species of earthworms do not affect soil formation in the same way because of very different behavior patterns. Some species consume mainly inorganic fractions of soil, whereas others feed almost exclusively on decaying organic matter. They can deposit their feces as casts either on the soil surface or in their burrows, depending on the species concerned, but all earthworm species contribute in different degrees to the comminution and mixing of the organic and inorganic components of soil and decrease the size of not only organic particles, but also mineral particles.

During passage through the earthworm gut, different kinds of mineral particles become mixed intimately with organic matter and form aggregates, which improve both the drainage and moisture-loading capacity of the soil. These aggregates are usually very water stable and improve many of the desirable characteristics of soils. There have been various suggestions as to the possible ways in which earthworms form aggregates, such as by production of gums or molecules. Earthworms move large amounts of soil from the deeper strata to the surface. The amounts moved in this way range from 2 to 250 tons/ha/annum, equivalent to bringing a layer of soil between 1 mm and 5 cm thick to the surface every year, creating a stone free layer on the soil surface.

Earthworms also affect soil structure in other ways. Some species make permanent burrows, whereas others move randomly through the soil, leaving cracks and crevices of different sizes. Both sorts of burrows are important in maintaining soil aeration, drainage, and porosity. Moreover, earthworm burrows are usually lined with a protein-based mucus that helps stabilize these channels, and many of the species with permanent burrows cast their feces around the lining of the burrows, with the cast material usually containing more plant nutrients in a readily available form than the surrounding soil.

Although all species of earthworms contribute to the breakdown of plant-derived organic matter, they differ greatly in the ways in which they break down organic matter and incorporate it into the soil. Their activities can be of three kinds, each associated with a different group of species. Some species are limited mainly to the plant-litter layer on the soil surface, decaying organic matter or wood, and seldom penetrate soil more than superficially. The main role of these species seems to be comminution of the organic matter into fine particles, which facilitates microbial activity. Other species live just below the soil surface most of the year, except when the weather is very cold or very dry; do not have permanent burrows; and ingest both organic matter and inorganic materials. These species produce organically enriched soil materials in the form of casts, which they deposit either randomly in the surface layers of soil or as distinct casts on the soil surface.

Finally, there are the truly soil-inhabiting species with permanent burrows that penetrate deep into the soil. These species feed primarily on organic matter but also ingest considerable quantities of inorganic materials and mix these thoroughly through the soil profile. These last species are of primary importance in pedogenesis. All species depend on consuming organic matter in some form and play an important role in the final stages of organic matter decomposition, which is humification into complex amorphous colloids containing phenolic materials, probably by promoting microbial activity.

Range of Nutrients in Vermicompost

Organic carbon (%)	9.15.17.98
Total nitrogen (%)	0.50.1.50
Available phosphorus (%)	0.10.0.30
Available potassium (%)	0.15.0.56
Available sodium (%)	0.06.0.30
Calcium and magnesium (meq/100 g)	22.67.47.60
Copper (ppm)	2.00.9.50
Iron (ppm)	2.00.9.30
Zinc (ppm)	5.70.11.50
Available sulfur (ppm)	128.00.548.00

During feeding by earthworms, the carbon:nitrogen ratio in the organic matter falls progressively; moreover, most of the nitrogen is converted into the ammonium or nitrate form. At the same time, the other nutrients, phosphorus and potassium, are converted into a form available to plants. Soils that have poor populations of earthworms often develop a structure with a mat of decomposed organic matter at the soil surface.

Earthworms in Waste Management

The importance of earthworms in the breakdown of organic matter and the release of the nutrients that it contains has been known for a long time. It has been demonstrated clearly that some species of earthworms are specialized to live specifically in decaying organic matter and can degrade it into fine particulate materials, rich in available nutrients, with considerable commercial potential as plant growth media soil amendments.

For instance, earthworms are able to process sewage sludges and solids from wastewater; brewery wastes; processed potato wastes; waste from the paper industries; wastes from supermarkets and restaurants; animal wastes from poultry, pigs, cattle, sheep, goats, horses, and rabbits; as well as horticultural residues from dead plants, yard wastes, and wastes from the mushroom industry.

Wastes Used for Vermicomposting

Source of Waste Generation	Utilizable Waste for Vermicomposting
Agricultural wastes	Stubble, weeds, husk, straw, and farm-yard manure
Agricultural fields	Stems, leaf matter, fruit rind, pulp, and stubble
Plantations	Dung, urine, and biogas slurry
Animal wastes	
Urban solid waste	Kitchen waste from households and restaurants, waste from market yards and places of worship, and sludge from sewage treatment plants
Agroindustries wastes	Peel, rind, and unused pulp of fruits and vegetables
Food processing units	Press mud and seed husk
Vegetable oil refineries	Press mud, fine bagasse, and boiler ash
Sugar factories	Spent wash, barley waste, and yeast sludge
Breweries and distilleries	Core of fruits, paper, and date-expired seeds
Seed production units	Stems, leaves, and flowers after extraction of oil
Aromatic oil extraction units	Coir pith
Coir industries	

Most organic wastes can be broken down by these species of earthworms, but some organic wastes have to be pretreated in various ways to make them acceptable to the earthworms, and not all organic wastes will grow earthworms equally well. The characteristics of different wastes that have been tested are as follows:

1. *Cattle manure solids* are the easiest animal wastes in which to grow earthworms successfully. Except when they are very fresh, they usually contain no materials unfavorable to the growth of earthworms. Solids can be used, but it is much easier to produce good vermicomposts in solids separated mechanically from slurries of cattle manures, which are rich in nutrients for

plant growth; moreover, the liquids can be added back to the solids at a later stage in vermicompost production.

2. *Horse manure* is an excellent material for growing earthworms and needs very little modification other than maintenance of good environmental factors in the waste. However, the earthworms do not grow as rapidly in horse manure as in pig or cattle wastes.

3. *Pig manure solids* are probably the most productive organic waste for growing earthworms. If the waste is in the form of a slurry, it is better if the solids are separated in mechanical presses or by sedimentation. Pig wastes tend to contain relatively large amounts of ammonia and inorganic salts, and unless these are washed out, the waste may have to be composted thermophilically for about 2 weeks or longer before to inoculation with earthworms. Pig wastes sometimes have a content of heavy metals, particularly copper. The processed pig waste vermicomposts are high in nutrients for plant growth.

4. *Poultry wastes*, including chicken, duck, and turkey manures, contain significant amounts of inorganic salts, urea, and ammonia, which may kill earthworms in freshly deposited wastes. However, after removal of these materials through composting, washing, or aging, earthworms grow well in them, and the vermicomposts produced are high in nutrients for the growth of greenhouse or field plants.

5. *Potato wastes*, usually in the form of peel from the processed potato and frozen potato industries, make ideal growth media for earthworms and need few modifications in terms of moisture content or other kinds of pre-processing.

6. *Food wastes* or *restaurant wastes* are readily available wastes that are expensive to dispose in landfills or elsewhere and provide a good medium for growing earthworms if some form of bulk material such as paper waste or compost is mixed with them to lower the moisture contents.

7. *Paper pulp solids* are produced by mechanical separation, pressing, or sedimentation of solids from the production washings. These solids are really excellent material for any growth of earthworms, and there is no need for pre-processing or additives. They are particularly suitable for outdoor windrow processing systems because they form a crust over the bed that minimizes water loss, and at the same time, it allows water to be added.

8. *Brewery wastes* need no modification, in terms of moisture content, to grow earthworms. Earthworms can process brewery wastes very quickly and grow and multiply rapidly in them. The vermicompost produced from brewery wastes has a good structure and nutrient content.

9. *Spent mushroom compost* is a good medium for growing earthworms, which are able to break down the straw it contains into small fragments and produce a finely structured material. However, the vermicompost produced may be low in plant nutrients.

10. *Urban garden wastes,* including grass clippings and tree leaves, are good growth media for earthworms, particularly when they are first macerated and thoroughly mixed before use. However, their production and availability tends to be seasonal, which makes production of a standardized vermicompost year round difficult.

BREAKDOWN OF SEWAGE WASTES BY EARTHWORMS

Aerobic sewage sludges can be readily ingested by earthworms such as *Eisenia fetida* and egested as finely divided casts; in the process, the sludge is decomposed and stabilized (i.e., rendered innocuous) much faster than non-earthworm ingested sludge, probably because of the dramatic increases of microbial populations in the casts resulting from earthworm activity. During this process, relative to non-earthworm-ingested sludge, objectionable odors disappear from the wastes very quickly, and there is a rapid reduction in populations of human pathogenic microorganisms such as *Salmonella enteritidis, Escherichia coli,* and other Enterobacteriaceae, human viruses, and even helminth ova.

It has been found that mixing sewage sludge with other bulking materials (e.g., garden wastes, paper pulp sludge, or other lignin-rich wastes) before composting the mixtures using earthworms can accelerate the rates of decomposition. During passage through the earthworm gut, there is maceration and mixing of such materials and finely divided materials with high microbial activity in the casts.

EARTHWORM SPECIES USED IN VERMICOMPOSTING

Five main earthworm species were identified as potentially useful species to break down organic wastes. These were *E. fetida* (or the closely related *Eisenia andrei*), *Dendrobaena veneta,* and *Lumbricus rubellus* from temperate regions and *Eudrilus eugeniae* and *Perionyx excavatus* from the tropics. The survival, growth, mortality, and reproduction of these species were studied in detail in the laboratory, in a range of organic wastes, including pig, cattle, duck, turkey, poultry, potato, brewery, paper, and activated sewage sludge. All the species tested could grow and survive in a wide range of different organic wastes, but some were much more prolific, others grew rapidly, and yet others attained a large biomass quickly; these were all characters contributing in different ways to the practical usefulness of the earthworms in producing vermicomposts or animal feed protein.

Eisenia fetida (Savigny) and Eisenia andrei, Bouché

The species most commonly used for breaking down organic wastes is *E. fetida* or the very closely related species *E. andrei.* There are a number of reasons why these species are preferred in vermicomposting all over the world. They are peregrine species that are very common, and many organic wastes became colonized naturally by these species. They have a wide temperature tolerance and can live in organic wastes with a range of moisture contents. They are tough earthworms, readily handled, and in mixed species cultures, they usually become dominant, so that even when vermicomposting systems begin with other species, they often end dominated by *E. fetida* (or *E. andrei*).

Eudrilus eugeniae (Kinberg)

This is a large earthworm native to Africa but commonly found in many other countries, including the United States. It is commonly known as the African night crawler, grows extremely rapidly, and is quite prolific. It is cultured extensively for fish bait in the United States, and under optimum conditions, it would seem to be an ideal species for earthworm protein production. Its main disadvantages are its relatively poor temperature tolerance and poor handling capabilities because it is easily damaged and can be difficult to harvest. *E. eugeniae* has very high rates of reproduction. This species is used extensively in the tropics, especially India.

Perionyx excavatus (Perrier)

This tropical species of earthworm is extremely prolific; it is almost as easy to handle as *E. fetida* and very easy to harvest. Its main drawback for use under temperate conditions is its inability to withstand temperature conditions below 5°C for long periods, but for tropical conditions it is an ideal species. It has an extremely high reproductive rate.

Dendrobaena veneta (Rosa)

Dendrobaena veneta is a large earthworm with good potential for use in vermiculture, and unlike *E. fetida* and *P. excavatius*, it can also survive in soil but is not very prolific although it grows quite rapidly. However, it is a suitable species for organic matter breakdown and is specifically useful for transfer and use in agricultural soils to improve soil fertility.

Polypheretima elongata (Erseus)

Polypheretima elongata has been tested for use in the breakdown of organic solids, including municipal and slaughterhouse wastes; human, poultry, and dairy manures; and mushroom compost in India. *P. elongata* appears to be restricted to tropical regions and may not survive temperate winters.

Lumbricus rubellus (Hoffmeister)

Lumbricus rubellus is a species of earthworm common in moister soils as well as in organic matter, particularly those to which animal manure or sewage solids have been applied. It can be used for organic waste breakdown, but it is a relatively slow-growing species. It has potential for breeding in organic wastes and transfer to soils to improve soil fertility.

METHODS OF PROCESSING ORGANIC WASTES WITH EARTHWORMS

Processing systems of earthworms range from very simple methods involving low technology such as windrows, waste heaps, or containers, through moderately complex to completely automated continuous flow reactors. The basic principle of all successful processing systems is to add the wastes at frequent intervals in small, thin layers to the surface of the system and allow the earthworms to move into this and process successive aerobic layers of wastes. The earthworms will always move up and concentrate themselves in the upper 15 cm of waste and continue to move upward as each successive waste layer is added. Many of the operations involved in vermicomposting can be mechanized; a suitable balance is needed between the costs of mechanization and the savings in labor that result. The key to combining maximum productivity of vermicompost with the greatest rates of earthworm growth is to maintain aerobicity and optimal moisture and temperature conditions in the waste and to avoid wastes with excessive amounts of ammonia and salts. The addition of organic wastes in thin layers avoids overheating through thermophilic composting, but enough usually occurs to maintain suitable temperatures for earthworm growth during cold winter periods. Hence, for year-round production to maintain a reasonable temperature in temperate climates, the processing should always be done under cover, although heating is not usually necessary if the waste additions are managed well with addition of thicker layers during cold periods to provide some thermophilic composting.

Windrow Vermicomposting Can Be Carried Out in a Number of Different Ways

Static Pile Windrows (Batch)

Static pile windrows are simply piles of mixed bedding and feed (or bedding with feed layered on top) that are inoculated with worms and allowed to stand until the processing is complete. These piles are usually elongated in a windrow style but can also be squares, rectangles, or any other shape that makes sense for the person building them. They should not exceed one meter in height (before settling). Care must be taken to provide a good environment for the worms, so the selection of bedding type and amount is important.

Top-Fed Windrows (Continuous Flow)

Top-fed windrows are similar to the windrows described above, except that they are not mixed and placed as a batch, but are set up as a continuous-flow operation. This means that the bedding is placed first, then inoculated with worms, and then covered repeatedly with thin (less than 10 cm) layers of food. The worms tend to consume the food at the food/bedding interface, then drop their castings near the bottom of the windrow. A layered windrow is created over time, with the finished product on the bottom, partially consumed bedding in the middle, and the fresher food on top. Layers of new bedding should be added periodically to replace the bedding material gradually consumed by the worms.

Wedges (Continuous Flow)

The vermicomposting wedge is an interesting variation on the top-fed windrow. An initial stock of worms in bedding is placed inside a corral-type structure (three-sided) 15 of no more than 3 ft or 1 m in height. The sides of the corral can be concrete, wood, or even bales of hay or straw. Fresh material is added on a regular feeding schedule through the open side, usually by bucket loader. The worms follow the fresh food over time, leaving the processed material behind. When the material has reached the open end of the corral, the finished material is harvested by removing the back of the corral and scooping the material out with a loader. A fourth side is then put in place and the direction is reversed. Using this system, the worms do not need to be separated from the vermicompost and the process can be continued indefinitely. During the coldest months, a layer of insulating hay or straw can be placed over the active part of the wedges. The corrals can be any width at all, the only constraint being access to the interior of the piles for monitoring and corrective actions, such as adjustment of moisture content or pH level. A corral width of about 6 ft, with space between adequate for foot travel, would be ideal. The ideal length will depend on the material being processed, the size of the worm population, and other factors affecting processing times.

BEDS OR BINS

Top-fed beds (continuous flow): A top-fed bed works like a top-fed windrow. The main difference is that the bed, unlike a windrow, is contained within four walls and (usually) a floor, and is protected to some degree from the elements, often within an unheated building such as a barn. The beds can be built with insulated sides, or bales of straw can be used to insulate them in the winter. If the bins are fairly large, they are sheltered from the wind and precipitation, and the feedstock is reasonably high in nitrogen, the only insulation required may be an insulating "pillow" or layer on top. These can be as simple as bags or bales of straw.

Stacked bins (batch or continuous flow): One of the major disadvantages of the bed or bin system is the amount of surface area required. While this is also true of the windrow and wedge systems, they are outdoors, where space is not as expensive as it is under cover. Growing worms indoors or even within an unheated shelter is an expensive proposition if nothing is done to address this issue. Stacked bins address the issue of space by adding the vertical dimension to vermicomposting. The bins must be small enough to be lifted, either by hand or with a forklift, when they are full of wet material. They can be fed continuously, but this involves handling them on a regular basis. The more economical route to take is to use a batch process, where the material is pre-mixed and placed in the bin, worms are added, and the bin is stacked for a pre-determined length of time and then emptied.

FLOW-THROUGH REACTORS

The term "flow-through" refers to the fact that the worms are never disturbed in their beds—the material goes in the top, flows through the reactor (and the worms' guts), and comes out of the bottom (*E. fetida* tends to eat at the surface and drop castings near the bottom of the bedding). The method for pushing the materials out the bottom is usually a set of hydraulically powered "breaker bars" that move along the bottom grate, loosening the material so that it falls through.

METHODS OF HARVESTING WORMS

Expanding the operation (new beds) can be accomplished by splitting the beds, that is, removing a portion of the bed to start a new one and replacing the material with new bedding and feed. When worms are sold, however, they are usually separated, weighed, and then transported in a relatively sterile medium, such as peat moss. To accomplish this, the worms must first be separated from the bedding and vermicompost. There are three basic categories of methods used by growers to harvest worms: manual, migration, and mechanical.

Manual Methods

Manual methods are the ones used by hobbyists and smaller-scale growers, particularly those who sell worms to the home-vermicomposting or bait market. In essence, manual harvesting involves hand-sorting or picking the worms directly from the compost by hand. This process can be facilitated by taking advantage of the fact that worms avoid light. If material containing worms is dumped in a pile on a flat surface with a light above, the worms will quickly dive below the surface. The harvester can then remove a layer of compost, stopping when worms become visible again. This process is repeated several times until there is nothing left on the table except a huddled mass of worms under a thin covering of compost.

Self-Harvesting (Migration) Methods

These methods, like some of the methods used in vermicomposting, are based on the worms' tendency to migrate to new regions, either to find new food or to avoid undesirable conditions, such as dryness or light. Unlike the manual methods described above, however, they often make use of simple mechanisms, such as screens or onion bags. The screen method is very common and easy to use. A box is constructed with a screen bottom. The mesh is usually 1/4″, although 1/8″ can be used as well. There are two different approaches. The downward-migration system is similar to the manual system, in that the worms are forced downward by strong light. The difference with the screen system is that the worms go down through the screen into a prepared, pre-weighed container of moist peat moss. Once the worms have all gone through, the compost in the box is removed and a new batch of worm-rich compost is put in. The process is repeated until the box with the peat moss has reached the desired weight. Like the manual method, this system can be set up in a number of locations at once, so that the worm harvester can move from one box to the next, with no time wasted waiting for the worms to migrate. The upward-migration system is similar, except that the box with the mesh bottom is placed directly on the worm bed. It has been filled with a few centimeters of damp peat moss and then sprinkled with a food attractive to worms, such as chicken mash, coffee grounds, or fresh cattle manure. The box is removed and weighed after visual inspection indicates that sufficient worms have moved up into the material. The difference in this system that large onion bags are used instead of boxes. The advantage of this system is that the worm beds are not disturbed. The main disadvantage is that the harvested worms are in material that contains a fair amount of unprocessed food, making the material messier and opening up the possibility of heating inside the package if the worms are shipped. The latter problem can be avoided by removing any obvious food and allowing a bit of time for the worms to consume what is left before packaging.

Mechanical Methods

Mechanical harvesters are the quickest and easiest method for separating worms from vermicompost. The mechanical harvester is a trommel device, a rotating cylinder about 8–10 ft in length and 2–3 ft in diameter. The cylinder walls are composed of screen material of different mesh sizes. The cylinder is rotated by a small electric motor mounted on one end of the cylinder. The trommel is set at an angle; at the upper end of the rotating trommel, worms and their bedding including castings are added.

As the cylinder rotates, the castings fall through the screen. The worms "ride" the entire distance of the trommel and pass through the lower end into a wheelbarrow.

VERMICOMPOST TEA

The vermicompost tea is a mixture of aerobic microorganisms and vermicompost extracted in highly aerated water. This liquid contains beneficial bacteria and fungi which help to enrich the soil, which may be poor of microorganism in result of pesticide and inorganic fertilizer application, with these microorganisms. The aerobic microorganisms also are disease-suppressive for plant. It most noted that the leachate of vermicompost during vermicomposting process is not tea it is just vermicompost leachate and may contain significant amount of not decomposed organic material.

FURTHER READING

Dominguez J, Edwards CA, Webster M. Vermicomposting of sewage sludge: effect of bulking materials on the growth and reproduction of the earthworm *Eisenia andrei*. *Pedobiologia* 2000;44(1):24–32.

Edwards CA, Neuhauser EF (eds.) *Earthworms in Waste and Environmental Management.* The Hague: SPB Academic; 1988.

Ndegwa PM, Thompson SA. Integrating composting and vermicomposting in the treatment and bioconversion of biosolids. *Bioresour. Technol.* 2001;76:107–112.

Singh A, Sharma S. Composting of a crop residue through treatment with microorganisms and subsequent vermicomposting. *Bioresour. Technol.* 2002;85(2):107–111.

22 Biogas Production

The current global energy supply is highly dependent on fossil sources (crude oil, lignite, hard coal, natural gas). These are fossilized remains of dead plants and animals, which have been exposed to heat and pressure in the Earth's crust over hundreds of millions of years. For this reason, fossil fuels are non-renewable resources which reserves are being depleted much faster than new ones are being formed. The production and utilization of biogas from anaerobic digestion (AD) provides environmental and socioeconomic benefits for the society as a whole as well as for the involved farmers. Utilization of the internal value chain of biogas production enhances local economic capabilities, safeguards jobs in rural areas, and increases regional purchasing power. It improves living standards and contributes to economic and social development.

The World's economies are dependent today on crude oil. Unlike fossil fuels, biogas from AD is permanently renewable, as it is produced on biomass, which is actually a living storage of solar energy through photosynthesis. Biogas from AD will not only improve the energy balance of a country but also make an important contribution to the preservation of the natural resources and to environmental protection.

ADVANTAGES OF BIOGAS

Utilization of fossil fuels such as lignite, hard coal, crude oil, and natural gas converts carbon, stored for millions of years in the Earth's crust, and releases it as carbon dioxide (CO_2) into the atmosphere. An increase of the current CO_2 concentration in the atmosphere causes global warming as carbon dioxide is a greenhouse gas (GHG). The combustion of biogas also releases CO_2. However, the main difference, when compared to fossil fuels, is that the carbon in biogas was recently taken up from the atmosphere by photosynthetic activity of the plants. The carbon cycle of biogas is thus closed within a very short time (between one and several years). Biogas production by AD reduces also emissions of methane (CH_4) and nitrous oxide (N_2O) from storage and utilization of untreated animal manure as fertilizer. The GHG potential of methane is higher than that of carbon dioxide by 23-fold and of nitrous oxide by 296-fold. When biogas displaces fossil fuels from energy production and transport, a reduction of emissions of CO_2, CH_4, and N_2O will occur, contributing to mitigate global warming.

Fossil fuels are limited resources, concentrated in few geographical areas of our planet. This creates, for the countries outside this area, a permanent and insecure status of dependency on import of energy. Developing and implementing renewable energy systems such as biogas from AD, based on national and regional biomass resources, will increase security of national energy supply and diminish dependency on imported fuels.

One of the main advantages of biogas production is the ability to transform waste material into a valuable resource by using it as a substrate for AD. Biogas production

DOI: 10.1201/9781003272618-25

is an excellent way to utilize organic wastes for energy production, followed by recycling of the digested substrate as fertilizer. AD can also contribute to reducing the volume of waste and of costs for waste disposal.

A biogas plant is not only a supplier of energy. The digested substrate, usually named digestate, is a valuable soil fertilizer, rich in nitrogen, phosphorus, potassium, and micronutrients, which can be applied on soils with the usual equipment for application of liquid manure. Compared to raw animal manure, digestate has improved fertilizer efficiency due to higher homogeneity and nutrient availability, better C/N ratio, and significantly reduced odors.

ANAEROBIC DIGESTION (AD)

AD is a biochemical process during which complex organic matter is decomposed in absence of oxygen by various types of anaerobic microorganisms. The process of AD is common to many natural environments such as the marine water sediments, the stomach of ruminants, or the peat bogs. In a biogas installation, the result of the AD process is the *biogas* and the *digestate*. If the substrate for AD is a homogenous mixture of two or more feedstock types (e.g., animal slurries and organic wastes from food industries), the process is called "co-digestion" and is common to most biogas applications today.

The formation of methane is a biological process that occurs naturally when organic material (biomass) decomposes in a humid atmosphere in the absence of air but in the presence of a group of natural microorganisms which are metabolically active, i.e., methane bacteria. By nature, methane is formed as marsh gas (or swamp gas) in the digestive tract of ruminants, in plants for wet composting, and in flooded rice fields. Biogas consists mainly of methane and carbon dioxide, but also contains several impurities.

Composition of biogas

Compound	Content (%)
Methane	50–75
Carbon dioxide	25–45
Water vapor	2 (20°C)–7 (40°C)
Oxygen	<2
Nitrogen	<2
Ammonia	<1
Hydrogen	<1
Hydrogen sulfide	<1

SUBSTRATES FOR BIOGAS PRODUCTION

In general, all types of biomass can be used as substrates as long as they contain carbohydrates, proteins, fats, cellulose, and hemicellulose as main components. It is important that the following points are taken into consideration when selecting the biomass:

- The content of organic substance should be appropriate for the selected fermentation process.
- The nutritional value of the organic substance, hence the potential for gas formation, should be as high as possible.
- The substrate should be free of pathogens and other organisms which would need to be made innocuous prior to the fermentation process.
- The content of harmful substances and trash should be low to allow the fermentation process to take place smoothly.
- The composition of the biogas should be appropriate for further application.
- The composition of the fermentation residue should be such that it can be used, e.g., as fertilizer.
- Like natural gas, biogas has a wide variety of uses, but, as it is derived from biomass, it is a renewable energy source. There are many other benefits to be derived from the process of converting substrates in a biogas plant.
- The economic pressure on conventional agricultural products always continues to rise. Many farmers are forced to give up their occupation, since their land no longer brings sufficient yield. However, the production of biogas is subsidized in many countries, giving the farmer an additional income. For the farmer, biogas production does not mean major reorientation, because microorganisms for methanation require similar care to that needed for livestock in the stable.
- With the present tendency for farms to become large-scale enterprises and with the widespread abandonment of agricultural areas, the cultural landscape is changing. Biogas production from corn or grass could contribute to the maintenance of the structure of the landscape with small farmyards.
- Biomasses that are not needed are often left to natural deterioration, but energy can be generated from these biomasses.

With aerobic degradation, the low-energy compounds CO_2 and H_2O are formed at least, i.e., much energy is lost to air—about 20 times as much as with an anaerobic process. In the case of the anaerobic degradation metabolism, products of high energy (e.g., alcohols, organic acids, and, in the long run, methane) result, which serve other organisms as nutrients (alcohols, organic acids) or are energetically used (biogas).

- Reduction of landfill area and the protection of the groundwater: the quantity of organic waste materials can be reduced down to 4% sludge when the residue is squeezed off and the wastewater from the biogas plant is recycled into the wastewater treatment plant.
- Substantial reduction of the disposal costs of organic wastes, even including meaningful reuse (e.g., as fertilizers), because the quantity of biomass decreases so significantly.
- If plants are used as co-substrates for biogas production and the residues are recycled to agriculture, no mineral fertilizer need be bought. A cycle of nutrients is reached. Nitrate leaching is reduced. Plant compatibility and plant health are improved.
- When storing the residue from the fermentation plant, much less odor—methane and nitrous oxide emissions (N_2O)—is released with unfermented

substrates: odor-active substances and the organic acids are reduced, giving a volume contraction of $c = 0.5\,g/L$. The ammonium content, which is high compared with untreated liquid manure, and the higher pH value can lead, however, to higher ammonia emissions.

- CO_2-neutral production of energy (especially electrical power and heat) is achieved. The climatic protection goal agreed upon in the Kyoto minutes is thereby effectively supported.

Methane yields of different feedstock materials

Feedstock	Methane Yield (%)	Biogas Yield (m³/t Fresh Feedstock)
Liquid cattle manure	60	25
Liquid pig manure	65	28
Distillers grains with solubles	61	40
Cattle manure	60	45
Pig manure	60	60
Poultry manure	60	80
Beet	53	88
Organic waste	61	100
Sweet sorghum	54	108
Forage beet	51	111
Grass silage	54	172
Corn silage	52	202

BIOGAS PLANT COMPONENTS

A biogas plant is a complex installation, consisting of a variety of elements. The layout of such a plant depends to a large extent on the types and amounts of feedstock supplied. As there are many different feedstock types suitable for digestion in biogas plants, there are, correspondingly, various techniques for treating these feedstock types and different digester constructions and systems of operation. Furthermore, depending on the type, size, and operational conditions of each biogas plant, various technologies for conditioning, storage, and utilization of biogas are possible to implement. As for storage and utilization of digestate, this is primarily oriented toward its utilization as fertilizer and the necessary environmental protection measures related to it.

The core component of a biogas plant is the digester (AD reactor tank), which is accompanied by a number of other components (Figure 22.1).

Agricultural biogas plants operate with four different process stages:

1. Transport, delivery, storage, and pretreatment of feedstock;
2. biogas production (AD);
3. storage of digestate, eventual conditioning, and utilization;
4. storage of biogas, conditioning, and utilization.

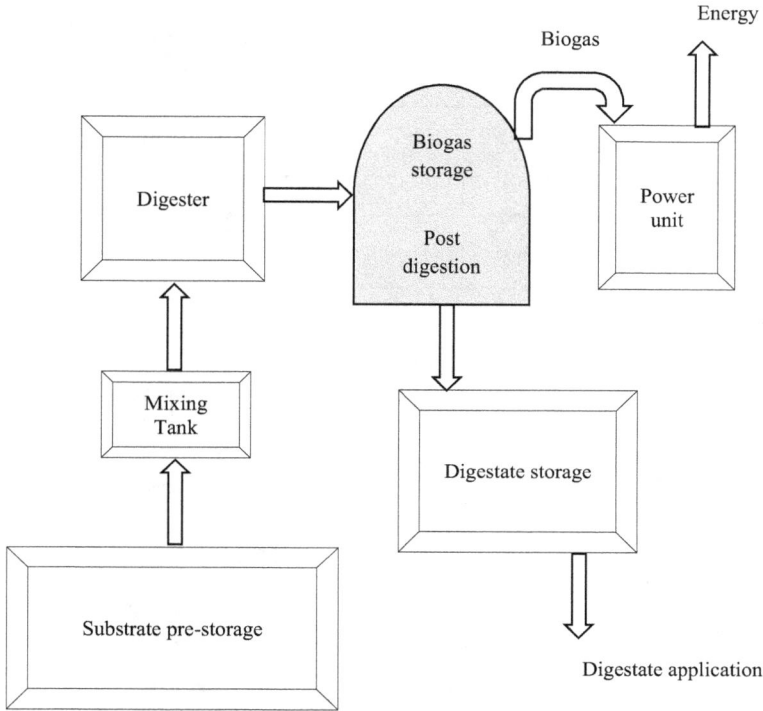

FIGURE 22.1 Main components and general process flow for biogas production.

1. The first process stage (storage, conditioning, transport, and insertion of feedstock) includes the storage tank for manure (2), the collection bins (3), the sanitation tank (4), the drive-in storage tanks (5), and the solid feedstock feeding system (6).
2. The second process stage includes the biogas production in the biogas reactor (7), also referred to as the digester.
3. The third process stage is represented by the storage tank for digestate (10) and the utilization of digestate as fertilizer on the fields (11).
4. The fourth process stage (biogas storage, conditioning, and utilization) consists of the gas storage tank (8) and the combined heat and power (CHP) production unit (9).

These four process stages are closely linked to each other (e.g., stage 4 provides the necessary process heating for stage 2) (Figure 22.2).

When building a biogas plant, the choice of type and the design of the plant are mainly determined by the amount and type of available feedstock. The amount of feedstock determines the dimensioning of the digester size, storage capacities, and CHP production unit. The feedstock types and quality (DM-content, structure, origin, etc.) determine the process technology.

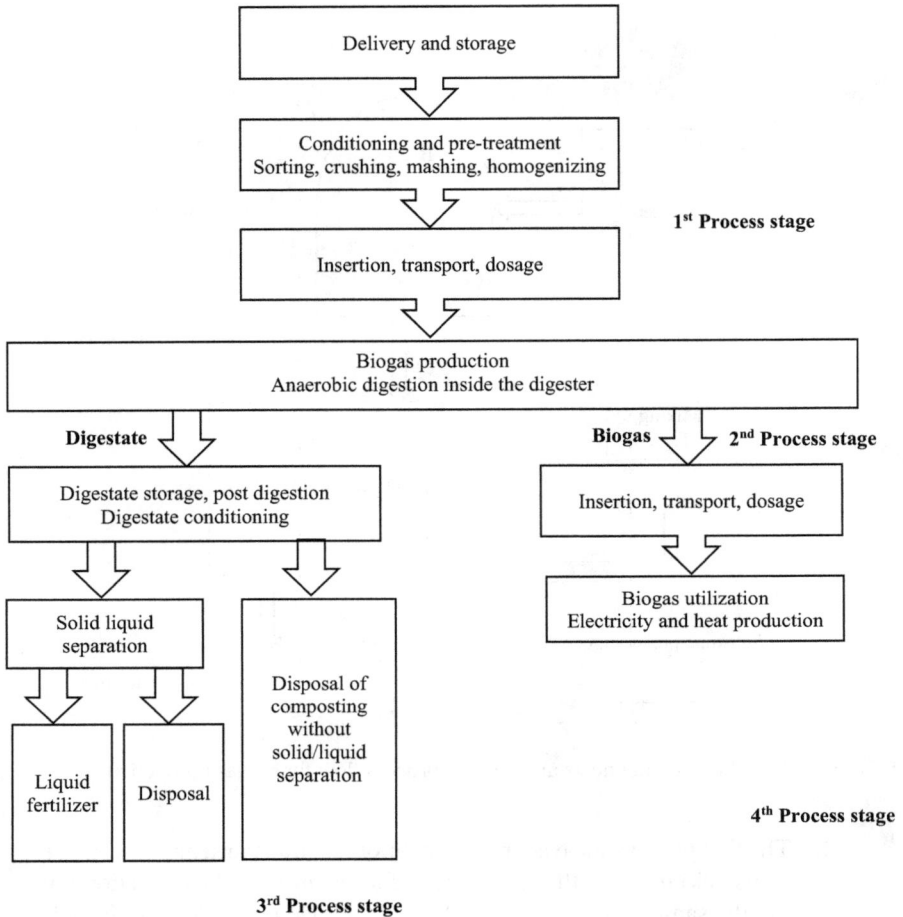

FIGURE 22.2 Process stages of agricultural biogas plants.

Depending on the composition of the feedstock, it may be necessary to separate problematic materials, to mash the feedstock or even to add water, in order to convert it into a pumpable mixture. If the supplied feedstock is prone to contamination, it is necessary to include a pre-sanitation step in the overall design of the future plant. In the case of wet digestion, single-stage AD plants operating with flow-through process are usually used. In the two-stage process, a pre-digester is placed before the main digester. The pre-digester creates the optimal conditions for the first two process steps of the AD process (hydrolysis and acid formation). After pre-digester, the feedstock enters the main digester, where the subsequent AD steps take place. The digested substrate (digestate) is pumped out of the digester and stored in storage tanks.

These storage tanks should be provided with covers of gas proof membranes, to facilitate collection of the biogas production which can take place inside these tanks, at ambient temperature (post-digestion). Alternatively, digestate can be stored in open digestate containers, with natural or artificial floating layer aimed to minimize surface emissions. The produced biogas is stored, conditioned, and used for energy generation. The actual standard use of biogas is for CHP production in, e.g., block-type thermal plants, for the simultaneous production of electricity and heat.

BIOGAS FORMATION

The formation of methane from biomass follows in general the equation:

$$C_cH_hO_oN_nS_s + yH_2O \rightarrow xCH_4 + nNH_3 + sH_2S + (c-x)CO_2$$

BIOLOGY OF BIOGAS PRODUCTION

Methane fermentation is a complex process, which can be divided up into four phases of degradation, named hydrolysis, acidogenesis, acetogenesis, and methanation, according to the main process of decomposition in this phase. The individual phases are carried out by different groups of microorganisms, which partly stand in syntrophic interrelation and place different requirements on the environment.

If the first stage runs too fast, the CO_2 portion in the biogas increases, the acid concentration rises, and the pH value drops below 7.0. Acidic fermentation is then also carried out in the second stage. If the second stage runs too fast, methane production is reduced. There are still many bacteria of the first stage in the substrate. The bacteria of the second stage must be inoculated. With biologically difficult degradable products, the hydrolytic stage limits the rate of degradation. In the second stage, the acetogenesis possibly limits the rate of decomposition (Figure 22.3).

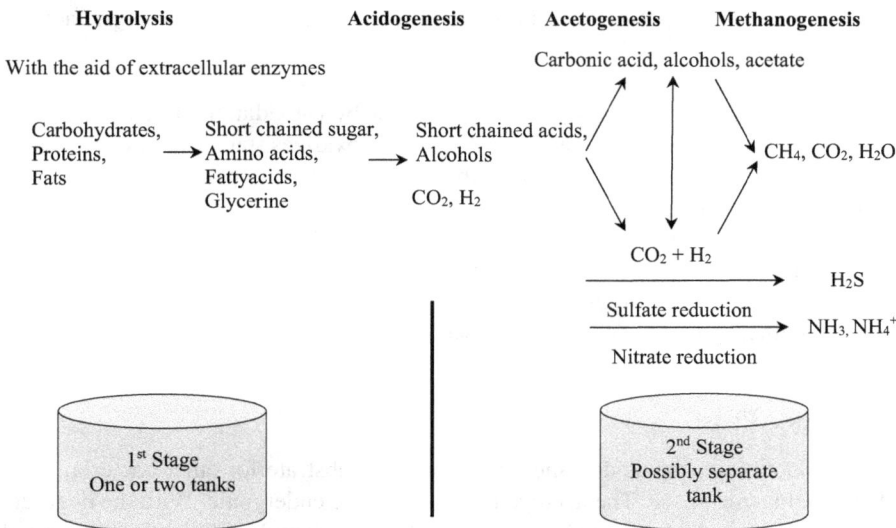

FIGURE 22.3 Biochemistry of methane gas production.

HYDROLYSIS

In the first phase (the hydrolysis), undissolved compounds, like cellulose, proteins, and fats, are cracked into monomers (water-soluble fragments) by exoenzymes (hydrolase) of facultative and obligatorily anaerobic bacteria. Actually, the covalent bonds are split in a chemical reaction with water. The hydrolysis of carbohydrates takes place within a few hours, and the hydrolysis of proteins and lipids within few days. Lignocellulose and lignin are degraded only slowly and incompletely. The facultative anaerobic microorganisms take the oxygen dissolved in the water and thus cause the low redox potential necessary for obligatorily anaerobic microorganisms.

ACIDOGENIC PHASE

The monomers formed in the hydrolytic phase are taken up by different facultative and obligatorily anaerobic bacteria and are degraded in the second, the acidogenic phase, to short-chain organic acids, C1–C5 molecules (e.g., butyric acid, propionic acid, acetate, acetic acid), alcohols, hydrogen, and carbon dioxide. The concentration of the intermediately formed hydrogen ions affects the kind of the products of fermentation. The higher the partial pressure of the hydrogen, the fewer reduced compounds, like acetate, are formed.

The pathways of degradation are as follows:

a. Carbohydrates:
   Formation of propionic acid by *Propionibacterium* via the succinate pathway and the acrylic pathway.
   Formation of butyric acid (butyric acid pathway) above all by *Clostridium*.
   Acetic acid → 2-hydroxy butyrate → trans-2-butenic acid → butyric acid → butanol
b. Fatty acids:
   These are degraded, e.g., from acetobacter by β-oxidation. Therefore, the fatty acid is bound on Coenzyme A and then oxidizes stepwise, as with each step two C atoms are separated, which are set free as acetate.
c. Amino acids:
   These are degraded by the Stickland reaction by *Clostridium botulinum* taking two amino acids at the same time—one as hydrogen donor, the other as acceptor—in coupling to acetate, ammonia, and CO_2. During splitting of cysteine, hydrogen sulfide is released.

ACETOGENIC PHASE

The products from the acidogenic phase serve as substrate for other bacteria, those of the acetogenic phase. The acetogenic reactions are endergonic. With the degradation of propionic acid are needed $\Delta G_f = +76.11$ kJ/mol, and with the degradation of ethanol, $\Delta G_f' = +9.6$ kJ/mol.

In the acetogenic phase, homoacetogenic microorganisms constantly reduce exergonic H_2 and CO_2 to acetic acid:

$$2CO_2 + 4H_2 \rightarrow CH_3COOH + 2H_2O$$

Acetogenic bacteria are obligatory H_2 producers. The acetate formation by oxidation of long-chain fatty acids (e.g., propionic or butyric acid) runs on its own and is thus thermodynamically possible only with very low hydrogen partial pressure.

Acetogenic bacteria can get the energy necessary for their survival and growth, therefore, only at very low H_2 concentration. Acetogenic and methane-producing microorganisms must therefore live in symbiosis. Methanogenic organisms can survive only with higher hydrogen partial pressure. They constantly remove the products of metabolism of the acetogenic bacteria from the substrate and so keep the hydrogen partial pressure, at a low level suitable for the acetogenic bacteria.

When the hydrogen partial pressure is low, H_2, CO_2, and acetate are predominantly formed by the acetogenic bacteria. When the hydrogen partial pressure is higher, predominantly butyric, capronic, propionic, and valeric acids and ethanol are formed. From these products, the methanogenic microorganisms can process only acetate, H_2, and CO_2. About 30% of the entire CH_4 production in the anaerobic sludge can be attributed to the reduction of CO_2 by H_2, but only 5%–6% of the entire methane formation can be attributed to the dissolved hydrogen. This is to be explained by the "interspecies hydrogen transfer," by which the hydrogen moves directly from the acetogenic microorganisms to the methanogenics, without being dissolved in the substrate.

The anaerobic conversion of fatty acids and alcohols goes energetically at the expense of the methanogenics, where these, however, in return, receive the substrates (H_2, CO_2, acetic acid) needed for growth from the acetogenic bacteria. The acetogenic phase limits the rate of degradation in the final stage. From the quantity and the composition of the biogas, a conclusion can be drawn about the activity of the acetogenic bacteria.

At the same time, organic nitrogen and sulfur compounds can be mineralized to hydrogenic sulfur by producing ammonia. The reduction of sulfate follows, for example, the stoichiometric equations below. Sulfate-reducing bacteria such as *Desulfovibrio*, *Desulfuromonas*, *Desulfobulbus*, *Desulfobacter*, *Desulfococcus*, *Desulfosarcina*, *Desulfonema*, and *Desulfotomaculum* participate in the process, which uses the energy released by the exergonic reaction:

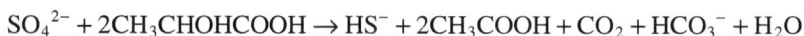

$$SO_4^{2-} + CH_3COOH \rightarrow HS^- + CO_2 + HCO_3^- + H_2O$$

$$SO_4^{2-} + 2CH_3CHOHCOOH \rightarrow HS^- + 2CH_3COOH + CO_2 + HCO_3^- + H_2O$$

METHANOGENIC PHASE

In the fourth stage, the methane formation takes place under strictly anaerobic conditions. This reaction is categorically exergonic. As follows from the description of the methanogenic microorganisms, all methanogenic species do not degrade all substrates.

During the formation of methane from acetate and/or CO_2 in microorganisms, long-chain hydrocarbons are involved such as methanofurans (e.g., $R-C_{24}H_{26}N_4O_8$) and H_4TMP (tetrahydromethanopterin) as cofactors. Corrinoids are molecules which have four reduced pyrrole rings in a large ring and can be represented by the empirical formula $C_{19}H_{22}N_4$.

When the methane formation works, the acetogenic phase also works without problems. When the methane formation is disturbed, over-acidification occurs.

Problems can occur when the acetogenic bacteria live in symbiosis instead of with a methanogenic species with other organisms using H_2. In wastewater technology, symbioses can occur with microorganisms which reduce sulfate to hydrogen sulfide. Therefore, they need hydrogen and compete with the methanogenics.

The methanogenics get less feed and form less methane. Additionally, hydrogen sulfide affects the methanogenics toxically. All methane-forming reactions have different energy yields. The oxidation of acetic acid is, in comparison to the reduction of $CO_2 + H_2$, only a little exergonic:

$$CH_3COOH \leftrightarrow CH_4 + CO \text{ at } \Delta G^0 = 31\,kJ/kmol$$

$$CO_2 + 4\,BADH/H^+ \leftrightarrow CH_4 + 2H_2O + 4NAD + \text{ at } \Delta G^0 = 136\,kJ/kmol$$

Nevertheless, only 27%–30% of the methane arises from the reduction, while 70% arises from acetate during methanation. This also is the case in sea sediments. Acetate-using methanogenics like *Methanosarcina barkeri, Methanobacterium sohngenii*, and *Methanobacterium thermoautotrophicum* grow in acetate very slowly, theoretically with a regeneration time of at least 100 hours, whereas CO_2 has turned out to be essential for the growth.

FACTORS INFLUENCING ANAEROBIC DIGESTION PROCESS

The efficiency of AD is influenced by some critical parameters, thus it is crucial that appropriate conditions for anaerobic microorganisms are provided. The growth and activity of anaerobic microorganisms is significantly influenced by conditions such as exclusion of oxygen, constant temperature, pH value, nutrient supply, stirring intensity as well as presence and amount of inhibitors (e.g., ammonia). The methane bacteria are fastidious anaerobes, so that the presence of oxygen into the digestion process must be strictly avoided.

TEMPERATURE

The AD process can take place at different temperatures, divided into three temperature ranges: psychrophilic (below 25°C), mesophilic (25°C–45°C), and thermophilic (45°C–70°C). The temperature stability is decisive for AD. In practice, the operation temperature is chosen with consideration to the feedstock used and the necessary process temperature is usually provided by floor or wall heating systems inside the digester. The solubility of various compounds (NH_3, H_2, CH_4, H_2S, and volatile fatty acids (VFAs)) also depends on the temperature. This can be of great significance for materials which have an inhibiting effect on the process.

pH VALUES AND OPTIMUM INTERVALS

The pH value is the measure of acidity/alkalinity of a solution (respectively of substrate mixture, in the case of AD) and is expressed in *parts per million* (ppm). The pH value of the AD substrate influences the growth of methanogenic microorganisms and affects the dissociation of some compounds of importance for the AD process (ammonia, sulfide, organic acids). Experience shows that methane formation takes place within a relatively narrow pH interval, from about 5.5 to 8.5, with an optimum interval between 7.0 and 8.0 for most methanogens. Acidogenic microorganisms usually have lower value of optimum pH. The optimum pH interval for mesophilic digestion is between 6.5 and 8.0, and the process is severely inhibited if the pH value decreases below 6.0 or rises above 8.3. The solubility of carbon dioxide in water decreases at increasing temperature. The pH value in thermophilic digesters is therefore higher than that in mesophilic ones, as dissolved carbon dioxide forms carbonic acid by reaction with water.

The value of pH can be increased by ammonia, produced during degradation of proteins or by the presence of ammonia in the feed stream, while the accumulation of VFA decreases the pH value. The value of pH in anaerobic reactors is mainly controlled by the bicarbonate buffer system. Therefore, the pH value inside digesters depends on the partial pressure of CO_2 and on the concentration of alkaline and acid components in the liquid phase. If accumulation of base or acid occurs, the buffer capacity counteracts these changes in pH up to a certain level. When the buffer capacity of the system is exceeded, drastic changes in pH values occur, completely inhibiting the AD process. For this reason, the pH value is not recommended as a stand-alone process monitoring parameter.

VOLATILE FATTY ACIDS (VFAS)

The stability of the AD process is reflected by the concentration of intermediate products like VFA. The VFAs are intermediate compounds (acetate, propionate, butyrate, lactate), produced during acidogenesis, with a carbon chain of up to six atoms. In most cases, AD process instability will lead to accumulation of VFA inside the digester, which can lead furthermore to a drop of pH value. However, the accumulation of VFA will not always be expressed by a drop of pH value due to the buffer capacity of the digester through the biomass types contained in it. Animal manure, e.g., has a surplus of alkalinity, which means that the VFA accumulation should exceed a certain level, before this can be detected due to significant decrease of pH value. At such point, the VFA concentration in the digester would be so high that the AD process will be already severely inhibited.

AMMONIA

Ammonia (NH_3) is an important compound, with a significant function for the AD process. NH_3 is an important nutrient, serving as a precursor to foodstuffs and fertilizers and is normally encountered as a gas with the characteristic pungent smell. Proteins are the main source of ammonia for the AD process. Too high ammonia concentration inside the digester, especially free ammonia (the unionized form of ammonia), is considered to be responsible for process inhibition. This is common to AD of animal slurries, due to their high ammonia concentration, originating from urine. For its inhibitory effect, ammonia concentration should be kept below 80 mg/L. Methanogenic bacteria are especially sensitive to ammonia inhibition. The concentration of free ammonia is direct proportional to temperature, so there is an increased risk of ammonia inhibition of AD processes operated at thermophilic temperatures, compared to mesophilic ones.

MACRO- AND MICRONUTRIENTS (TRACE ELEMENTS) AND TOXIC COMPOUNDS

Microelements (trace elements) like iron, nickel, cobalt, selenium, molybdenum, or tungsten are equally important for the growth and survival of the AD microorganisms as the macronutrients carbon, nitrogen, phosphorus, and sulfur. The optimal ratio of the macronutrients carbon, nitrogen, phosphorus, and sulfur (C:N:P:S) is considered 600:15:5:1. Insufficient provision of nutrients and trace elements, as well as too high digestibility of the substrate, can cause inhibition and disturbances in the AD process. Another factor, influencing the activity of anaerobic microorganisms, is the presence of toxic compounds. They can be brought into the AD system together with the feedstock or are generated during the process. The application of threshold values for toxic compounds is difficult, on one hand, because these kinds of materials are often bound by chemical processes and, on the other hand, because of the capacity of anaerobic microorganisms to adapt, within some limits, to environmental conditions, herewith to the presence of toxic compounds.

FURTHER READING

Mudhoo A. *Biogas Production: Pretreatment Methods in Anaerobic Digestion.* Scrivener Publishing LLC. John Wiley & Sons; 2012.
Speight JG. *Biogas: Production and Properties.* Nova Science Publishers, Inc; 2019.
Treichel H, Fongaro G. *Improving Biogas Production: Technological Challenges, Alternative Sources, Future Developments.* Springer; 2019.
Vico A, Artemio N. *Biogas: Production, Applications and Global Developments.* Nova Science Publishers, Inc; 2017.
Wellinger A, Murphy J, Baxter D. *The Biogas Handbook. Science, Production and Applications.* Woodhead Publishing Series in Energy; 2013.

Index

For Product Safety Concerns and Information please contact our EU
representative GPSR@taylorandfrancis.com
Taylor & Francis Verlag GmbH, Kaufingerstraße 24, 80331 München, Germany

9 781032 224503